Tourism, Hospitality & Event Management

This book series covers all topics relevant to the tourism, hospitality, and event industries. It includes destination management and related aspects of the travel and mobility industries as well as effects from developments in information and communication technologies. "Tourism, Hospitality & Event Management" embraces books both for professionals and scholars, and explicitly includes undergraduate and advanced texts for students. In this setting, the book series reflects the close connection between research, teaching, and practice in tourism research and tourism management and the related fields.

This series is indexed in SCOPUS.

Marco Valeri
Editor

Contemporary Religious Tourism

Multidisciplinary insights, Environmental Engagement, and Community Impact

Springer

Editor
Marco Valeri
Faculty of Economics
Niccolò Cusano University
Rome, Italy

ISSN 2510-4993 ISSN 2510-5000 (electronic)
Tourism, Hospitality & Event Management
ISBN 978-3-031-79073-7 ISBN 978-3-031-79074-4 (eBook)
https://doi.org/10.1007/978-3-031-79074-4

This Springer imprint is published by the registered company Springer Nature Switzerland AG
The registered company address is: Gewerbestrasse 11, 6330 Cham, Switzerland

Contents

About the Editor and Contributors

About the Editor

Marco Valeri is Associate Professor of Organizational Behavior, Faculty of Economics, Niccolò Cusano University, Italy. He is Lecturer in Applied Organizational Behaviour, Xenophon College, UK. He is Visiting Professor at the University of Information Science and Technology (UIST) "St. Paul The Apostle," Macedonia. He is Adjunct Professor at the Faculty of Social Sciences and Leisure Management, School of Hospitality, Tourism and Events, Taylors University, Malaysia. He is Adjunct Professor at the Faculty of Economics, Lovely Professional University (LPU), Phagwara, India. He is Associate Researcher in Strategy, Magellan Research Center, School of Management, Iaelyon Business School, Jean Monnet University, France. He is Honorary Associate Professor, University of Pannonia, Hungary. His teaching and consultancy fields include strategic management, leadership development, cross-cultural management, and international hospitality management. He is a member of several Editorial Boards of international tourism journals and reviewer and editor of several handbooks on entrepreneurship, tourism, and hospitality management (Emerald Publishing, Springer Publishing, Routledge Publishing, Edward Publishing, De Gruyter, and IGI Global Publishing).

Contributors

Madalena Abreu Polytechnic Institute of Coimbra, Coimbra, CEOS.PP, Coimbra, Portugal

Julia N. Albrecht Department of Tourism, University of Otago, Dunedin, New Zealand

Vincenzo Asero Department of Political and Social Sciences (DPSS), University of Catania, Catania, Italy

António Azevedo School of Economics and Management of University of Minho, iBMS, Braga, Portugal

Ayan Banik Department of Tourism and Hospitality Management, Gopalganj Science and Technology University, Gopalganj, Bangladesh

Jean-Gabriel Bela Faculty of Social Sciences, Pontifical Gregorian University, Rome, Italy

Ana María Campón-Cerro Facultad de Empresa, Finanzas y Turismo, Universidad de Extremadura, Cáceres, Spain

Universidade da Beira Interior & NECE–Research Centre for Business Sciences, Covilhã, Portugal

Jingjing Dai Office of Domestic and Foreign Affairs, Ningxia University, Ningxia, China

Monalisha Dash Department of Hospitality & Tourism, Galgotias University, Greater Noida, Uttar Pradesh, India

Bob Yi-Chen Duan School of Vocational and Technical Education, Yunnan Normal University, Yunnan, China

Griffith Institute for Tourism, Griffith University, Brisbane, QLD, Australia

Long Hai Duong School of Social Sciences and Humanities, Can Tho University, Can Tho, Vietnam

Margarida Franca Polytechnic Institute of Leiria, Leiria, UCP-CITER, Lisboa, Portugal

Anthi Gousiou Department of Business Administration, University of the Aegean, Mytilene, Greece

Sandeep Guleria Department of Tourism, Chandigarh University, Mohali, Punjab, India

Da Van Huynh School of Social Sciences and Humanities, Can Tho University, Can Tho, Vietnam

Md Nurnobi Islam Department of Tourism, University of Otago, Dunedin, New Zealand

Jyoti Department of Hospitality & Tourism, Galgotias University, Greater Noida, Uttar Pradesh, India

Svetoslav Kaleychev Faculty of Economics of Infrastructure, University of National and World Economy, Sofia, Bulgaria

Evangelia Kasimati Economic Analysis and Research Department, Bank of Greece, Athens, Greece

Tilemachos K. Koliopoulos Telegeco Research and Development. Collaborator University of West Attica, Faculty of Health, Athens, Greece

Ljiljana Kostic Faculty of Education in Uzice, University of Kragujevac, Uzice, Serbia

Suneel Kumar Department of Commerce, Shaheed Bhagat Singh College, University of Delhi, New Delhi, India

Madhu Kumari Department of Tourism and Hospitality Management, Chandigarh University, Mohali, Punjab, India

Bapon Chandra Kuri Department of Tourism and Hospitality Management, Gopalganj Science and Technology University, Gopalganj, Bangladesh

Maria Lagou Department of Accounting and Finance Administration, University of West Attica, Athens, Greece

Peter Lah Faculty of Social Sciences, Pontifical Gregorian University, Rome, Italy

Brent Lovelock Department of Tourism, University of Otago, Dunedin, New Zealand

Emily Ma Department of Hospitality and Events, School of Hospitality and Tourism Management, University of Surrey, Surrey, UK

Cernicova-Buca Mariana Faculty of Communication Sciences, Politehnica University of Timisoara, Timisoara, Romania

Kamarun Muhsina Department of Tourism and Hospitality Management, Daffodil Institute of IT, Dhaka, Bangladesh

Noel M. Muniz Department of Statistics, Econometrics, Operational Research, Business Organization, and Applied Economics, University of Cordoba, Cordoba, Spain

Manisha Paliwal R&D cell at Sri Balaji University, Pune, Maharashtra, India

Bárbara Sofía Pasaco-González Institutos Universitarios de Investigación, Universidad de Extremadura, Cáceres, Spain

Leonor M. Pérez Department of Statistics, Econometrics, Operational Research, Business Organization, and Applied Economics, University of Cordoba, Cordoba, Spain

Panoraia Poulaki Department of Economics and Management of Tourism, University of the Aegean, Chios, Greece

Chandra Prakash R&D cell at Sri Balaji University, Pune, Maharashtra, India

Daliborka Puric Faculty of Education in Uzice, University of Kragujevac, Uzice, Serbia

Theodoros Rachiotis Department of Physical Education and Sport Science, National and Kapodistrian University of Athens, Athens, Greece

Ricardo Gouveia Rodrigues Universidade da Beira Interior & NECE–Research Centre for Business Sciences, Covilhã, Portugal

Maribel Rodríguez Department of Statistics, Econometrics, Operational Research, Business Organization, and Applied Economics, University of Cordoba, Cordoba, Spain

Elena Mirela Samfira Teacher Training Department, University of Life Sciences from Timisoara, Timisoara, Romania

Francisco Sánchez-Cubo Department of Applied Economics (Economic Structure), University of Málaga, Málaga, Spain

Mariya Stankova Faculty of Economics, South-West University, Blagoevgrad, Bulgaria

Paraskevas F. Tsorvas Department of Business Administration, University of the Aegean, Mytilene, Greece

Nikolaos Vasilakis Ionian University, Kerkira, Greece

María J. Vázquez-García Department of Statistics, Econometrics, Operational Research, Business Organization, and Applied Economics, University of Cordoba, Cordoba, Spain

Yuetian Zhang School of Hospitality and Tourism Management, Macau University of Science and Technology, Macau, China

Introduction

Marco Valeri

Abstract Contemporary Religious Tourism. Multidisciplinary insights, environmental engagement, and community impact aims to analyze the impact of religious tourism phenomena on touristic destinations. The presence of research gaps in so many areas necessitate drawing attention of the academia toward this impactful research topic. Tourism has largely flourished due to various psychosociological factors, with religious and spiritual motives taking a significant role. The impact of religion and spirituality on tourism is considerable and remains strong today. Specifically, the desire for spiritual experiences is often a key incentive for journeys, as many religions encourage their followers to attend sacred sites and participate in religious events and festivals, which assist in alleviating both spiritual and material challenges while simultaneously enriching their spiritual lives. The concept of a religious motive is multifaceted, carrying various meanings and varying in intensity based on personal belief and societal context, from outright disbelief to extreme devotion.

The aim of this book is to analyze the impact of religious tourism phenomena on touristic destinations. The presence of research gaps in so many areas necessitate drawing attention of the academia towards this impactful research topic. Tourism has largely flourished due to various psychosociological factors, with religious and spiritual motives taking a significant role. The impact of religion and spirituality on tourism is considerable and remains strong today. Specifically, the desire for spiritual experiences is often a key incentive for journeys, as many religions encourage their followers to attend sacred sites and participate in religious events and festivals, which assist in alleviating both spiritual and material challenges while simultaneously enriching their spiritual lives. The concept of a religious motive is multifaceted, carrying various meanings and varying in intensity based on personal belief

M. Valeri (✉)
Faculty of Economics, Niccolò Cusano University, Rome, Italy

Faculty of Social Sciences and Leisure Management, School of Hospitality, Tourism and Events, Taylor's University, Subang Jaya, Selangor, Malaysia
e-mail: marco.valeri@unicusano.it

M. Valeri (ed.), *Contemporary Religious Tourism*, Tourism, Hospitality & Event Management, https://doi.org/10.1007/978-3-031-79074-4_1

and societal context, from outright disbelief to extreme devotion. Travelers with religious motivations display a range of motivations and expectations. The extent of their faith influences their willingness to embark on journeys of differing lengths, sometimes spanning vast distances, to fulfill their spiritual needs.

The concept of religious motive is intricate and varies in intensity based on personal beliefs and social circumstances, spanning from complete disbelief to extreme religious fervor. Travelers motivated by religion also show varied motivations and expectations. The extent of one's religious beliefs influences how far individuals are willing to travel—sometimes over considerable distances—to fulfill their spiritual needs and convictions. The aim of these journeys, so-called pilgrimages, is for believers to follow their religious feelings and needs or perform an act designated by their religion, and the destinations of such pilgrimages are various holy places. The reasons for embarking on a spiritual journey are complex and varied at the same time. Even when religion is the primary driver, additional motives may also influence the decision. Additionally, due to the strong motivation required for a spiritual journey, expectations tend to be greater compared to those of regular religious tourists. One of the most apparent links between tourism and religion is the multitude of sacred sites that attract and draw in visitors. Basically, in views of the tourists, the main attraction is to be found in the cultural content of historical value of those sacred buildings.

It is thought that the role of religious elements in tourism is influenced by their purpose in religion. The religious functions of these elements can be divided into two main categories: the first category consists of sacred sites visited by followers to carry out religious ceremonies and fulfill their spiritual needs, including churches, mosques, and temples; the second category encompasses various institutions that do not traditionally possess a religious identity but are utilized by practitioners and often belong to religious bodies, such as monasteries, convents, seminaries, and religious schools. Nonetheless, the first category should also encompass all holy sites, whether small or large, where devotees gather for their religious schools. However, the definition of the first group shall include all holy places, small or greater areas, places, and regions where believers come because of their religious content or religious character.

Contemporary Religious Tourism: Multidisciplinary Insights, Environmental Engagement, and Community Impact aims to contribute to the literature on religious tourism phenomena and its impact on touristic destinations by uncovering multidisciplinary aspects such as marketing, economics, finance, psychology, social development, sustainability, and operations.

1 Structure of the Book

The book involves research studies of different nationalities. The book contains an introduction and 19 chapters written by 47 authors located in 15 different countries (Australia, Bangladesh, Bulgaria, China, Ecuador, Greece, India, Italy, New Zealand, Portugal, Romania, Serbia, Spain, UK, and Vietnam) and affiliated with 24 different international universities. The book is structured in two parts. The first part focuses on *tourism, religion, and spiritual journeys.* This part collects chapters that explore the topic of religious tourism and pilgrimage, examining it from a tourism industry perspective, reflecting on definitions and motivations. The second part focuses on *case studies: intersections between tourism and religion.* This part collects chapters drawing together understandings from different religious traditions, presenting data and motivations on a variety of pilgrimage types. As the chapters demonstrate, this niche product is indeed important and stimulates international travels.

Chapter 2 titled "*Religious tourism: Unveiling the role of authenticity and emotional solidarity in enhancing tourists' memorable experiences and destination loyalty*" is written by Da Van Huynh and Long Hai Duong. This chapter investigates how objective, constructive, and existential authenticity at religious sites influences tourists' emotional solidarity and, subsequently, how this solidarity contributes to memorable experiences and loyalty to religious destinations. A total of 430 valid responses (239 domestic tourists and 191 international tourists) were used for hypothesis testing and multigroup analysis. The findings highlight significant impacts of all three aspects of authenticity on emotional solidarity. Additionally, the findings indicate a critical role of emotional solidarity in promoting tourists' memorable experiences and destination loyalty. Notably, memorable experiences were found to have a significant effect on tourists' loyalty to religious tourism destinations. Importantly, the multigroup analysis revealed a significant difference only in the effect of emotional solidarity on memorable experiences between the two data samples. Theoretical contributions include advancing understanding of how perceived authenticity predicts emotional solidarity and exploring the intricate relationships among emotional solidarity, memorable experiences, and destination loyalty in religious tourism. Practical implications suggest strategies for religious tourism marketers and destination managers to enhance emotional connections, thereby enriching the memorability and loyalty of visitors to these sacred destinations.

Chapter 3 titled "*Role of religion in engaging religious tourists in pro-environmental behavior*" is written by Nurnobi Islam, Julia N. Albrecht, and Brent Lovelock. This chapter argues religions contain environmental teachings that align with modern sustainability teachings. However, a lack of knowledge about environmental teachings in religion and the separation of religions from our lives are responsible for tourists' irresponsible behavior. To change tourists' behavior in religious tourism destinations, ignorance must be removed. Moreover, coordination in individual, social, and inter-religion is important. The chapter is structured to first discuss the role of religion in shaping environmental ideologies. Second, the method is briefly discussed, followed by a discussion on human-nature relationships, causes of the ecological crisis from religious perspectives, human behavior as guardian of nature, and, finally, a few recommendations for bringing tourists' behavioral changes.

Chapter 4 titled "*Immersive and transformative spiritual experiences: from the inner self to the sense of belonging to a community*" is written by Margarida Franca, Madalena Abreu, and António Azevedo. This chapter aims to discuss the new trends in the design of spiritual, transformative, and religious experiences. These new itineraries are now being positioned as spiritual tourism. This is a new form of tourism, which is becoming increasingly important for academics and organizations alike. Why is religion important? Often faith, religiosity, and spirituality are used interchangeably and inconsistently by scholars. The rise of secularism created a distinction between religion and spirituality. The level of religiosity is the intensity with which an individual believes in a religion and thus affiliates and participates in it, whereas spirituality is being described in contrast, as an individual expression that speaks to the greatest of human capacities. To analyze some experiences of spiritual tourism, particularly on the supply side of this tourism product, this chapter presents several case studies in Portugal of institutions that offer spiritual retreat experiences or other types of offers associated with spirituality.

Chapter 5 titled "*Religious tourism as an educational experience*" is written by Daliborka Puric, Ljiljana Kostic, and Elena Mirela Samfira. The purpose of this study was to examine the possibilities of acquiring educational experience through the developed model of religious tourism. The goals of the study were related to theoretical foundation of the possibility of religious tourism for educational purposes; developing a model of educational-religious tourism that includes historical, artistic, and cultural elements; and pointing out the gap in the literature and creating recommendations for future research. The study design was a theoretical research, focus groups, and survey. The methodology first identified the basic question of acquiring an educational experience as a benefit of religious tourism and then its contribution to the educational experience was examined through the creation of the religious tourism model. The sample in empirical part of the study consists of in-service teachers and students—future teachers from Serbia. The research findings show that organized religious tourism can be applied for educational purposes. The developed model of educational-religious tourism can contribute to the achievement of teaching goals and subject outcomes, but also to the realization of educational intentions of a wider scope, which go beyond subject, class, and even institutional frameworks. At the same time, this model represents only the initial step in achieving indicated goals, and its content and implementation possibilities could be further improved.

Chapter 6 titled "*Exploring the Multifaceted Impact of Religious Tourism: A Multidisciplinary Perspective*" is written by Bapon Chandra Kuri, Ayan Banik, and Marco Valeri. This chapter aims to comprehensively analyze the multifaceted impact of religious tourism from a multidisciplinary perspective. It seeks to explore the diverse dimensions of religious tourism phenomena and their implications for destinations by examining the intersections of marketing, economics, finance, psychology, social development, sustainability, and operations. By delving into the underlying motivations of religious travelers, ranging from spiritual quests to cultural exploration, this research endeavors to uncover patterns of travel behavior and their effects on destination management strategies. Furthermore, it examines sacred sites, considering their roles as religious pilgrimage destinations and tourist attractions. By exploring religious tourism's economic contributions and challenges, such as sustainability and over-tourism, this chapter seeks to provide insights into the management and preservation of religious destinations. Additionally, it explores the psychological aspects of religious tourism, including its impact on individual well-being and spiritual fulfillment, and explore opportunities for social development and community engagement within religious tourism contexts. The chapter adopts a qualitative approach, employing a literature review and case studies to explore the multifaceted impact of religious tourism. It synthesizes findings from diverse disciplines such as marketing, economics, psychology, and sustainability to comprehensively understand the topic. The analysis reveals that religious tourism significantly impacts destinations economically, socially, and culturally. It highlights the importance of understanding the motivations of religious travelers and their implications for destination management. On the other hand, the study identifies opportunities for sustainable development and community engagement within religious tourism contexts, emphasizing the need for multidisciplinary approaches to address this phenomenon's complex challenges and opportunities.

Chapter 7 titled "*Exploring the Dynamics of Religious Tourism: A Bibliometric Analysis of Pilgrimage Sites and Modern Trends*" is written by Chandra Prakash and Manisha Paliwal. The study aims to explore the evolving landscape of religious travel within the global tourist economy, focusing on pilgrimage sites and the motivations behind modern society's inclination towards religious tourism. Utilizing bibliometric analysis from 1989 to 2023, drawing from the Scopus database, the study examines 1368 papers, refining the selection to 838 articles. VOSviewer and Biblioshiny tools are employed to address four research objectives: publication patterns, citations, authorship, and collaborations. The findings reveal diverse academic approaches and multidisciplinary perspectives on religious tourism, covering various methodologies, subjects, applications, geographical focuses, and manifestations of the phenomenon. Notably, recent research extends beyond tourist motives to investigate identities, religious connections, and religiosity. This study contributes to understanding religious tourism by recognizing the significance of structures and activities surrounding religious sites and identifying viable research domains and theme clusters. It sheds light on the expanding quantity of publications in the field, primarily driven by European nations, and highlights areas of underrepresented religious traditions such as Judaism, Hinduism, and Buddhism.

Chapter 8 titled "*A profile analysis of travellers participating in a mass gathering religious event*" is written by Bárbara Sofía Pasaco-González, Francisco Sánchez-Cubo, Ana María Campón-Cerro, and Ricardo Gouveia Rodrigues. Religious events have acquired great relevance among religious tourism, given that they constitute a form of cultural expression that reflects how religion has retained a significant place within society. Consequently, literature has emphasized the need to identify their characteristics. Therefore, from an event management approach, this study aims to define the profile of tourists participating in religious events. Using data from a questionnaire distributed by the authors at 2023 World Youth Day in Lisbon (n = 225), descriptive and quantitative—latent class analysis (LCA)—analyses were performed. The findings reveal that religious events attract visitors with a high level of education who are primarily interested in having a unique religious experience and growing spiritually. Plus, attendees also consider their participation in the event as an opportunity to learn about the destination and the local culture, improving their knowledge about themselves and the world. These results position religious events as a holistic experience framed by the search for benefits derived from both participation in the event and knowledge of the destination.

Chapter 9 titled "*Religious Tourism Phenomena Evolution and Future Research Perspectives*" is written by Kamarun Muhsina. The objective of this study is to examine the historical development, present situation, and future potential of religious tourism, with a focus on its economic conditions, various phenomena, and strategies for progress. The research develops a thorough literature review methodology to track the historical development of religious tourism, starting from ancient pilgrimages to contemporary organized tours. The investigation of current trends and challenges in religious tourism is conducted through the analysis of scholarly articles, reports, and case studies. The study emphasizes notable advancements in religious tourism, such as the influence of technology on visitor experiences, the trend of integrating spirituality with recreational pursuits, and the economic consequences for destination regions. It identifies significant obstacles, such as the impact of global politics and concerns regarding authenticity and commercialization. This study enhances the comprehension of religious tourism by combining historical viewpoints with current patterns and difficulties. This study offers valuable insights to policymakers, tourism stakeholders, and scholars, enabling them to improve the sustainability and cultural authenticity of religious tourism.

Chapter 10 titled "*Tourist destination placement and supported architectural digital tools for cultural pilgrimage tourism—Lefkada island*" is written by Tilemachos Koliopoulos and Marco Valeri. This chapter describes the importance of digital tools for description of safe cultural parts of historical religion monuments with visitors coming to explore several historical sites, heritage sites, and local goods at traditional villages. A useful framework for promotion of local religious tourism cultural architectural elements based on 3D digital representation for associative heritage elements within religious tourism at Lefkada island making useful conclusions for the growth of associative alternative types of tourism including the sustainable religious tourism. In a post-COVID world, cultural religion tourism is likely to adapt and evolve in several ways. Digital architectural designs are presented for 3D historical representations and user useful digital experiences applying proper tools and techniques in 3D simulation world with avatars. The investigated material is to promote safe sustainable architectural designs at outdoors in terms of health and safety, sanitation at the post-COVID-19 era and promotion of local cultural goods, old historical cultural monuments, and alternative tourism activities on Lefkada island. Useful solutions are presented for people with several disabilities promoting old buildings with their history based on maintenance projects for several religious regional building architectural travel destinations. Useful results and sanitary digital drawings are analyzed supporting safe combination between religious tourism, qualitative cultural tourism, green architectural educational tourism facilities, and landscape upgrade on Leukada island. It is also known as Lefkada, which is a popular tourist destination located in the Ionian Sea in Greece.

Chapter 11 titled *"Religious Tourism and Local Development: case study of Samos Island" is written by* Panoraia Poulaki, Theodoros Rachiotis, and Nikolaos Vasilakis. The objective of this research is to investigate the opinions and perceptions of tourist experts regarding the prospects for the development of religious tourism on the island of Samos. The main objective of the research is to make possible the requested utilization of the findings of the empirical research regarding the development of religious tourism in the northern Aegean region. As is well known, the tourist development of island regions is the subject of intense concern due to the geographical peculiarity that characterizes them. The purpose of this research is to develop religious tourism on the island of Samos in the context of local tourism development by highlighting a special region with a rich historical and religious heritage, considering the problems it faces due to insularity and geographical isolation. Religious tourism can be a driver of local economic development, since as an activity it utilizes the cultural capital of the past. The empirical field of the research includes qualitative research using interviews with public and private sector stakeholders of the island of Samos and the North Aegean region, in order to examine the research hypotheses related to the development of religious tourism in the case study. In this research, experts related to the development of religious tourism were selected from the church, the state, and the tourist agencies, to highlight the prospects for the development of this special form.

Chapter 12 titled "*Sustainable Development Practices in Tourism and Hospitality Industry: A Case study of Garhwal Himalayan region of Uttarakhand, India*" is written by Madhu Kumari, Sandeep Guleria, and Suneel Kumar. This chapter aims to assess the sustainable development practices adopted by the tourism and hospitality industry in the study area and to analyze their effectiveness of corporate social responsibility (CSR) strategies for nature conservation and preservation as well as their role in the sustainable integrated development of the area for host communities. The Garhwal Himalayan region of Uttarakhand has long attracted curious tourists and pilgrims from all over the world due to its towering peaks, majestic landscapes, and rich sociocultural resources. Due to these factors, the Garhwal region's tourism and hospitality sector now serves as a major engine for socioeconomic growth. The Garhwal Himalayan region draws visitors who are looking for fun, excitement, spiritual tranquilly, and the numerous natural and cultural benefits of mountains. Tourism and hospitality provide significant economic, business, and employment prospects for the host towns, and they also generate income and profits for the state government and private businesses. The Garhwal Himalayan region is one of the most delicate and sensitive human-environment systems, making ecological carrying capacity a significant barrier to some types of tourism. Tourism and hospitality industry development and promotion in the Garhwal Himalayan region should, therefore, be built around the principles of sustainable development. Issues often linked to current tourism and hospitality industry models in the Garhwal Himalayan region can create environmental harm and pollution, threaten the sociocultural legacy, improperly utilize scarce resources, and even have a detrimental impact on society's externalities. These development dynamics, which include the effects of climate change and natural disasters, are significant change drivers that have an impact on the Garhwal Himalayan region's ability to develop sustainably.

Chapter 13 titled "*A heavenly trail: The Cammino Celeste as a non-place of personal religion for a self-healing religious walk experience*" is written by Jean-Gabriel Bela and Peter Lah. The spiritual component of tourism appears in the great ensemble of cultural tourism, one of the goals of which, in this regard, is only the preservation of places of worship that are part of the landscape and of cultural heritage. It follows that the motivations of the religious tourist can be described in terms of to learn, discover, experience, and consume the tangible and intangible cultural attractions/products in a tourism destination. These attractions/products relate to a set of distinctive material, intellectual, spiritual, and emotional features of a society that encompasses arts and architecture, historical and cultural heritage, culinary heritage, literature, music, creative industries, and the living cultures with their lifestyles, value systems, beliefs, and traditions. Religious tourism is therefore first and foremost cultural tourism. On the booking sites of hotels, asked to indicate the reason for the trip, one's choice is limited to "leisure" or "business." The religious and the spiritual would then fit into the "leisure" box. The problem with this understanding of tourism, namely, presenting the religious and the spiritual only as consumer products, is that it negatively sanctions individuals' quest for meaning. This is therefore the first non-place of religious tourism. It is reduced to leisure as consumer product, in the lexicon of what Marc Augé calls *supermodernity*.

Chapter 14 titled "*Determinants of the development of Religious Tourism in Greece*" is written by Paraskevas F. Tsorvas, Maria Lagou, and Anthi Gousiou. The aim of this chapter is to examine the factors that influence religious tourism in Greece. An empirical study was conducted to identify these factors, which, if properly utilized, can attract tourists, and promote the development of religious tourism in Greece. In addition, a framework for a tourism policy on religious tourism is proposed. This framework includes selected actions for planning, management, and administration. The aim is to emphasize the cultural and economic value of religious monuments in Greece. The methodological framework used in the research is factor analysis, which was based on selected questions which are the variables. For the factor analysis to be effective and reliable, it is tested using the technique of principal component analysis (PCA). For the final solution of determining the component factors, the rotated component matrix (PCM) was chosen to show the two factors with their respective landings on the variable after orthogonal rotation. Finally, linear regression was performed to determine the mathematical model that describes, predicts, and interprets the values of the selected variables in relation to the values of several other variables that influence them. The factor analysis showed that certain variables serve as both facilitators and barriers to the development of religious tourism in Greece. The multiple regression analysis identified independent variables that demonstrate correlations, which can be utilized to create mathematical models for tourism forecasting. The results of the study can also be of use in the formulation of tourism policy for religious tourism. The originality of the research lies in the use of two quantitative methods (factor analysis and regression analysis) to identify the factors affecting the development of religious tourism in Greece.

Chapter 15 titled "*Religion and tourism: management of cognitive, emotional, and spiritual experiences in destination Bulgaria*" is written by Mariya Stankova and Svetoslav Kaleychev. The aim of the research is to explore the potential of selected religious sites and practices in Bulgaria and to arrive at specific recommendations concerning the management of cognitive, emotional, and spiritual experiences in destination Bulgaria. The focus of the present proposal is the tourist destination Bulgaria. For its 1343-year history, the country guards the treasures of previous civilizations from the Balkan region, as well as witnesses to the progress of Christianity in Europe. But also, visiting it, tourists can get to know special examples of Islamic religious art and construction, as well as those related to Judaism. Destination Bulgaria attracts tourists with an interest in visiting religious sites. The research methodology involves a survey of attitudes among visitors to selected sites, an interview with tour guides and representatives at the sites, and an analysis of the results obtained. Formalization as an approach to transforming substantive knowledge is expected to remove the ambiguity of observed phenomena and processes to reach a new understanding of them. The results of the research present the extremely important role of religious tourism in terms of the spiritual development of modern society. The presented research confirms the role of religion as one of the leading motives for carrying out tourist trips. Personal satisfaction, spiritual development, and personal self-improvement are only some of the factors that motivate a huge number of religious tourists to visit religious sites, turning them into centers of world cultural and religious heritage.

Chapter 16 titled "*Religious tourism in Greece: Current status and future development*" is written by Evangelia Kasimati and Vincenzo Asero. The purpose of this chapter is to highlight the importance of religious tourism, as a special form of tourism, in the tourist development of Greece, to identify its current status, to determine the possibilities that exist for its further development and to provide policy proposals for its sustainable growth. The human movement, with the aim of searching for the sacred or the divine, constitutes an important psychological need for every individual, regardless of race, nationality or religion, which entails multiple economic, cultural and social consequences in the places of pilgrimage or religious destination. In the academic literature, this phenomenon is called religious tourism. The main motive that pushes people to make trips of this type is the religious element, which is often combined with other motives that cover their tourist needs. Sacred places or places of divine worship and religious rituals have been destinations for travelers since ancient times. The methodological approach is based on a qualitative analysis, considering secondary data and research work. Our findings contribute to the Greek tourism literature by providing valuable insights for the Greek religious tourism, especially since the body of such academic research for Greece is not adequate.

Chapter 17 titled "*Strategic management in Catholic tourist destinations and their territorial network of stakeholders: Examining their life cycle for the present and future of religious tourism*" is written by Noel M. Muniz, María Jesús Vázquez García, Maribel Rodríguez Zapatero, and Leonor M. Pérez Naranjo. The objective of this work is to analyze how, within the life cycle model for tourism planning, strategic management in Catholic tourist destinations is inherently related to its territorial network of key actors (stakeholders), being crucial for the optimization of the present and future management of religious tourism in a region, and, therefore, of the stages of growth, development, consolidation, and subsequent evolution of its initiatives. To this end, the recent history of strategic planning and networking management in religious tourism is examined. Similarly, a case analysis of catholic religious tourism sites, located in both Spain and Latin America, is carried out, identifying through stakeholder maps the intervention times and potential value generation in territorial destination management while basing on the life cycle model of tourism planning. Finally, together with the current trends in global religious tourism, the capacity of territorial networking management to act as a competitive driver for change in the sustainable and inclusive development of this tourism model has been analyzed. The analysis exposes how the weight of certain key actors in catholic religious tourism sites may accelerate certain stages of the development of tourist destination, while other stakeholders, traditionally considered as pivotal, could come to play a less decisive role given the social transitions experienced by region communities and society in the face of more inclusive and sustainable territorial management.

Chapter 18 titled "*Identity deconstruction through pilgrimage in Lhasa*" is written by Bob Yi-Chen Duan, Yuetian Zhang, Jingjing Dai, and Emily Ma. Most religious tourism involves sightseeing and relaxation, which has not differentiated it from mass tourism or normal holidays. Only a segment of the market and visitors are purely for religious purposes, such as spiritual uplifting, to undertake a pilgrimage trip. Hence, this feature is closely linked to one's identity. For example, self-identity, social identity, and self-congruity theory. Thus, a social psychology approach matches religious tourism. However, identity's role in religious tourism has few observations. Further, in this study context (Tibet), existing studies have also indicated that religion is not a motivation for religious tourism. Cui found Han Chinese seeing the local nature of a pilgrimage but not the religious. Hence, a focus on religious believers (Han Chinese) visiting Tibet for pilgrimage is missing in the literature. This chapter investigates, from an identity perspective, whether religious tourism influences the visitors' identities. Specially, the research questions (RQ) are as follows. RQ1: How do visitors see their identities after the trip? RQ2: How do visitors see local people's identities? In-depth interviews were carried out using snowball and purposive sampling. Data was collected from 12 interviewees, and data analysis was guided by a pre-framework based on identity theory. Findings reveal two aspects of identity. First, tourists' identity and local people's identity seen by tourists (short for local identity after) are two different concepts of identity. Second, a religious trip has deconstructed the inherent identities of tourists as well as locals. First, for tourists, identity appears at the self-identity level, emphasizing how tourists understand themselves. Thus, tourists are more focused on individual-level identities, such as "I am a good person." A social identity is perceived by tourists regarding the local people. This identity highlights how tourists see the local people as different from them (intergroup compassion). In other words, tourists and locals both belong to Buddhism, but the tourists see the locals as different from them. Tourists perceive the local people as having a collective identity (Tibetan) rather than an individual local; A person is an individual, and B is another. Second, the identities of the tourists and the local identities are changed after the trip. Tourists found the Buddhist identity to be hard and not easy to maintain. Some visitors have not reached their ideal identity after the trip (uplifting). For locals, tourists see them as not as pure as they think. For example, tourists think local people are careless about material life and only interested in religious things.

Chapter 19 titled "*Micro-pilgrimage as a ritual of commemorating fighters for freedom in Romania*" is written by Cernicova-Buca Mariana. This chapter is dedicated to examining the attempt of Timisoara activists to establish micro-pilgrimage traditions to commemorate the events leading to the upturning of the communist regime in Romania. After the ban on religious gatherings and pilgrimages during the communist rule, the spiritual life in Romania regained momentum, also reviving the religious pilgrimages. Religious or religious-like manifestations responded to the need of Romanians for solemnity and rituals, to "re-enchant" the public life. In this logic, from 1990 onwards, the places where the opponents to Ceausescu regime clashed with the repressive forces and resulted in deaths became places of remembrance and parts of a ritualistic heritagization of the sites. For more than three decades these places were developed through monument-building and media promotion, rehearsed, and imposed in a route entitled "The route of freedom," offered on demand to tourists but made obligatory as part of the celebrations dedicated to the memory of the 1989 Revolution. The participants, initially thousands, in the late years, hundreds, reunite for prayer, flower laying, and storytelling on a route that encompasses not Timisoara only sites (which give substance to the micro-pilgrimage), but also to other cities in Romania declared martyr cities of the 1989 Revolution, especially to the Crematory in Bucharest, where 47 bodies of 1989 victims were cremated by the repressive regime. This practice is challenged by a city-hall initiative to secularize the route and offer it as a touristic experience without the religious rituals that typically accompany the 1989 remembrance ceremonies, or the storytelling performed by the participants in the 1989 events. The research is based on direct observation of the ceremonial unfolding of the commemorative rituals of the described micro-pilgrimages, supplemented with insights drawn from interviews and with media monitoring. The analysis of media coverage highlights the religious vocabulary used to describe what is in fact a secular initiative: the development of thanatourism or micro-pilgrimage to celebrate places where modern Romanian history encounters losses of human lives in the clash with Ceausescu regime. The history of almost three decades and a half of developing micro-pilgrimage routes in Timisoara shows uneven success of the sites included in the "long" route, some of which are left aside in the city-hall project to establish "the route of Revolution." This is happening despite the efforts to develop such sites with visual objects that can help the heritagization of the places and despite the storytelling that it itself seems compelling.

Chapter 20 titled "*An Overview of Religious Tourism with Special Reference to Shakti Peeths in Odisha, India*" is written by Monalisha Dash· Suneel Kumar and Jyoti. This chapter has tried to arrange the exploration of various Shakti Peeth in a critical manner with significance and strong belief associated to sacred Shakti Peeth in Odisha while analyzing the significance and potentiality of those holy abodes. Suggesting how the community could support towards the religious tourism in contributing revenue generation to the nation economy along with the cultural exchange. Qualitative and quantitative approaches have been initiated to delve into the research finding. It has been found that Odisha has more than 10 numbers of Shakti Peeth and each Shakti Peeth has a unique supernatural event associated with it. Religious tourism refers to the practice of faith-based individuals or groups travelling for spiritual or religious purposes in search of a deeper meaning in life. Therefore, the research's primary goal is to promote these locations and establish them as important tourist attractions. India has venerated the Goddess from the Vedic era. Everything comes from her, according to the Bible. Shakti originates from her. Starting and finishing, she contains the entire creation. In Hinduism, Shakti—sometimes known as "the Great Divine Mother"—is the idea or embodiment of the divine feminine creative force. The Shakti Peethas are among one sacred places also called Shakti Peethas, are important pilgrimage sites, and are holy abode and hallowed shrines where people go to worship Goddess Durga in her many Adi Shakti guises. In order to promote those spiritual sites, accessibility to specific religious locations has been developed, and extensive research on the diverse religious tourist locations across Odisha's regions, along with a clear examination of statistical data, has proven extremely helpful in revealing some remarkable facts regarding landscape of religious tourism in Odisha, with a particular focus on the popularity of Shakti Peeth which stimulate to market complete travel packages.

Part I
Tourism, Religion, and Spiritual Journeys

Religious Tourism: Unveiling the Role of Authenticity and Emotional Solidarity in Enhancing Tourists' Memorable Experiences and Destination Loyalty

Da Van Huynh and Long Hai Duong

Abstract This study explores how objective, constructive, and existential authenticity at religious sites influence tourists' emotional solidarity, which, in turn, enhances memorable experiences and destination loyalty. An analysis of 430 responses (239 domestic and 191 international tourists) showed that each authenticity type significantly impacts emotional solidarity. The findings indicate emotional connection further strengthens both memorable experiences and destination loyalty, with memorable experiences proving essential for loyalty. The multi-group analysis revealed a significant difference in the impact of emotional solidarity on memorable experiences between Vietnamese and international tourists. Theoretically, the findings advance understanding of how perceived authenticity fosters emotional solidarity and loyalty-building experiences. Practical implications are also offered for religious tourism destination marketers and managers to enhance tourists' emotional connections, memorable experiences, and loyalty to religious sites.

1 Introduction

Religious tourism has seen substantial changes over the past few decades and continues to evolve (Collins-Kreiner 2020). This tourism genre is likely to gain increasing popularity post-COVID-19 due to a renewed quest for spiritual meaning, emotional healing, and reconnection with cultural and religious traditions (van Iwaarden and Nawijn 2024; Santus et al. 2023). In other words, many seek the peace and solitude of religious sites in natural settings after prolonged social distancing (Graburn 2024). In this sense, the common aspect of religion can offer a sense of solidarity and shared experience, appealing to those who experienced isolation during the long-lasting pandemic crisis (Jiang et al. 2023). Therefore, understanding

D. V. Huynh · L. H. Duong (✉)
School of Social Sciences and Humanities, Can Tho University, Can Tho, Vietnam
e-mail: hvda@ctu.edu.vn; hailong@ctu.edu.vn

M. Valeri (ed.), *Contemporary Religious Tourism*, Tourism, Hospitality & Event Management, https://doi.org/10.1007/978-3-031-79074-4_2

the needs of visitors to holy places after the pandemic recovery is crucial for destinations to meet their experience expectations (Liutikas 2023).

Authenticity is crucial in religious tourism as it profoundly enriches the spiritual journey, ensuring that tourists' experiences are deeply meaningful, culturally resonant, and emotionally fulfilling (Moufahim and Lichrou 2019). Preserving the historical and cultural integrity of religious sites allows visitors to connect deeply with the sacred traditions and significance of these places (Séraphin and Jarraud 2022). Additionally, genuine interactions with local communities and religious practitioners offer richer, more meaningful insights, enhancing the spiritual journey and fostering a sense of connection and respect (Ruttanavisanon and Agmapisarn 2022). Participating in authentic practices and rituals provides a sense of immersion and devotion, making the religion more impactful (Daskin and Pala 2023). By creating lasting positive impressions, authentic religious experiences help build a loyal community of visitors who are committed to preserving and honouring the spiritual and cultural significance of the site (Bhandari et al. 2022). Therefore, these authentic elements ensure that tourists have a profound and transformative experience, fostering spiritual growth, cultural appreciation, and memorable experiences.

Emotional solidarity, defined as the sense of connection and shared identity between tourists and residents, is essential for creating genuine and enriching experiences (Lan et al. 2021). Although many studies have examined the social connection between locals and visitors and forecasted locals' attitudes towards expansion of tourism, few have focused on the closeness and intimacy of this connection (Hasani et al. 2016). According to Chen et al. (2022), perceived authenticity at a destination may greatly influence tourists' emotional bonds to the locale and their behavioural intentions. However, there is insufficient understanding on the important connection between perceived authenticity and ES in the religious tourism context. As such, more research is needed to explore how religious tourism, with its deep cultural and historical contexts, can nurture these emotional connections and provide memorable experiences for tourists.

Moreover, memorable experiences in religious tourism are essential as they deeply enrich the spiritual journey, fostering a profound sense of fulfilment and connection (Bhandari et al. 2022). Tourists seek not only physical travel but also meaningful encounters that resonate with their faith and personal beliefs (Ribeiro et al. 2022; Valeri 2022). These experiences can enhance spiritual growth, provide a sense of accomplishment, and strengthen religious devotion (Bhandari et al. 2022). Moreover, memorable religious encounters can inspire revisitation and positive recommendations, benefiting destinations economically and culturally (Chen et al. 2020). Although various research has looked into memorable encounters in the setting of pilgrimage travel (Bhandari et al. 2022), continued investigation is critical to grasp the complex nature of spiritual experiences. This understanding can help religious sites meet tourists' expectations, support sustainable tourism, and preserve religious heritage.

Prior literature suggests that understanding the diverse spiritual motivations, travel behaviours, and expectations of various tourist types can optimize site management, increase visitor contentment and support sustainable practices (Bhandari et al. 2022; Ribeiro et al. 2022; Kumar et al. 2024). Despite the growing popularity of religious tourism, there is limited research on this specific experience in the post-COVID-19 pandemic context. Therefore, exploring religious tourism experiences is crucial for effectively tailoring services fulfil the adapting requirements of tourists. To address these gaps, this study attempts to investigate how objective, constructive, and existential authenticity can enhance tourists' emotional solidarity during their visits. Additionally, it seeks to understand the critical social connection between tourists and residents, exploring its impact on shaping memorable experiences and fostering long-term commitment to these sacred places. Importantly, the study investigates the link between memorable experiences and destination loyalty to religious sites. Ultimately, it aims to explore whether different types of tourists (domestic and international) may derive varying outcomes from their religious experiences. This research will provide significant theoretical and practical implications for managing religious destinations in a more sustainable way.

2 Key Aspects of Perceived Authenticity

The notion of objective authenticity, derived from Boorstin's 'pseudo-events', examines the inherent inauthenticity of modern life and tourism. Boorstin (1992) claims this inauthenticity is pervasive in contemporary experiences. On the other hand, MacCannell (1973) argues that people look for genuine travel encounters to compensate for its absence in daily life, often finding only staged authenticity. Despite facing criticisms, MacCannell regards tourism as a meaningful endeavour akin to a religious site (Uriely 1997). Both scholars have been criticized for oversimplifying the motivations of tourists (Mkono 2013). Objective authenticity concerns the genuine nature of tourism objects, such as activities, artifacts, and historical events (Wang 1999). It is determined by the originality or historical accuracy of these objects (Chronis and Hampton 2008) and is typically assessed by experts, as tourists often lack the necessary expertise (Rickly-Boyd 2012). The objectivist view, which sees authenticity as an inherent quality, faces challenges in the ever-evolving modern world (Reisinger and Steiner 2006). Despite its complexity and contentious nature, the concept remains essential in tourism studies, particularly regarding staged authenticity and its influence on tourist experiences and motivations (Mkono 2013).

Constructivists argue that tourists seek symbolic rather than objective authenticity, shaped by societal norms (Wang 1999). They believe tourism objects gain authenticity from their symbolic significance or social construction, rather than their originality (Cohen 1988). This perspective implies that what people perceive as truth is a collective construct, shaped and defined by prevailing ideologies and societal conventions (Chronis and Hampton 2008). Constructive authenticity emphasizes the interaction between tourists and hosts, with authenticity closely tied to the local context (Reisinger and Steiner 2006). It views authenticity and inauthenticity not as opposites but as points on a continuum (Belhassen et al. 2008).

Commodification can positively impact the preservation and revitalization of culture, with traditionally crafted local products often seen as markers of authenticity (Halewood and Hannam 2001). Authenticity is thus seen as the result of a mutual construction process through interactions between tourists and locals, deeply embedded in the local cultural context (Reisinger and Steiner 2006).

Existential authenticity, proposed by Wang (1999), emphasizes personal fulfilment through tourism activities rather than the authenticity of objects. This concept focuses on the individual's authenticity, assessed through personal perception (Reisinger and Steiner 2006). It suggests that even fabricated tourist objects can provide authentic experiences (Wang 1999). Rooted in philosophical ideas about existence and happiness, existential authenticity addresses the emptiness of modern life, with tourism offering paths to self-discovery by shedding everyday roles (Mkono 2013). Wang (1999) categorizes it into intrapersonal authenticity (self-discovery and escaping daily life) and interpersonal authenticity (deepening emotional connections and creating a sense of community, or tourist communitas). However, existential authenticity is transient, context-dependent, and often challenging to integrate into daily life, offering limited and not universally applicable impacts (Knudsen et al. 2016; Reisinger and Steiner 2006).

3 Emotional Solidarity

Emotional solidarity involves a profound sense of mutual understanding, empathy, and support in interpersonal relationships (Joo and Woosnam 2020). This concept traces back to Emile Durkheim's presentation in 1912, 'The Elementary Forms of Religious Life', where he studied aboriginal religious communities (Durkheim 1995). Durkheim's examination of Australian Aboriginal tribes led him to hypothesize that emotional connections among individuals are foundational for uniting them. He emphasized that shared sacred beliefs and ritualistic practices foster deeper emotional bonds within religious groups. Later, Collins (1975) introduced connection as a crucial component towards the development of ES, expanding on Durkheim's findings. This concept encompasses mutual understanding, empathy, support, and community connectedness (Lai and Wong 2022). The framework of ES, initially proposed by Durkheim (1995), was initiated to tourism-related inquiries by Woosnam et al. (2009). They challenged the traditional perspective of tourist-resident connection as fleeting and transactional (Harrill 2004), highlighting that tourists and residents can form enduring and intimate connections. This integration of foundational works led to the contemporary emotional solidarity theory, which is now central to tourism research. Numerous studies have explored emotional solidarity in various tourism contexts, focusing on resident-tourist interactions (Hasani et al. 2016; Lai and Wong 2022).

Prior research suggests that perceived authenticity may significantly impact emotional solidarity in resident-tourist interactions through genuine connections and shared experiences (Chen et al. 2022). According to Taheri et al. (2020), objective authenticity, grounded in tangible and verifiable cultural elements, can provide tourists with a general representation of a destination, enhancing their respect and appreciation for the local community. This mutual understanding and empathy are crucial for developing emotional solidarity. Additionally, constructive authenticity, which stems from the personal meanings tourists attach to their experiences, facilitates meaningful engagement with residents by immersing tourists in local customs and traditions (Nguyen and Cheung 2016). Wu et al. (2019) argue that the engagement with residents might nurture empathy and mutual support. According to Nguyen and Cheung (2016), existential authenticity which is centred on personal fulfilment and self-discovery may promote deeper, more reflective interactions with residents, which may lead to profound emotional connections between tourists and residents. For example, when tourists seek authentic encounters, they tend to be more involved with local cultures and people in a respectful and meaningful way, which may foster a sense of belonging and support for the destination. Therefore, the three key aspects of authenticity have the potential to create fertile ground for emotional solidarity by encouraging genuine, empathetic, and supportive interactions that strengthen the bonds between tourists and residents. As such, the below assumptions are suggested:

Hypothesis (H1): Objective authenticity of a religious tourism destination has positive direct effect on the emotional solidarity between tourists and residents.

Hypothesis (H2): Constructive authenticity of a religious tourism destination has positive direct effect on emotional solidarity between tourists and residents.

Hypothesis (H3): Existential authenticity of a religious tourism destination has positive direct effect on emotional solidarity between tourists and residents.

4 Memorable Tourism Experience (MTE) and Destination Loyalty

The MTE notion has attracted considerable academic interest, shifting the focus from conventional economic transactions to the experiential and emotional dimensions of tourism. (Hosany et al. 2022; Jiang et al. 2023). Researchers have explored MTE in various contexts, including cultural, adventure, ecotourism, and religious tourism (Chen and Rahman 2018; Murad et al. 2022; Albayrak et al. 2018). Key factors influencing MTE include authenticity, novelty, emotional engagement, and personal relevance (Dagustani et al. 2017; Hosany et al. 2022). MTEs are linked to higher satisfaction, increased loyalty, and a greater propensity for future tourism (Chen and Rahman 2018; Kim 2018). While MTE focuses on positive outcomes, negative experiences can also have a lasting impact (Tung and Ritchie 2011). Current research emphasizes the positive dimensions of MTE, based on Kim et al.'s (2012) framework, which includes hedonism, novelty, meaningfulness, knowledge, social interaction, involvement, and refreshment. These elements collectively enhance the overall tourism experience.

Existing literature indicates that there is critical association between MTE and ES (Jiang et al. 2022). In religious tourism context, shared spiritual journeys can foster profound emotional bonds and mutual understanding between visitors and local people (Bhandari et al. 2022). Furthermore, pilgrims who perceive a warm welcome and hospitality from residents and religious sites are likely to strengthen their emotional bonds and form lasting memories during their travels (Jiang et al. 2022). These meaningful and unforgettable connections may foster a sense of community and belonging, vital for maintaining resident-tourist relationships and emotional ties to religious (An et al. 2024) destinations. Therefore, the following hypotheses are posited:

Hypothesis (H4): Emotional solidarity has significant direct impact on tourists' memorable experience during their visit at a religious destination.

Hypothesis (H5): Emotional solidarity has significant direct impact on tourists' loyalty to a religious tourism destination.

Previous literature also indicates that memorable experiences contribute to heightened levels of satisfaction and enjoyment during travel (Hosany et al. 2022). These positive emotions are closely linked to loyalty behaviours such as repeat visitation and positive feedback (Kim 2018). Furthermore, memorable experiences often involve aspects such as authenticity, novelty, and meaningful engagement, which contributes to enhance experience outcomes (Dagustani et al. 2017). In religious tourism, memorable experiences foster emotional bonds and attachment to sacred destinations. These meaningful connections enhance tourists' satisfaction and desire to revisit, contributing to destination loyalty through profound spiritual and personal significance. Therefore, we propose that the memorable experiences at religious destinations may contribute to the enhancement of tourists' loyalty as follows:

Hypothesis (H6): Tourists' memorable experience has significant effect on their loyalty to a religious tourism place.

Building on the theoretical overview, we introduce a theoretical framework and associated hypotheses (refer to Fig. 1):

5 Research Method

The southern region of Vietnam was selected for this research due to its abundance of religious tourism destinations that attract thousands of both domestic and international visitors. The region hosts various spiritual activities, including festivals, rituals, and annual events, drawing numerous participants, religious believers and non-believers who visit sacred sites for authentic and distinctive experiences. This investigation employed a quantitative approach for gathering data. Participants were recruited on-site during religious festivals, ceremonies, and rituals at religious destinations. Data was collected by using both English and Vietnamese questionnaire

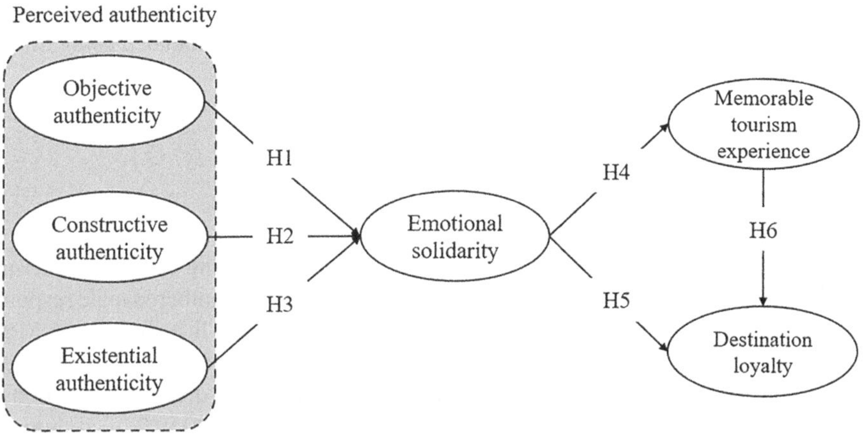

Fig. 1 Research model

for both categories of respondents. To guarantee the accuracy of the translation from English to Vietnamese, two professionals fluent in English and knowledgeable about religious tourism were involved in the process. Only minor revisions were needed to adapt some religious tourism terms to the research context. Then, a pilot study involving 20 tourists who had experienced religious visits to sacred sites was conducted. This step was crucial for improving content validity and preparing the questionnaire for the formal data collection stage. The pilot study resulted in minor feedback, leading to further refinement of the final questionnaire.

The formal questionnaire comprises three parts using a 5-point Likert scale. The first section addresses tourists' motivations for their religious tourism experience, while the third part collects respondents' basic demographic information. Importantly, the second section assesses key variables of perceived authenticity using 20 measurement items adapted from (Zhang and Yin 2020), emotional solidarity using 8 measurement items adapted from An et al. (2024), memorable experience using 4 adapted items from Vada et al. (2022), and destination loyalty using 4 items borrowed from An et al. (2024).

The data was gathered from January to August 2023, coinciding with large gatherings of tourists at annual religious events in the southern parts of Vietnam. These events provided ample opportunities for researchers to gather sufficient data. Respondents in this study were recruited voluntarily, emphasizing transparency and informed consent throughout the process. Each participant received comprehensive information about the project's objectives, the types of data collected, and how their information would be utilized. They were also informed of their rights, including the option to withdraw from participation and have their data removed from the study before a predetermined deadline. This approach ensured that participants could make informed decisions about their involvement and felt confident in the handling of their personal information. After filtering out disqualified responses due to missing data, lack of engagement, or extreme multivariate outliers, the final 430 valid responses were utilized for subsequent analyses. The hypotheses and MGA were tested by Smart PLS while demographic statistics were calculated using SPSS 23 software.

6 Results

This study analysed 430 valid responses from tourists visiting religious tourism destinations in Mekong delta region. Female participants outnumbered male respondents, comprising 70.9% compared to 29.1%. In terms of age distribution, 46.51% were aged 20–30, 25% were aged 30–40, 15% were aged 40–50, 6.6% were under 20, and 6.89% were aged 50–60 or over 60. Participants had diverse occupations: 30.23% worked in office jobs, 23.25% were in service roles, 20.93% were housewives, 18.6% worked in manufacturing, and 6.99% had other jobs. Educationally, 59.3% held a university degree, 20.9% had vocational training or a college education, and 19.8% had graduate degrees. Regarding annual income, more than half (65.11%) earned less than USD$5000, 22.15% earned between USD$5000–10,000, and 12.74% earned over USD$10000.

The study investigated the convergent and discriminant validity of all indicators by examining six key factors and 31 measurement items using component-based PLS-SEM. Additionally, the measurement focused on evaluating construct reliability, validity, and discriminant validity (Table 1). Assessing the outer model is vital for predicting the correlation between each latent construct and the observed variables. Leguina (2015) highlighted the importance of a strong relationship between indicators to guarantee consistency and validate the reliability of a construct. Therefore, the composite reliability (CR) was used to test internal reliability and validity, with a recommended range of 0.6 to 0.7. After removing one measurement item for each perceived authenticity construct that fell below the recommended threshold, the majority of factor loading values (λ) for remaining items exceeded 0.7, thus confirming both convergent and discriminant validity. Additionally, all the values of average variance extracted (AVE) were over 0.5, showing that the constructs were both reliable and valid.

The study also evaluated the discriminant validity by analysing the distinctiveness of the constructs using cross-loading of indicators and the HTMT ratio (Hair et al. 2017). Initially, it was observed that the factor loading indicators for each designated construct were higher than those for other constructs. Moreover, the study found that α- and CR values, as shown in Table 1, are all above 0.7, which is a widely accepted benchmark in social science research (Rönkkö and Cho 2022).

Significantly, the results suggest that this study satisfied the minimum requirement for convergent validity, as all AVE coefficient values were above 0.5. The study further assessed discriminant validity using the HTMT ratio, finding that all values ranged between 0 and 1, thereby meeting the validity criteria (Table 2).

The study also conducted the structural path analysis. According to Hair et al. (2017), in social science research, path coefficients are considered significant if

Table 1 Measurement model

Construct	Item	Loading	Skewness	Kurtosis	α	CR	AVE
OA	OA1	0.797	−1.051	1.173	0.755	0.845	0.576
	OA2	0.726	−0.604	−0.263			
	OA3	0.753	−0.821	0.443			
	0A4	0.759	−1.253	1.294			
CA	CA1	0.798	−1.216	1.241	0.789	0.863	0.612
	CA2	0.791	−1.241	1.275			
	CA3	0.762	−1.418	2.013			
	CA4	0.779	−1.334	1.670			
EA	EA1	0822	−1.293	1.263	0.811	0.877	0.642
	EA2	0.835	−1.286	1.459			
	EA3	0.847	−1.370	1.863			
	EA4	0.700	−1.428	2.187			
MTE	MTE1	0.824	−1.291	1.685	0.859	0.905	0.703
	MTE2	0.858	−1.437	2.125			
	MTE3	0.832	−1.404	2.088			
	MTE4	0.841	−1.318	1.735			
ES	ES1	0.704	−1.335	1.884	0.874	0.901	0.553
	ES2	0.754	−1.381	2.015			
	ES3	0.768	−1.334	1.783			
	ES4	0.782	−1.351	1.905			
	ES5	0.702	−1.396	2.009			
	ES6	0.700	−1.524	2.519			
	ES7	0.701	−1.336	1.657			
	ES8	0.729	−1.384	1.787			
DL	DL1	0.912	−1.375	1.641	0.942	0.958	0.851
	DL2	0.923	−1.269	1.251			
	DL3	0.933	−1.282	1.319			
	DL4	0.923	−1.267	1.250			

Note: *OA* Objective authenticity, *CA* constructive authenticity, *EA* existential authenticity, *MTE* memorable tourism experience, *ES* emotional solidarity, *DL* destination loyalty

Table 2 Heterotrait-monotrait ratio of correlations (HTMT)

Construct	CA	(DA	ES	EA	MTE	OA
CA						
DA	0.730					
ES	0.809	0.645				
EA	0.764	0.571	0.716			
MTE	0.734	0.572	0.621	0.734		
OA	0.835	0.613	0.724	0.623	0.620	
Q^2	–	0.335	0.283	–	0.201	–

t-values exceed 1.65 (at a significance level of 10%), 1.96 (at 5%), and 2.57 (at 1%). In this study, all hypotheses were confirmed (Fig. 2). Specifically, CA exhibited the strongest direct influence on ES, with a standardized coefficient of 0.367, a t-value of 6.142, and a significance level of $p < 0.001$. Similarly, OA significantly influenced ES with a standardized coefficient of 0.222, a t-value of 3.927, and a significance level of $p < 0.001$. Additionally, EA had a significant impact on ES, with a standardized coefficient of 0.271, a t-value of 4.908, and a significance level of $p < 0.001$. Notably, ES significantly influenced both MTE with a standardized coefficient of 0.541, a t-value of 10.597, and a significance level of $p < 0.001$, and DL with a standardized coefficient of 0.437, a t-value of 8.317, and a significance level of $p < 0.001$. Finally, MTE played a significant role in enhancing tourists' destination loyalty, with a standardized coefficient of 0.279, a t-value of 5.913, and a significance level of $p < 0.001$.

The predictive efficacy of the research model was assessed using two key measures: predictive accuracy and predictive relevance. Predictive accuracy evaluates precision within the dataset, while predictive relevance assesses the model's ability to generalize to new data. The coefficient of determination (R^2) was used to measure predictive accuracy, showing that objective, constructive, and existential authenticity collectively explained 54.4% of emotional solidarity, while emotional solidarity explained 29.3% of memorable experience. Both ES and MTE together explained 40% of DL. Predictive relevance was evaluated using blindfolding, with results indicating positive values ($Q^2 > 0$), demonstrating significant predictive capability.

Hair et al. (2010) suggest that relationship between two variables is significant when t-value is above 1.960 and p-values below 0.05. Consequently, the multi-group analysis (MGA) (Fig. 3) indicated no significant differences between international and domestic tourists regarding the effects of objective, constructive, and existential authenticity on emotional solidarity ($t < 1.960$, $p > 0.05$). In a similar

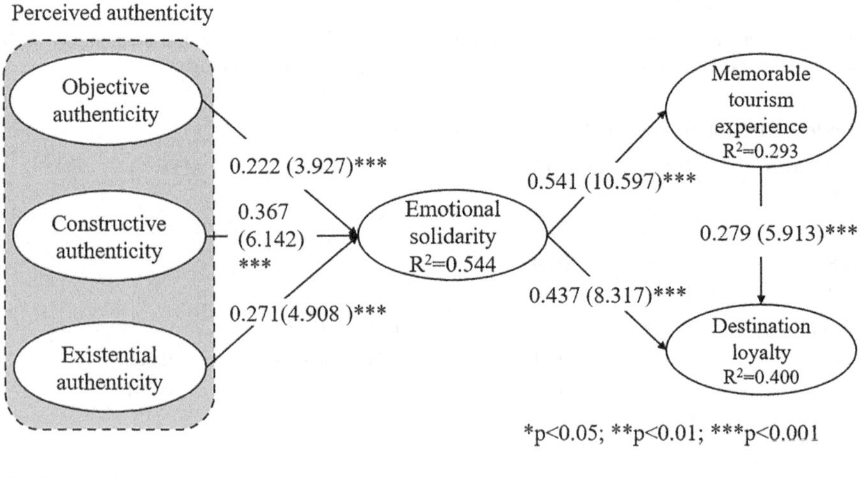

Fig. 2 Results of path analysis

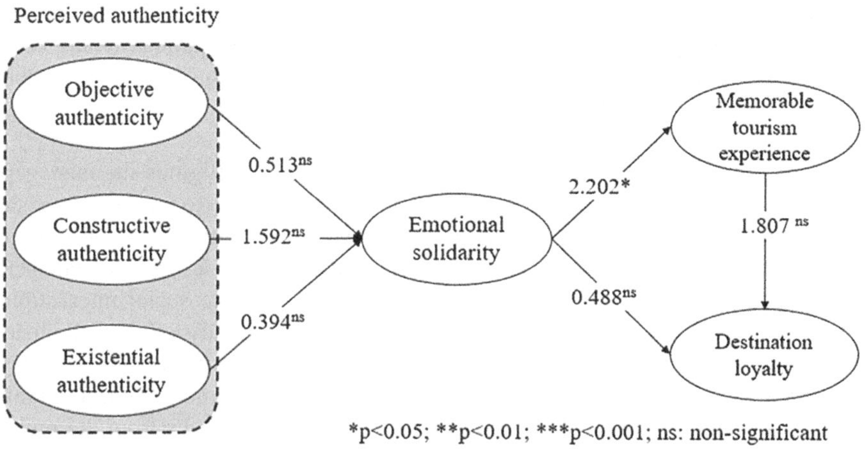

Fig. 3 MGA results (the value on each path represents the *t*-value of the difference between the path coefficients of the models for the international and domestic data sample)

vein, no differences were observed in the impact of ES on DL ($t = 0.488, p > 0.05$) or in the effect of MTE on DL ($t = 1.807, p > 0.05$) across the data samples. Nonetheless, a notable disparity was observed in the influence of ES on MTE across the two data samples ($t = 2.202, p < 0.05$).

7 Discussions and Implications

Despite extensive literature on perceived authenticity in religious tourism (Nguyen and Cheung 2016; Wu et al. 2019), there is limited investigation focusing on its relationship with MTE, ES, and DL, particularly comparing the different perspectives of domestic and international markets. This study confirms the significance of perceived authenticity in shaping tourists' experiences at cultural heritage sites. It highlights how objective, constructive, and existential authenticity significantly influence emotional solidarity. Moreover, findings also confirm that ES enhances MTE and increases DL. Furthermore, the study shows that tourists' memorable experiences during their visit at religious sites are crucial in fostering their destination loyalty.

Theoretically, this study provides further insight into the perceived authenticity within the realm of religious tourism. The findings corroborate previous research, emphasizing the roles of OA, CA, and EA (Nguyen and Cheung 2016; Wu et al. 2019). Notably, this research identifies a crucial link between these types of authenticity and emotional solidarity, thereby expanding on prior literature (Chen et al. 2022). This insight into perceived authenticity offers a deeper understanding of its effect on promoting memorable tourism experiences. Despite limited prior research on how emotional solidarity connects to memorable experiences, the findings in this study highlight that tourists' social interactions with residents contribute significantly to meaningful and lasting memories, expanding prior literature (Jiang et al. 2022). Furthermore, the research enhances our understanding of how memorable experiences at religious sites foster destination loyalty, confirming earlier findings (Ribeiro et al. 2022; Stavrianea and Kamenidou 2022).

Practically, this investigation offers valuable insights for religious site managers and destination marketers aiming to tailor religious experiences to meet the expectations of tourists. The findings reveal that perceived authenticity significantly influences emotional solidarity. Therefore, it is recommended that destination managers and tour operators design religious experiences that promote social interactions between residents and tourists. This approach ensures that tourists feel welcomed, fairly treated, and emotionally connected to the local community (Lan et al. 2021). The findings in this investigation stress the role of ES in promoting unforgettable encounters and promoting destination loyalty. Notably, the results from the multi-group analysis imply a meaningful difference observed in the association between ES and MTE between the two tourist markets. In this sense, it is crucial to enhance the memorability of their experiences to attract more tourists. As such, religious site managers can organize community events and cultural exchanges, such as guided tours, shared meals, and local ceremonies, to deepen emotional engagement between tourists and residents at religious destinations. In other words, marketing strategies should focus on highlighting authentic and interactive experiences. Regularly assessing tourists' needs through feedback mechanisms, such as surveys, is crucial for understanding the different tourist markets and improving their religious experiences. Collaborating with local artisans to create unique souvenirs and training staff at religious sites to foster positive interactions between tourists and locals can further enhance emotional solidarity. By implementing these strategies, destination managers and marketers can create a more engaging, memorable, and loyalty-inspiring religious experience tailored to the diverse preferences of each market.

While this research holds considerable significance, it does have its limitations. Importantly, the sample size of respondents visiting the religious destinations was limited. Prospective studies may increase the sample sizes to better represent the wider population of religious tourism. Additionally, this study employed a quantitative method which could not investigate all the facets of perceived authenticity. Therefore, prospective research should consider using different methods (e.g. mixed methods) to obtain deeper understandings of the complex essence of religious tourism experiences. Furthermore, the study's focus on a few popular religious sites might not be representative of all religious tourism destinations in the southern parts of Vietnam. Future studies should examine religious tourism experiences in various locations and different religions. Exploring different genres of religious experiences can provide a deeper understanding of how perceived authenticity and emotional solidarity vary across different cultural and religious contexts.

References

Albayrak T, Herstein R, Caber M, Drori N, Bideci M, Berger R (2018). Exploring religious tourist experiences in Jerusalem: The intersection of Abrahamic religions. Tourism Management, 69, 285–296

An S, Eck T, Woosnam KM, Jiang L (2024) Applying mindfulness theory to enhance voluntourism experiences. Tour Manag Perspect 50:101203

Belhassen Y, Caton K, Stewart WP (2008) The search for authenticity in the pilgrim experience. Ann Tour Res 35(3):668–689

Bhandari H, Mittal A, Aggarwal A (2022) Memorable religious tourism experience, attitude towards pilgrimage, behavioral intention and the moderating role of religiosity. Co-Editors 76

Boorstin DJ (1992) The image: a guide to pseudo-events in America. Vintage

Chen H, Rahman I (2018). Cultural tourism: An analysis of engagement, cultural contact, memorable tourism experience and destination loyalty. Tourism Management Perspectives, 26, 153–163

Chen X, Cheng Z, Kim G-B (2020) Make it memorable: tourism experience, fun, recommendation and revisit intentions of Chinese outbound tourists. Sustainability 12(5):1904

Chen J, Liu C, Si Y, Law R, Zhang M (2022) A study on the mediating role of emotional solidarity between authenticity perception mechanism and tourism support behavior intentions within rural homestay inn tourism. Behav Sci 12(9):341

Chronis A, Hampton RD (2008) Consuming the authentic Gettysburg: how a tourist landscape becomes an authentic experience. J Cons Behav: An Int Res Rev 7(2):111–126

Cohen E (1988) Authenticity and commoditization in tourism. Ann Tour Res 15(3):371–386

Collins R (1975) Conflict sociology: toward an explanatory science. Academic Press, New York

Collins-Kreiner N (2020) Pilgrimage tourism-past, present and future rejuvenation: a perspective article. Tour Rev 75(1):145–148

Dagustani D, Kartini D, Oesman YM, Kaltum U (2017) Memorable tourism experience. Antecedents and destination image outcome in Indonesia. J Environ Manag Tourism 8(8 (24)):1482–1493

Daskin M, Pala K (2023) Linking emotional solidarity with residents' support for religious tourism development: An empirical research from a cultural heritage context. Kybernetes

Durkheim E (1995) The elementary forms ofreligious life. *Trans. Karen E. Fields.* Free Press, New York

Graburn N (2024) Tourism, pilgrimage and the sacred: at home or away. Ann Tour Res 104:103719

Hair JF, Anderson RE, Babin BJ, Black WC (2010) Multivariate data analysis: a global perspective, vol 7. Pearson, Upper Saddle River, NJ

Hair J, Hollingsworth CL, Randolph AB, Chong AYL (2017) An updated and expanded assessment of PLS-SEM in information systems research. Ind Manag Data Syst 117(3):442–458

Halewood C, Hannam K (2001) Viking heritage tourism: authenticity and commodification. Ann Tour Res 28(3):565–580

Harrill R (2004) Residents' attitudes toward tourism development: a literature review with implications for tourism planning. J Plan Lit 18(3):251–266

Hasani A, Moghavvemi S, Hamzah A (2016) The impact of emotional solidarity on residents' attitude and tourism development. PLoS One 11(6):e0157624

Hosany S, Sthapit E, Björk P (2022) Memorable tourism experience: a review and research agenda. Psychol Mark 39(8):1467–1486

Jiang L, Eck T, An S (2022) A study on the effect of emotional solidarity on memorable tourism experience and destination loyalty in volunteer tourism. SAGE Open 12(1):21582440221087264

Jiang T, Gao J, Zheng X, Liao Y (2023) Engendering prosociality through tourist-to-tourist interaction in religious tourism: an existential transformative perspective. J Sustain Tour:1–17

Joo D, Woosnam KM (2020) Measuring tourists' emotional solidarity with one another—a modification of the emotional solidarity scale. J Travel Res 59(7):1186–1203

Kim JH, Ritchie JRB, McCormick B (2012). Development of a scale to measure memorable tourism experiences. Journal of Travel Research, 51(1), 12–25

Kim J-H (2018) The impact of memorable tourism experiences on loyalty behaviors: the mediating effects of destination image and satisfaction. J Travel Res 57(7):856–870
Knudsen DC, Rickly JM, Vidon ES (2016) The fantasy of authenticity: touring with Lacan. Ann Tour Res 58:33–45
Kumar J, Bayram GE, Sharma R, Valeri M, Basheer S (2024) Talent management practices and the motivation of female employees of five star Hotels in Antalya, Turkey. Int J Tour Policy 14(3):290–309. https://doi.org/10.1504/IJTP.2024.138401
Lai IKW, Wong JWC (2022) Comparing the effects of tourists' perceptions of residents' emotional solidarity and tourists' emotional solidarity on trip satisfaction and word-of-mouth intentions. J Travel Res:00472875221143504
Lan T, Zheng Z, Tian D, Zhang R, Law R, Zhang M (2021) Resident-tourist value co-creation in the intangible cultural heritage tourism context: the role of residents' perception of tourism development and emotional solidarity. Sustainability 13(3):1369
Leguina A (2015) A primer on partial least squares structural equation modeling (PLS-SEM), vol 38. Taylor & Francis, p 220
Liutikas D (2023) Post-COVID-19 tourism: transformations of travelling experience. In: COVID-19, tourist destinations and prospects for recovery: volume one: a global perspective. Springer, pp 277–301
MacCannell D (1973) Staged authenticity: arrangements of social space in tourist settings. Am J Sociol 79(3):589–603
Mkono M (2013) African and Western tourists: object authenticity quest? Ann Tour Res 41:195–214
Moufahim M, Lichrou M (2019) Pilgrimage, consumption and rituals: spiritual authenticity in a Shia Muslim pilgrimage. Tour Manag 70:322–332
Murad NAC, Rahim NA, Ghazali NANM, Khairuddin R, Bakar NA (2022). A Cross-Ecotourism Activities Comparison of Memorable Tourism Experience Among Penang National Park Visitors
Nguyen THH, Cheung C (2016) Chinese heritage tourists to heritage sites: what are the effects of heritage motivation and perceived authenticity on satisfaction? Asia Pacific J Tour Res 21(11):1155–1168
Reisinger Y, Steiner CJ (2006) Reconceptualizing object authenticity. Ann Tour Res 33(1):65–86
Ribeiro MA, Patwardhan V, Woosnam KM, Thirugnanasambantham K (2022) Antecedents to religious festival visitors loyalty: the role of memorable experiences, religious faith, emotion and satisfaction. In: Handbook on the tourist experience. Edward Elgar Publishing, pp 350–368
Rickly-Boyd JM (2012) Authenticity & aura: a Benjaminian approach to tourism. Ann Tour Res 39(1):269–289
Rönkkö M, Cho E (2022) An updated guideline for assessing discriminant validity. Organ Res Methods 25(1):6–14
Ruttanavisanon W, Agmapisarn C (2022) Place attachment and destination loyalty of religious tourists in Chiang Mai, Thailand: examining the role of emotional solidarity and risk perception. Thammasat Rev 25(2):1–25
Santus K, Nafi S, Mallik N, Valeri M (2023) Mediating effect of emotional intelligence on the relationship between employee job satisfaction and firm performance of small business. Eur Bus Rev 35:624. https://doi.org/10.1108/EBR-12-2022-0249
Séraphin H, Jarraud N (2022) COVID-19: impacts and perspectives for religious tourism events. The case of Lourdes pilgrimages. J Conv Event Tour 23(1):15–40
Stavrianea A, Kamenidou I (2022) Memorable tourism experiences, destination image, satisfaction, and loyalty: an empirical study of Santorini Island. EuroMed J Bus 17(1):1–20
Taheri B, Gannon MJ, Kesgin M (2020) Visitors' perceived trust in sincere, authentic, and memorable heritage experiences. Serv Ind J 40(9–10):705–725
Tung VWS, Ritchie JRB (2011). Exploring the essence of memorable tourism experiences. Annals of Tourism Research, 38(4), 1367–1386
Uriely N (1997) Theories of modern and postmodern tourism. Ann Tour Res 24(4):982–985

Vada S, Prentice C, Filep S, King B (2022) The influence of travel companionships on memorable tourism experiences, Well-being, and behavioural intentions. Int J Tour Res 24(5):714–724

Valeri M (2022) New governance and management in tourist destinations. IGI Global Publishing, Hershey PA, USA

van Iwaarden M, Nawijn J (2024) Eudaimonic benefits of tourism: the pilgrimage experience. Tour Recreat Res 49(1):37–47

Wang N (1999) Rethinking authenticity in tourism experience. Ann Tour Res *26:349*

Woosnam KM, Norman WC, Ying T (2009) Exploring the theoretical framework of emotional solidarity between residents and tourists. J Travel Res 48(2):245–258

Wu D, Shen C, Wang E, Hou Y, Yang J (2019) Impact of the perceived authenticity of heritage sites on subjective Well-being: a study of the mediating role of place attachment and satisfaction. Sustainability 11(21):6148

Zhang T, Yin P (2020) Testing the structural relationships of tourism authenticities. J Destin Mark Manag 18:100485

Role of Religion in Engaging Religious Tourists in Pro-Environmental Behavior

Md Nurnobi Islam, Julia N. Albrecht, and Brent Lovelock

Abstract Motivating all kinds of tourists, including pilgrims, to engage in pro-environmental behavior is essential to minimize the adverse impacts of tourism on the environment and progress toward sustainability goals. Religion, despite its massive impacts on human lives, decision-making, and behavior, has not yet received much attention from tourism scholars in this regard. This chapter argues that religions contain environmental teachings, which can be used in promoting tourists' pro-environmental behavior. This chapter has discussed the role of religion in shaping environmental ideologies and the human-nature relationship in religions. Moreover, it has highlighted the importance of overcoming ignorance, sanctification of nature, promoting eco-spirituality, empowering and engaging root-level religious leaders in environmental education, and inter-religion coordination for reducing the environmental footprints of tourism through engaging religious tourists in pro-environmental behavior.

1 Introduction

Tourism is recognized for its substantial socio-economic benefits to a destination, a country, and the global economy. However, these benefits come with significant environmental costs. Tourism is one of the biggest contributors to GHG emissions, accelerating climate change and making the world warmer and disaster-prone. This sector contributes to 8%–11% of GHG emissions globally by creating pressure on water, energy, food, and land (WTTC 2021). If its efforts to outpace volume growth continue as usual, the tourism sector will double its GHG emissions in just another 25 years (Girma 2022). Tourism, in general, is associated with negative environmental outcomes for all kinds of destinations. Environmental impacts of tourism

M. N. Islam (✉) · J. N. Albrecht · B. Lovelock
Department of Tourism, University of Otago, Dunedin, New Zealand
e-mail: nurnobi.islam@postgrad.otago.ac.nz; julia.albrecht@otago.ac.nz; brent.lovelock@otago.ac.nz

M. Valeri (ed.), *Contemporary Religious Tourism*, Tourism, Hospitality & Event Management, https://doi.org/10.1007/978-3-031-79074-4_3

include pollution, waste generation, lack of cleanliness, destruction of natural areas, loss of biodiversity, soil contamination, and overcrowding, affecting destinations worldwide. Religious tourism destinations are no different from other tourism destinations regarding their environmental footprint. During the Hajj pilgrimage, Macca faces adverse environmental consequences, including extensive waste generation, a high level of GHG emissions, pollution, and degraded air quality (Abonomi et al. 2022; El Hanandeh 2013). Over 42,000 tons of waste are generated by more than 3 million pilgrims in a few days, draining natural and economic resources and burdening the authorities to dispose of the wastes (Osra et al. 2021; Qarjouli 2023). Each Hajj pilgrim to Saudi Arabia contributes 60.5 kg of CO2 daily, with air travel being the biggest contributor, and food, lodging, and other activities contribute to around 40% of emissions (El Hanandeh 2013). The degradation poses significant health risks to the pilgrims and the residents (Ahmed et al. 2006). The Saudi Government has a robust awareness campaign to encourage behavioral change in pilgrims, in addition to many greening projects to minimize the environmental footprints of this religious event. The awareness campaigns highlight the religious teachings on environmental stewardship to encourage the pilgrims to behave responsibly (Qarjouli 2023). Similarly, Vatican City, another prominent cultural and religious tourism destination, aims to achieve zero carbon emissions by 2050 through sustainable policies (Pullella 2020). These sustainable initiatives are praiseworthy, but they will bring the desired results only if the tourists cooperate by behaving responsibly. In developing countries, like India, environmental degradation in pilgrimage destinations was reported (Shinde 2011); however, initiatives to minimize the environmental footprints were less common.

Nearly all major religions contain sacred texts and teachings to educate followers on the intrinsic value of nature, environmental stewardship, emotional solidarity, ethical judgments, frugality consciousness, and other ego-driven restrictive behaviors that consider other humans, nonhumans, and the environment (Arbuckle and Konisky 2015; Chavan and Sharma 2024; El-Gammal and Abozaid 2021; Thathong 2012). In sacred places, where tourists come for sacred purposes, their religious commitments are expected to engage them in more obedient and environment-friendly behavior. However, this is not always the case. Like other forms of mass tourism, religious tourism also contributes significantly to environmental degradation (El Hanandeh 2013; Shinde 2011). One of the major causes of this degradation is reported to be tourists' unsustainable behavior (Apollo et al. 2020). Like mass tourists, religious tourists seek high-quality hedonic experiences and engage in secular and hedonic activities (Lochrie et al. 2018). Environmental degradation from tourist behavior in religious tourism destinations poses a pressing challenge to the assumption that tourists' religiosity can contribute to pro-environmental behavior.

Scholars stressed the importance of finding solutions within religions to mitigate environmental crises (Hoffman and Sandelands 2005). The tourism sector itself needs to be conscious of the adverse impacts, as climate change will ultimately impact it economically (Belias et al. 2022). The religious tourism market is expanding rapidly, with more than 330 million tourists visiting spiritual sites annually (UNWTO 2014), highlighting the need for sustainable practices. This chapter argues religions contain environmental teachings that align with modern sustainability teachings. However, a lack of knowledge about environmental teachings in religion and the separation of religions from our lives are responsible for tourists' irresponsible behavior. To change tourists' behavior in religious tourism destinations, ignorance must be removed. Moreover, coordination in individual, social, and inter-religion is important. The chapter is structured to first discuss the role of religion in shaping environmental ideologies. Second, the method is briefly discussed; followed by a discussion on human-nature relationships, causes of the ecological crisis from religious perspectives, and human behavior as guardian of nature; and finally, a few recommendations for bringing tourists' behavioral changes.

2 The Role of Religion in Shaping Environmental Ideologies

Anthropocentrism and ecocentrism are contrasting ideologies that help us understand the relationship between humans and nature. Anthropocentrism views nature as existing for human benefit and supports unlimited human progress by exploiting nature (Mylius 2018). Contrarily, ecocentrism acknowledges the intrinsic value of nature, emphasizing that all living objects are valuable for this ecosystem to function properly (Holden 2003). These ideologies are closely associated with the beliefs of "dominion over nature" and "environmental stewardship," respectively (Leary et al. 2016; Morrison et al. 2015). The dominion concept suggests that humans have the right to exploit nature for their needs, while the stewardship concept suggests that nature is a divine gift. As a vicegerent of God, every human is responsible for its care. Religious stewardship stresses that all creatures have some value for the universe, rejecting anthropocentrism, which prioritizes human interests over those of other species; claims the superiority of human values, capacities, and positions actively or passively; and makes decisions based on this claim (Mylius 2018). Stewardship supports "species egalitarianism," which argues that all species have equal moral standing and command equal respect (Taylor 1986).

The anthropocentrism and ecocentrism ideologies arose from a political debate in the twentieth century in the United States over water needs in San Francisco and a proposal to construct a dam in Hetch-Hetchy Valley in Yosemite National Park (Hoffman and Sandelands 2005). One group took an anthropocentric perspective, valuing nature for human use, progress, and economic gains. They argued for dam construction. The other group took an ecocentric perspective, arguing for nature's intrinsic and spiritual value. They opposed the dam construction. Some scholars criticized that anthropocentric ideology has its roots in religion; particularly Judaic-Christian beliefs are condemned for ecological damage (White Jr. 1967). White Jr. (1967) argued that Christian anti-environmental beliefs have deeply infused the Western culture, fostering "dominion over nature" beliefs and contributing to environmental degradation through Western technologies and egoistic thinking patterns. The thesis of Lynn White Jr. sparked substantial debate on the relationship between religion and nature. Cohen (1985) emphasized the need for substantive evidence to validate the claim. Moncrief Lewis (1970) criticized White Jr. for oversimplifying the relationship between religion and environmental attitude. He argued that capitalism and democratization led to urbanization, population growth, increased wealth in individual hands, and individual resource ownership, directly contributed to environmental degradation. Many subsequent scholars supported White Jr.'s assertion (Arbuckle and Konisky 2015; Taylor et al. 2016). A verse from the Bible, which is often used as evidence to blame Christianity, read:

> Be fruitful and multiply and fill the earth and subdue it and have dominion over the fish of the sea and over the birds of the air and over every living thing that moves upon the earth. (Genesis 1: 8)

This verse led to the belief that human beings were given superiority over other living and nonliving creatures (Kay 1989; Morrison et al. 2015). Bruce (1983) argued that the command to subdue the earth in the Bible does not mean to exploit it; rather, it is a command to cultivate the earth and transform the wild areas into usable lands. The author advocated a cultural-linguistic reading that situates biblical narratives within their original ritual and communal contexts. This approach will enable the biblical text to be understood as shaping moral communities and practices rather than just providing isolated proof texts. In response to White Jr.'s criticism, other scholars defended Christianity by advocating the biblical teachings on environmental stewardship (Van Dyke 2013). Environmental stewardship counters the interpretation of dominion. Jenkins (2009) questioned White Jr.'s legacy and asked for an alternative way of inquiry in connecting environmental problems to religious deliberations. Gulick (1991) argued that biblical texts and teachings inform and inspire eco-spirituality that fosters deep respect toward nature. Moreover, environmental beliefs vary across and within religions. Therefore, Christianity cannot be blamed indiscriminately. Many verses in the Bible support environmental stewardship. A biblical verse read:

> The righteous care for the needs of their animals. (Proverbs 12:10)

This verse indicates that the Bible commands people to be sympathetic to other creatures. Other religions, such as Islam, Hinduism, and Buddhism, advocate environmental stewardship, rejecting the dominion over nature views. Islamic view of Khalifa is aligned with stewardship responsibility (Abdelzaher and Abdelzaher 2015). Hinduism and Buddhism promote Ahimsa (nonviolence) for nature (Ottuh 2020). Taoism emphasizes the harmony of all human and nonhuman things (Holden 2003).

Despite their debates, scholars agreed that solutions to environmental problems should come from religion. White Jr. (1967) asserted that science and technology alone cannot resolve the present ecological crisis and suggested that the remedy must be religious. Other scholars also stress that by transforming mindsets and environmental beliefs and developing moral obligations and a respectful attitude toward nature, religion can contribute to sustainable behavior across individuals (Chavan and Sharma 2024; Gulick 1991; Hoffman and Sandelands 2005; Ives et al. 2023; Ottuh 2020; Rice 2006). This discussion of environmental ideologies is very relevant to tourism because anthropocentric ideologies have long influenced its development (Zhang et al. 2020). For example, the development of wildlife tourism, which restricts animals' freedom to prioritize human hedonic pleasure, reflects anthropocentric values. Although environmental ethics have been discussed in tourism, they have not achieved the desired environmental outcomes. This happened because the discourses took place in an anthropocentric framework where tourism stakeholders' economic and social interests were prioritized over nature's rights in the environmental discourses (Holden 2003).

3 Method of Study

This research used the content analysis method. Content analysis is a systematic approach to analyzing and interpreting various forms of communication, such as verbal and visual data (Harwood and Garry 2003). The data used for this research include verses (texts) from the Quran and the Bible. The relevant textual data were collected from research papers, websites of faith-based environmental organizations (e.g., Green Faith), news articles, blogs, and reports. The textual data collected from these sources were cross-checked through electronic versions of the Quran and the Bible before including them for the analysis. After collecting the sacred texts from different sources, they were analyzed and coded. The unit of analysis was sentences and phrases. The patterns and relationships among coded data were analyzed, and then the data were organized into sub-themes and main themes. A summary of the data analysis process is included in Appendix 1.

Appendix 1 A Summary of the Data Analysis Process

Theme	Sub-theme	Code	Verses from the Quran and Bible
Human-nature relationship	Stewardship responsibility	Creation of humans as an authority on the earth; be fruitful to nature; humans are tenants and will be made accountable; caring creatures is righteousness.	"And when the Lord said to the angels, "Indeed, I will make upon the earth a successive authority." They said, "Will you place upon it one who causes corruption therein and sheds blood, while we declare your praise and sanctify you?" Allah said, "Indeed, I know that which you do not know" (The Quran, 2:30). "Be fruitful and multiply and fill the earth and subdue it and have dominion over the fish of the sea and over the birds of the air and over every living thing that moves upon the earth" (Genesis 1:28). "The land cannot be sold permanently because the land is mine and you are foreigners—you're my tenants. You must provide for the right of redemption for any of the land that you own" (Leviticus 25:23–24). "The righteous care for the needs of their animals" (Proverbs 12:10).
	Sanctification of nature	God is the creator of the earth and the heavens; God created all living and non-living objects; all creatures pray to God.	"Your Lord is God, who created the heavens and earth in six days, then established Himself on the throne..." (The Quran, 7:54). "The earth is the Lord's, and everything in it, the world, and all who live in it; for he founded it upon the seas and established it upon the waters" (Psalm 24:1–2). "All the earth bows down to you; they sing praise to you; they sing the praises of your name" (Psalm 66:4).

(continued)

Theme	Sub-theme	Code	Verses from the Quran and Bible
Causes of ecological crisis	Human's unsustainable behavior	Humans are responsible for wrongdoings; greed causes anarchy, caution against greed, and temptation/greed/ desire misleads.	"Mischief (corruption) has spread on land and sea due to what people's hands have done" (Quran, 30:41). "And you devour others inheritance greedily, and love wealth fervently. Enough! When the earth is entirely crushed over and over" (Quran 89: 19–21). "But those who desire to be rich fall into temptation, into a snare, into many senseless and harmful desires that plunge people into ruin and destruction" (Timothy 6:9).
Prescribed human behavior as guardian of nature	Discouraged behavior	Avoid overconsumption; the creator hates wasteful behavior; wastefulness is the act of devils; discouraged corruption; discouraged pollution; discouraged arrogance; reminding humans about wrongdoings.	"O children of Adam! Eat and drink but be not excessive. Indeed, He likes not those who commit excess" (the Quran, 7:31). "Indeed, the wasteful are brothers of the devils, and ever has Satan been to his Lord ungrateful" (The Quran, 17:27). *"Eat and drink from the provision of Allah* and do not commit abuse on the earth, spreading corruption" (The Quran, 2:60) "Do not pollute the land where you live" (Numbers 35:33). "Do not strut arrogantly on the earth. You will never split the earth apart nor will you ever rival the mountains' stature" (The Quran, 17: 37). "I brought you into a fertile land to eat its fruit and rich produce. But you came and defiled my land and made my inheritance detestable." (Jeremiah 2:7).
	Encouraged behavior	Call for humbleness, compassion, and peacefulness.	"The 'true' servants of the most compassionate are those who walk on the earth humbly, and when the foolish address them 'improperly', they only respond with peace" (The Quran, 25:63).

4 Findings and Discussion

4.1 Human-Nature Relationship in the Sacred Books of Religion

Religion prescribes some code of conduct that informs us what type of behavior is expected or not by God, religious community or institutions, or society. Our perception of, attitude toward, and behavior in particular places and with objects (living and nonliving) are shaped by religious beliefs. People who believe in religions will be less likely to believe they are exempted from their religious duties when travelling, making religious teachings about nature significant for promoting pro-environmental behavior in tourism. Abrahamic religion (Islam, Christianity, and Judaism) approaches nature from a stewardship perspective. The Quran and the Bible explicitly mention the importance of nature for human existence. Hundreds of verses in the Quran are related to nature, specifically plants, animals, water, air, land, and other natural resources (Helfaya et al. 2016). The Quran and Bible sanctify nature, mentioning that the universe is the property of the almighty creator; everywhere in it, there are signs of the almighty creator. A verse from the Quran read:

> Your Lord is God, who created the heavens and earth in six Days, then established Himself on the throne…. (The Quran 7:54)

A similar verse from the Bible read:

> The earth is the Lord's, and everything in it, the world, and all who live in it. (Psalm 24:1)

A biblical verse suggests that other creatures pray to God as humans do. It calls for the moral responsibility of humans to take care of other worshippers and avoid exploiting them. A verse from the Bible read:

> All the earth bows down to you; they sing praise to you, they sing the praises of your name. (Psalm 66:4)

Religion entrusted humans with stewardship responsibilities. According to the Bible, God has given human beings intellectual superiority over other creatures (Genesis 1:26) and commanded human beings to take care of his creations (Genesis 2:15). Similarly, the Quranic verses entrust human beings with the responsibility of this earth:

> And when the Lord said to the angels, 'Indeed, I will make upon the earth a successive authority.' They said, 'Will You place upon it one who causes corruption therein and sheds blood, while we declare Your praise and sanctify You?' Allah said, 'Indeed, I know that which you do not know.' (The Quran 2:30)

The religious texts provide important motivations or ethical grounds for humans to be cautious in interacting with nature. Islam encourages humans to take responsibility for stewardship through commitment to heavenly rewards. The Bible also warned against the actions that keep other creatures at stake (Jeremiah 2:7; Numbers 35:33). Prophet Muhammad (PBUH) said,

> There is none amongst the Muslims who plants a tree or sows seeds, and then a bird, or a person or an animal eats from it but is regarded as a charitable gift for him. (Al-Bukhari 2320)

A few verses (Geneses 1:26–28) from the Bible that support the human's intellectual superiority over other creatures and stewardship responsibility are often misinterpreted and wrongly perceived by environmentalists, who claim that religions contribute to environmental degradation (White Jr. 1967). However, in the historical context, these verses did not mean to exploit nature but to reproduce and cultivate the land, take care of it, and responsively manage it (Bruce 1983; Kay 1989). The enlightened persons will hardly be convinced that the dominion mentioned in the Bible is the same as the dominion of a ruthless person who deals harshly with others to achieve his desires and selfishness (Bruce 1983). Other verses also prove that humans are not given the absolute power to exploit or misuse nature. Because humans are vicegerents, they will be made accountable for their actions on the earth. A verse reads:

> The land cannot be sold permanently because the land is mine and you are foreigners—you're my tenants. You must provide for the right of redemption for any of the land that you own. (Leviticus 25:23–24)

The Bible suggests that God will never destroy living creatures (Genesis 8:21). As God has taken responsibility for protecting nature, humans should do the same. Since God created the universe and everything within it, every creature within this universe has value to God. Thus, every creature should be treated with care and respect.

4.2 *The Causes of Ecological Crisis from Religious Perspectives*

In prehistoric times, it was taught that natural disasters were sent from God as a result of human misdeeds (Gulick 1991). The concept of Karma in Buddhism and Hinduism suggests that every human action, evil or good, has its consequences; so, an individual has to face the evil consequences of his misdeeds (Ottuh 2020; Sponsel and Natadecha-Sponsel 2003). Buddhism blames greed, aversion, and selfishness as the root causes of this spiritual crisis (Thathong 2012). Thus, religions hold human actions responsible for environmental degradation and imbalance in the ecosystem. A verse from the Quran read:

> Mischief (Corruption) has spread on land and sea due to what people's hands have done. (The Quran 30:41)

The Bible also considers greed as the root cause of all evils (Timothy 6:9). Wasteful behavior is considered a kind of corruption and a sinful act (Haram) in Islam (Abdelzaher and Abdelzaher 2015; Filimonau et al. 2022). Greed is a heinous in Islam as well. A few verses from the Quran read:

> And you devour others inheritance greedily, and love wealth fervently. Enough! when the earth is entirely crushed over and over. (The Quran 89: 19–21)

The above verse suggests that greed causes humans to disrespect others' rights (both human and natural), destroy others' properties, and create anarchy. Greed and other negative human qualities, like selfishness, aversion, and lack of self-control, are embedded in the materialistic lifestyle. Materialistic individuals lead luxurious, comfortable, and wasteful lives. They focus on egoism or self-interest, excessive consumption, and the practice of throwaway culture (Evers et al. 2018). They exploit natural resources unsustainably. Such behavior reflects a disregard for nature's intrinsic value and future generations' well-being. A verse from the Bible read:

> But those who desire to be rich fall into temptation, into a snare, into many senseless and harmful desires that plunge people into ruin and destruction. (Timothy 6:9)

Materialism also reduces subjective well-being (Csutora and Zsóka 2014) and may propel people into depression, aggression, violence and other mental health issues (Capstick et al. 2022), which can lead to harmful behavior toward nature. Opposite to materialism, frugality emphasizes simplicity, balance, and avoiding overconsumption. All major religions discourage materialism and encourage frugality in life. Religious teachings emphasizing compassion, love, care, simplicity, etc. are against materialistic values.

According to Buddhism, ignorance causes human suffering, including environmental problems. Studies show that not all religious people are conscious of environmental issues; 58% of highly religious Christians do not believe that climate change is very serious (Alper 2022). Another study suggests that religious consumers are less aware of the importance of environmentally protective behavior, such as recycling (Arli et al. 2021). Being a follower of religion does not guarantee that an individual is aware of his doctrinal teachings about the environment. For example, many non-Arabic-speaking Muslims recite the Quran in Arabic without understanding its meaning. They recite the Arabic Quran because it is sacred to them, but the translated version is not. The Quran is unlikely to change their behavior when they do not understand its meaning. For example, although Islam prohibits wasteful behavior, many Muslims are not sufficiently aware of this teaching and engage in wasteful behavior (Filimonau et al. 2022), indicating that awareness of doctrinal principles about environmental behavior is more important than ritual participation.

4.3 *Human Behavior as Guardian of Nature*

The Quran and Bible discourage excess consumption, waste, and activities such as crime or corruption that destroy resources. In Islam, humans are not allowed to pollute the environment, misuse natural resources, or adopt overconsumption (Abdelzaher and Abdelzaher 2015; Helfaya et al. 2016). Religious sacred texts compared the wasters with the evils and termed them ungrateful. The Bible also suggests that nature should not be abused or overused (Leviticus 25:1–7). Islam informs humankind that the Almighty does not love the wasters. A verse from the Quran read:

> O children of Adam! .... eat and drink, but be not excessive. Indeed, He likes not those who commit excess. (The Quran, 7:31)

According to Abdelzaher and Abdelzaher (2015), humans can fulfil their environmental responsibilities by seeking balance in activities, avoiding overconsumption, practicing accountability, and acknowledging interdependence with other creatures. Any negligence in fulfilling their responsibilities or human activities that endanger nature is synonymous with ignoring the responsibilities and disregarding the commands of Allah. Any ungrateful behavior is punishable in the life hereafter in the eyes of religion. A verse from the Quran read:

> Indeed, the wasteful are brothers of the devils, and ever has Satan been to his Lord ungrateful. (The Quran 17:27)

The Quran also informs human beings that the world has been created in proportion or measure (The Quran 54:49), that the beauty of nature (The Quran 6:99), that the similarities of creations (The Quran 6:38), and that nature is a blessing for human beings (The Quran 16:65) and instructed human beings not to commit any crime or abuse on the earth (The Quran 2:60). A few verses from Quran that instructed human beings not to create havoc on the earth and be arrogant on the earth read:

> Eat and drink from the provision of Allah and do not commit abuse on the earth, spreading corruption. (The Quran 2:60)

A similar verse from the Bible discourages activities that harm nature:

> Do not pollute the land where you live. (Numbers 35:33)

Religion prohibits arrogancy, which is one of the vital causes of corrupt or wasteful behavior, and encourages compassion, gratitude, peace, and humbleness. A verse from the Quran read:

> Do not strut arrogantly on the earth. You will never split the earth apart nor will you ever rival the mountains' stature. (The Quran 17:37)

Similar verses from the Bible read:

> I brought you into a fertile land to eat its fruit and rich produce. But you came and defiled my land and made my inheritance detestable. (Jeremiah 2:7)

Another verse from the Quran encouraging humbleness read:

> The 'true' servants of the most compassionate are those who walk on the earth humbly, and when the foolish address them 'improperly', they only respond with peace. (The Quran 25:63)

These religious teachings have implications for our consumption behavior both in home and tourism contexts. Religions direct us not to overuse but to conserve natural resources, including water, electricity, food, etc., and not to pollute, kill, or harm others, which are the central ideas of sustainability. The Quran and the Bible teach us to be humble and avoid arrogance.

Buddhism and Hinduism also acknowledge the intrinsic value of nature. The central teachings of Buddhism are simplicity, wisdom, compassion, nonviolence, mindfulness, morality, and selflessness (Sponsel and Natadecha-Sponsel 2003; Thathong 2012). Hinduism sanctifies nature, viewing it as inseparable from the divine. Nature-friendly actions such as planting trees have been encouraged in Hindu scriptures (Coward 2003). Other religions, such as Taoism, Confucianism, Sikhism, Jainism, and Indigenous traditions, also promote the harmonious relationship between nature and humans. In tourism, tourists' unsustainable behavior, such as food waste, pollution, overconsumption and degradation of natural resources, cruelty toward animals, high levels of GHG emissions from extravagant lifestyles, and other forms of moral disengagement, worsens ecological crises. Religious environmental teachings are powerful resources that can be used to deal with tourists' unsustainable behavior.

5 Effectiveness of Religious Teachings in Pro-Environmental Behavior

Research shows that religious teachings can contribute to environmental concerns and pro-environmental behavior. Gohary et al. (2022) found that activating the concept of God in tourists' minds increases their environmental concerns. This research found that benevolent views of God reduce environmental concern, and punitive views of God have the opposite effect. Care should be taken in positioning God in tourists' minds. Punitive views of God may risk weakening the relationship between tourists and religion. Religion encourages frugality in life and discourages a materialistic lifestyle. The research found that frugality consciousness is positively associated with the willingness to act in an environment-friendly way (Wang et al. 2020). Contrarily, materialism is negatively associated with pro-environmental behavior (Indaryani et al. 2023). Religious teachings of compassion, sympathy, and gratitude are found to be associated with a pro-environmental attitude and behavioral intention (Allen and Ferrand 1999; Robina-Ramírez and Pulido-Fernández 2019). Religions teach personal and social norms, which also contribute to religious tourists' pro-environmental behavior (Kala and Chaubey 2023; Ullah et al. 2024). In sum, the research findings suggest religious teachings, if followed in life, can help engage followers in pro-environmental behavior in both home and tourism contexts.

6 Recommendations

Effective engagement in pro-environmental behavior requires awareness of environmental teachings, the sanctification of nature, and adoption of eco-spirituality at the individual level, the empowerment of religious leaders and environmental education at the community level, and interfaith coordination on a larger scale.

6.1 Overcoming Ignorance

Getting free from ignorance is an important requirement of spirituality. Ignorance leads individuals to commit mistakes. A verse from the Hebrew Bible read: *"My people are destroyed from lack of knowledge. Because you have rejected knowledge, I also reject you"* (Hosea 4:6). Religions do not prohibit tourism, travelling, and leisure activities but prohibit behaviors or activities that can put other people, animals, plants, and the planet in trouble. Although tourists seek a break from daily monotonous life, they are still bound by religious duties, including respect for nature. A precondition to performing religious duties is the awareness that religions emphasize interdependence with nature, emotional solidarity, sustainable living, and stewardship. Religious awareness can help people differentiate between right and wrong behavior, develop a deeper understanding of their actions and potential impact on others, and change or weaken the effect of anthropocentric beliefs. A lack of awareness and environmental knowledge among adherents often leads to behaviors that do not reflect these values (Alper 2022). Climate change skepticism, which can have an association with religious fundamentalism, may stem from a lack of religious knowledge (Alper 2022; Muñoz-García 2014; Preston and Shin 2022). Raising awareness about environmental ethics and stewardship beliefs can correct this skepticism. Social media is an effective platform for raising environmental awareness and activating environmental norms among tourists. Environmental moral education in educational institutions and home can foster awareness of stewardship and environmental citizenship behavior among future generations (Liu et al. 2024; Siagian et al. 2023).

6.2 *Sanctification of Nature*

Almost all major religions consider nature as sacred. Reinforcing the concept of the sanctification of nature and environmental stewardship in individuals' minds is crucial to changing their dealings with nature. Individuals sacrifice their money, time, energy, and comfort for sacred matters (Pargament and Mahoney 2005). Sanctification of nature can elicit spiritual emotions such as love, gratitude, solidarity, etc. for nature. It may also trigger a feeling of eco-guilt when religious people feel that they cannot revere sacred nature sufficiently. Eco-guilt motivates individuals to correct their wrong behavior toward nature and engage more in ecologically responsible behavior (Ullah et al. 2024). Individuals who sanctify the environment to a greater degree are more likely to hold pro-environmental beliefs and engage in environmentally protective behaviors (Tarakeshwar et al. 2001). Religions promise divine rewards for environmental behavior. For example, Islam considers planting a tree an act of charity (Sadaqah). In Islam, divine reward is promised for charitable acts. Contrarily, divine retribution was promised for deviations. Religious beliefs can provide a moral motivation to reinforce a positive attitude toward nature. Thus, a motivation-based teaching approach, such as divine reward and retribution beliefs, may be fruitful in making religious people aware of their environmental footprints and behavior in tourism destinations (Teng et al. 2023). Environmental teachings in religions differ from secular teachings in that religions establish a link between the awareness of the divine, such as rebirth, divine rewards, punishments, etc., and worldly behavior. Therefore, a religious approach can change religious individuals' environmental attitudes and behavior in tourism destinations regardless of their spiritual or secular importance.

6.3 Promoting Eco-Spirituality

A life filled with simplicity, nonviolence, self-control, emotions, morality, accountability, wisdom, and responsibility is a religious pathway to solving environmental problems. These human qualities are embedded in a lifestyle called eco-spirituality. Eco-spirituality is a form of spirituality that takes a biocentric lifestyle rather than an ego-centric and extravagant lifestyle. Eco-spirituality emphasizes commitment to moral standards, stewardship responsibility, and inner power, including self-control and emotional solidarity with nature (Gulick 1991). Two important characteristics of eco-spirituality are that nature is considered sacred and worthy of respect and that connectedness with nature is emphasized (Taylor 2007). Spiritual or religious individuals take stewardship responsibility, which involves love, care, and gratitude toward nature; practice self-control, which involves nonviolence, frugality, simplicity, and repentance; and are committed to moral standards, which involves environmental ethics (Garfield et al. 2014; Omoyajowo et al. 2023). By enhancing connectedness with God, nature, and self, eco-spirituality can contribute to behavioral changes (Burkhardt 1989). Therefore, the adoption of eco-spirituality by individuals is paramount to enhancing deep respect toward nature and reducing environmental footprints at individual levels.

6.4 Empowering and Engaging Root-Level Religious Leaders

The root-level religious and community leaders are more socially attached to their followers than the top-level religious leaders. They are the first to contact their followers regarding religious or social issues. They have more influence on the masses in their communities, because of their social presence and close interactions. A study found that more than 60% of the respondents take environmentally conservative measures more seriously if their religious leaders give them spiritual teachings (Omoyajowo et al. 2023). Another study found that religious leaders are more effective than municipal bodies in engaging ethnic minority groups in recycling (Lakhan 2018). The involvement of local religious leaders in environmental discourses will make it easier to change individuals' behavior. Top-level religious leaders, such as Pope Francis, have been recently involved in environmental awareness campaigns. Coordination among different layers of religious leaders is important to prioritize environmental issues in all religious discourses. The local tourism organizations and environmentally concerned groups can work closely with the root-level religious leaders and engage them in disseminating religious teachings on environmental protection. The environmental crisis has reached such a high level that this issue should be a priority in day-to-day religious discussion in all layers of society.

6.5 *Inter-Religion Coordination*

Tourism is a global industry. Tourism destinations are places of interest to visitors from different religions. Some religious tourism destinations are sacred places to the followers of different faiths. For example, Jerusalem is a place of religious importance to the followers of Islam, Judaism, and Christianity. Coordination among different religions is important to achieve environmental sustainability in tourism, specifically religious tourism. One religious group believing in "dominion over nature" and another believing in "environmental stewardship" will not bring desired environmental outcomes in tourism. Rather, differences in environmental ideologies may put religious tourism destinations at risk of conflicts, vandalism, and degradation. Developing cultural sensitivity is also important to avoid conflicts. Different religions have recently taken initiatives to greenish their religions (Konisky 2017; Xie 2019). The greening process has occurred in two ways. First, traditional religions undergo the greening process. In this process, the environmental ethics that remained out of central importance in traditional religions are now being rediscovered, and, second, the emergence of new forms of green religions whose followers express solidarity, reverence, and respect toward nature without believing in the divinity (Xie 2019). The first form of greening process of traditional religions has not yielded the desired results yet (Taylor et al. 2016). Christians in America tend to show less concern about global warming and environmental pollution. Over the last two decades, their concern for environmental issues has declined (Konisky 2017). This can be a consequence of secularizing effects influenced by conservative politics, which is linked to the declining influence of religions on Christians in America (Perry 2024). An interdisciplinary coordination between religion and politics is also required. Without greening politics, attempts to greenish religions may not bring behavioral change in religious followers. Coordination among different religions will help enhance our respect for religions and share knowledge to minimize environmental footprints through behavioral change. Coordination is important to promote ecocentric environmental ethics that every religious individual will practice regardless of religious affiliation. The different faith-based organizations that promote environmental stewardship should also be involved in the process.

7 Conclusion

This research found that religious books discuss human-nature relationships and the causes of environmental crises. The sacred books also prescribe humans' responsible behavior that aligns with modern-day sustainability teachings. Our discussion reveals that religions sanctify nature and provide humans with the responsibility to protect it. Lynn White's criticism of Judeo-Christian religion is too generalized. Research suggests that environmental concerns vary across different denominations of Judeo-Christianity (Arbuckle and Konisky 2015). However, the merit of Lynn

White's criticism of religion is that it instigated thoughts about environmental issues among religious institutions and leaders. The subsequent efforts toward greening religions are positive outcomes of this criticism. This research found that religious sacred books do not teach people to dominate nature indiscriminately but promote stewardship. People who believe in stewardship show pro-environmental behavior (Eom et al. 2021; Leary et al. 2016). It is a misperception that religion promotes dominion beliefs. The biblical verse instructing humans to subdue nature might have motivated humans in prehistoric times to develop confidence, take stewardship, and coexist in the wild environment. Contrary to Lynn White's claim, we claim that ignoring and not understanding religions deeply, but not actively following them is responsible for irresponsible behavior toward nature.

According to religions, human behavior, specifically greed, animosity, corruption, lack of compassion, selfishness, wastefulness, ignorance, etc., causes environmental degradation. Attribution of responsibility (for climate change) to God is not the faith but contrary to religious faith. Modern-day sustainability teachings also emphasize overcoming these negative behaviors and traits that religions prohibit. Religion encourages empathy, compassion, connectedness, peacefulness, frugality, and other positive traits to deal with social and environmental crises. Implementing the teachings of sacred books in life is more important than merely reciting them. The main purpose of these sacred books is to guide human behavior in worldly life so that peace prevails. Unconsciously reciting these sacred books does not serve the purpose of God. Human's obedience, not unconscious reading of sacred texts, is associated with the rewards in the life hereafter, as religions promised. Practicing environmental teachings is possible if the followers understand and become aware of the environmental teachings. Indeed, religions could not sufficiently engage their followers in pro-environmental behavior, as evidenced by the growing rate of environmental degradation. One reason is the lack of knowledge about environmental teachings embedded in religions among religious individuals. Being religiously committed, such as joining religious rituals and believing in God, does not necessarily mean an individual is aware of religious environmental ethics. Misperception of religious teachings, such as dominion over nature beliefs, is another reason for an anthropocentric attitude toward nature. Such misperception is present among some, not all, Christian followers who believe that the Bible has given them the right to dominate nature (Kay 1989). Whether arising from misinterpretation or selfish interpretation, the anthropocentric attitude of the developed Christian society has spread to other parts of the world. Islamic societies are not different from Western societies in terms of their treatment of the environment. It is not religion, but the marginalization of religion, and consequently dilution of religion from individual minds, is responsible for such anti-environmental attitudes and behavior (Coward 2003; Saniotis 2011).

It is concluded that deeper awareness of religious teachings, not merely reciting religious books and participation in religious rituals, is important for taking stewardship responsibility and correcting anthropocentric attitudes. Since religions do not exempt people from mandatory responsibilities, religiously affiliated individuals who become conscious of environmental ethics will engage in pro-environmental behavior regardless of their purposes for visiting a destination. Religious tourism brings religious leaders and followers closer, creating an opportunity to know the religions in depth. By growing awareness about religious teachings about the environment or nature, stewardship responsibility can be fostered among religious individuals and pilgrims. Adopting a spiritual lifestyle at the individual level, empowering root-level religious leaders, and coordinating among different religions, religious denominations, political parties, religious and environmental organizations, and tourism stakeholders are important for behavioral change and sustainability in tourism.

References

Abdelzaher DM, Abdelzaher A (2015) Beyond environmental regulations: exploring the potential of "eco-Islam" in boosting environmental ethics within SMEs in Arab markets. J Bus Ethics 145(2):357–371

Abonomi A, De Lacy T, Pyke J (2022) Environmental impact of the hajj. Int J Relig Tour Pilgr 10(1):133–151

Ahmed QA, Arabi YM, Memish ZA (2006) Health risks at the hajj. Lancet 367(9515):1008–1015

Allen JB, Ferrand JL (1999) Environmental locus of control, sympathy, and proenvironmental behavior: a test of Geller's actively caring hypothesis. Environ Behav 31(3):338–353

Alper BA (2022) How religion intersects with Americans' views on the environment. https://www.pewresearch.org/religion/2022/11/17/how-religion-intersects-with-americans-views-on-the-environment/. Accessed 7 June 2024

Apollo M, Wengel Y, Schänzel H, Musa G (2020) Hinduism, ecological conservation, and public health: what are the health hazards for religious tourists at Hindu temples? Religions 11(8):416

Arbuckle MB, Konisky DM (2015) The role of religion in environmental attitudes. Soc Sci Q 96(5):1244–1263

Arli D, Pentecost R, Thaichon P (2021) Does religion make consumers more environmentally friendly? Mark Intell Plan 39(8):1024–1041

Belias D, Rossidis I, Valeri M (2022) Tourism in crisis: the impact of climate change on the tourism industry. In Tourism risk: crisis and recovery management. Emerald Publishing Limited, p 163–179

Bruce FF (1983) The bible and the environment. In: The living and active word of god: studies in Honor of Samuel J Schultz. Eisenbrauns, pp 15–29

Burkhardt MA (1989) Spirituality: an analysis of the concept. Holist Nurs Pract 3(3):69–77

Capstick S, Nash N, Whitmarsh L, Poortinga W, Haggar P, Brügger A (2022) The connection between subjective wellbeing and pro-environmental behaviour: individual and cross-national characteristics in a seven-country study. Environ Sci Pol 133:63–73

Chavan P, Sharma A (2024) Religiosity, spirituality or environmental consciousness? Analysing determinants of pro-environmental religious practices. J Human Values 30(2):160–187

Cohen J (1985) The bible, man, and nature in the history of Western thought: a call for reassessment. J Relig 65(2):155–172

Coward H (2003) Hindu views of nature and the environment. In: Nature across cultures: views of nature and the environment in non-western cultures. Springer, pp 411–419

Csutora M, Zsóka Á (2014) May spirituality lead to reduced ecological footprint? Conceptual framework and empirical analysis. World Rev Entrep, Manage Sustain Develop 10(1):88

El-Gammal M, Abozaid A (2021) Environmental protection and sustainable development: an Islamic perspective. In: Islamic finance and circular economy: Connecting impact and value creation. Springer, pp 27–40

El Hanandeh A (2013) Quantifying the carbon footprint of religious tourism: the case of hajj. J Clean Prod 52:53–60
Eom K, Tok TQH, Saad CS, Kim HS (2021) Religion, environmental guilt, and pro-environmental support: the opposing pathways of stewardship belief and belief in a controlling god. J Environ Psychol 78:101717. https://doi.org/10.1016/j.jenvp.2021.101717
Evers U, Gruner RL, Sneddon J, Lee JA (2018) Exploring materialism and frugality in determining product end-use consumption behaviors. Psychol Mark 35(12):948–956
Filimonau V, Kadum H, Mohammed NK, Algboory H, Qasem JM, Ermolaev VA, Muhialdin BJ (2022) Religiosity and food waste behavior at home and away. J Hosp Market Manag 31(7):797–818. https://doi.org/10.1080/19368623.2022.2080145
Garfield AM, Drwecki BB, Moore CF, Kortenkamp KV, Gracz MD (2014) The oneness beliefs scale: connecting spirituality with pro-environmental behavior. J Sci Study Relig 53(2):356–372
Girma LL (2022) This is how tourism must shift to actually address climate change. Bloomberg. https://www.bloomberg.com/news/articles/2022-11-14/how-tourism-can-hit-net-zero-by-2050-an-unrealistic-but-not-impossible-solution
Gohary A, Shah A, Hosseini SR, Chan EY, Madani F (2022) God salience and tourists' pro-environmental behavior. Ann Tour Res 93:103318. https://doi.org/10.1016/j.annals.2021.103318
Gulick WB (1991) The bible and ecological spirituality. Theol Today 48(2):182–194
Harwood TG, Garry T (2003) An overview of content analysis. Mark Rev 3(4):479–498
Helfaya A, Kotb A, Hanafi R (2016) Qur'anic ethics for environmental responsibility: implications for business practice. J Bus Ethics 150(4):1105–1128
Hoffman AJ, Sandelands LE (2005) Getting right with nature: anthropocentrism, ecocentrism, and theocentrism. Organ Environ 18(2):141–162
Holden A (2003) In need of new environmental ethics for tourism? Ann Tour Res 30(1):94–108
Indaryani M, Marliana E, Widjanarko M, Miranac AE, Montales CB (2023) The relationship between religiosity, materialism, and pro-environmental behavior among consumers in Central Java (Indonesia) and Bicol region (The Philippines). Kasetsart J Soc Sci 44(3)
Ives CD, Buys C, Ogunbode C, Palmer M, Rose A, Valerio R (2023) Activating faith: pro-environmental responses to a Christian text on sustainability. Sustain Sci 18(2):877–890
Jenkins W (2009) After Lynn White: religious ethics and environmental problems. J Relig Ethics 37(2):283–309
Kala D, Chaubey DS (2023) Pro-environmental behavior of religious tourists: moderating role of religious beliefs. Cornell Hosp Q 65(1):105–119
Kay J (1989) Human dominion over nature in the Hebrew bible. Ann Assoc Am Geogr 79(2):214–232
Konisky DM (2017) The greening of Christianity? A study of environmental attitudes over time. Environ Poli 27(2):267–291
Lakhan C (2018) The garbage gospel: using the theory of planned behavior to explain the role of religious institutions in affecting pro-environmental behavior among ethnic minorities. J Environ Educ 49(1):43–58
Leary RB, Minton EA, Mittelstaedt JD (2016) Thou shall not? The influence of religion on beliefs of stewardship and dominion, sustainable behaviors, and marketing systems. J Macromark 36(4):457–470
Liu J, Yuan L, Li M, Li SJ, Sun Y, Yuan J (2024) Ecological experiential learning and tourists' pro-environmental behavior intentions: the mediating roles of awe and nature connection. Heliyon 10(1):e23410
Lochrie S, Baxter IW, Collinson E, Curran R, Gannon MJ, Taheri B, Thompson J, Yalinay O (2018) Self-expression and play: can religious tourism be hedonistic? Tour Recreat Res 44(1):2–16
Moncrief Lewis W (1970) The cultural basis of our environmental crises. Science 170(30):508–512
Morrison M, Duncan R, Parton K (2015) Religion does matter for climate change attitudes and behavior. PLoS One 10(8):e0134868
Muñoz-García A (2014) Religion and environmental concern in Europe. Arch Psychol Relig 36(3):323–343

Mylius B (2018) Three types of anthropocentrism. Environ Philos 15(2):159–194

Omoyajowo K, Danjin M, Omoyajowo K, Odipe O, Mwadi B, May A, Amos O, Rabie M (2023) Exploring the interplay of environmental conservation within spirituality and multicultural perspective: insights from a cross-sectional study. Environ Develop Sustain 26:16957–16985

Osra FA, Alzahrani JS, Alsoufi MS, Osra OA, Mirza AZ (2021) Environmental and economic sustainability in the hajj system. Arab J Geosci 14:1–15

Ottuh PO (2020) Religious approach to non-anthropocentric ethics in environmental philosophy. Cogito-Multidiscip Res J 1:7–24

Pargament KI, Mahoney A (2005) THEORY: “sacred matters: sanctification as a vital topic for the psychology of religion”. Int J Psychol Relig 15(3):179–198

Perry SL (2024) America’s becoming less religious. Is politics to blame? Time. https://time.com/6693016/americas-less-religious-politics-to-blame/

Preston JL, Shin F (2022) Opposing effects of spirituality and religious fundamentalism on environmental attitudes. J Environ Psychol 80:101772. https://doi.org/10.1016/j.jenvp.2022.101772

Pullella P (2020) Pope commits Vatican to net zero carbon emissions by 2050. Reuters. https://www.reuters.com/article/idUSKBN28M0RO/

Qarjouli A (2023) Green hajj: how to preserve Islam’s environmental pillars in pilgrimage. Doha News. https://dohanews.co/green-hajj-how-to-preserve-islams-environmental-pillars-in-pilgrimage/

Rice G (2006) Pro-environmental behavior in Egypt: is there a role for Islamic environmental ethics? J Bus Ethics 65(4):373–390

Robina-Ramírez R, Pulido-Fernández M (2019) What role do religious belief and moral emotions play in pilgrimage with regards to respecting nature? Ann Leisure Res 24(4):492–512

Saniotis A (2011) Muslims and ecology: fostering Islamic environmental ethics. Contemp Islam 6(2):155–171

Shinde KA (2011) “This is religious environment”: sacred space, environmental discourses, and environmental behavior at a Hindu pilgrimage site in India. Space Cult 14(4):448–463

Siagian N, Ridayani A, Kamsinah ME, Fatmawati E, Adi Pramono S, Fajri I (2023) The effect of environmental citizenship and spiritual norms as mediators on students’ environmental behaviour. Int J Adolesc Youth 28(1)

Sponsel LE, Natadecha-Sponsel P (2003) Buddhist views of nature and the environment. In: Nature across Cultures: views of nature and the environment in non-western cultures. Springer, pp 351–371

Tarakeshwar N, Swank AB, Pargament KI, Mahoney A (2001) The sanctification of nature and theological conservatism: a study of opposing religious correlates of environmentalism. Rev Relig Res 42(4):387–404

Taylor B (2007) Focus introduction: aquatic nature religion. J Am Acad Relig 75(4):863–874

Taylor B, Van Wieren G, Zaleha BD (2016) Lynn White Jr. and the greening-of-religion hypothesis. Conserv Biol 30(5):1000–1009

Taylor P (1986) Respect for nature: a theory of environmental ethics. Princeton University Press

Teng CC, Wang YC, Cheng YJ, Wang SN (2023) Religious beliefs and food waste prevention practices: mechanisms of divine and environmental awareness. J Hosp Market Manag 32(4):530–554

Thathong K (2012) A spiritual dimension and environmental education: Buddhism and environmental crisis. Procedia Soc Behav Sci 46:5063–5068

Ullah S, Lyu B, Ahmad T, Sami A, Kukreti M (2024) A mediated moderation model of eco-guilt, personal and social norms and religiosity triggering pro-environmental behavior in tourists. Curr Psychol 43(8):6830–6839

UNWTO (2014) Tourism can protect and promote religious heritage. https://www.unwto.org/archive/europe/press-release/2014-12-10/tourism-can-protect-and-promote-religious-heritage. Accessed 09 Sep 2024

Van Dyke F (2013) Christian stewardship. In: Values in sustainable development. Routledge, pp 134–144

Wang S, Wang J, Li J, Zhou K (2020) How and when does religiosity contribute to tourists' intention to behave pro-environmentally in hotels? J Sustain Tour 28(8):1120–1137
White L Jr (1967) The historical roots of our ecologic crisis. Science 155(3767):1203–1207
WTTC (2021) A net zero roadmap for travel & tourism. https://wttc.org/Portals/0/Documents/Reports/2021/WTTC_Net_Zero_Roadmap.pdf. Accessed 30 July 2024
Xie S (2019) Green religion as a way of life: Thoreau and his ecocentric esthetics of existence. ANQ Q J Short Artic Notes Rev 33(4):252–262
Zhang G, Higham JE, Albrecht JN (2020) New territory for environmental sociology: environmental philosophy and nature-based tourism. In: The Cambridge handbook of environmental sociology, pp 262–282

Immersive and Transformative Spiritual Experiences: From the Inner Self to the Sense of Belonging to a Community

Margarida Franca, Madalena Abreu, and António Azevedo

Abstract This chapter examines the emergence of novel trends in spiritual experiences, which are now designated as "spiritual tourism". The study encompasses six locations in Portugal, four of which facilitate the implementation of St. Ignatius' spiritual exercises. In-depth interviews with the individuals responsible for managing these locations elucidate the factors that motivate people to engage in spiritual tourism. Spiritual tourism in Portugal offers a connection to the Catholic heritage, facilitating well-being, prayer, and contemplation in natural environments. These spaces foster individual and small-group spiritual journeys, tailored to modern religious behaviour. Religious institutions shape and promote these experiences, aligning with tourists' desire for individual spiritual retreats.

1 Introduction

This chapter aims to discuss the new trends in the design of spiritual, transformative, and religious experiences. These new itineraries are now being positioned as spiritual tourism. This is a new form of tourism, which is becoming increasingly important for academics and organisations alike.

Victor and Treschuk (2020, pp. 109–110) write "According to White et al. (2011, p. 50), religion is defined as a "sentiment of learned behaviors and social expressions that reflect cultural values." Religion is more restricted than spirituality and is

M. Franca (✉)
Polytechnic Institute of Leiria, Leiria, UCP-CITER, Lisboa, Portugal
e-mail: margarida.franca@ipleiria.pt

M. Abreu
Polytechnic Institute of Coimbra, Coimbra, CEOS.PP, Coimbra, Portugal
e-mail: mabreu@iscac.pt

A. Azevedo
School of Economics and Management of University of Minho, iBMS, Braga, Portugal
e-mail: antonioa@eeg.uminho.pt

M. Valeri (ed.), *Contemporary Religious Tourism*, Tourism, Hospitality & Event Management, https://doi.org/10.1007/978-3-031-79074-4_4

often referred to as prayer, religious activities, and seeking spiritual guidance (Lepherd 2014; Taylor 2007; Tzeng and Yin 2008; Wachholtz et al. 2007). According to Booth (2012), religious institutions were constructed by men who spoke for God. However, spiritual experiences cannot be compared with religion. Spirituality requires personal experience and changes in the heart. Religion involves coding and conceptualizing that experience. Religious identification requires that members of the group are impelled to follow a certain set of rules, which control and enable group behavior (Tomass 2012)".

Why is religion important? Often faith, religiosity, and spirituality are used interchangeably and inconsistently by scholars (Paul Victor and Treschuk 2020; Zinnbauer et al. 1997; Zrinščak 2004). The rise of secularism created a distinction between religion and spirituality. The level of religiosity is "the intensity with which an individual believes in a religion, and thus affiliates and participates in it" (Sardana et al. 2018, p. 724), whereas spirituality is being described in contrast, as "an individual expression that speaks to the greatest of human capacities" (Pargament 1999). This is particularly visible in the West, where people often declare themselves to be "spiritual, but not religious" (Huss 2014; Wixwat and Saucier 2021).

For Willson et al. (2013), spirituality is the pursuit of a harmonious relationship or 'oneness' with self, 'others' (including other people, animals, the earth, nature), and/or God/Higher Power. People who engage in spiritual tourism usually feel 'energized', 'inspired', 'healed', and 'uplifted' (Cheer et al. 2017; Kujawa 2017).

To analyse some experiences of spiritual tourism, particularly on the supply side of this tourism product, this chapter presents several case studies in Portugal of institutions that offer spiritual retreat experiences or other types of offers associated with spirituality.

2 Literature Review

2.1 Intersection Between Religious/Spiritual Activities and Tourism Experiences

The concept of spirituality has itself broadened from concerning metaphysical and higher consciousness to include various aspects of mind and body well-being (Buzinde 2020; Huss 2014; Dillette et al. 2021). It now encompasses a very long list of products and services, such as yoga and meditation sessions, retreats (Mourtazina 2020), martial arts, spiritual and self-help discourses, spiritual counselling and mysticism, vegetarian and organic diets for spiritual cleansing, traditional massage techniques as spiritual healing, rural tourism (Sharpley and Jepson 2011), nature-based tourism (Wang et al. 2023), and spiritual tourism (Sardana et al. 2018).

The relationship between religion and tourism has been increasingly discussed by several authors from a multiplicity of disciplinary perspectives (Cheer et al. 2017; Christou et al. 2023; Kim et al. 2020; Shekhar and M. 2023).

Starting from a sample of 776 records, Collins-Kreiner (2020) made a meta-analysis of 40 papers and outlined a knowledge map reflecting the key areas of academic insight into the religion-tourism nexus. The boundaries between religion and tourism are blurred and de-differentiation between the two realms is a trend. Contemporary research adopts the approach of 'religion as a product' or 'transformative experience' (Chhabra and Kim 2024; Sheldon 2020). Collins-Kreiner (2020) did not provide a clear definition of religious tourism (RT) because there many different types of tourists and not only religiously motivated visitors. There are several types of motivations/purposes such as seeking spiritual well-being, enlightenment, knowledge, and social bonds. The concept of secular visitation has been developed to emphasize journeys to non-religious sacred sites. These approaches have emerged to provide a more holistic understanding of the intersections between tourism and religion and/or spirituality (Kirillova 2019) (Fig. 1).

Nevertheless, Iliev (2020) and, more recently, the bibliometric analysis of Das et al. (2024) compiled several definitions of religious tourism:

(a) "refers to a trip that is closely tied to the adherents of a specific religion, and those locations are sacred religious sites" (Lin and Fu 2019)
(b) "is described as travel to certain tourist places to experience religious activities or sites or the associated attractions, such as art, culture, customs and architecture" (Blackwell 2007; Heydari Chianeh et al. 2018)
(c) "a range of spiritual sites and associated services, which are visited for both secular and religious reasons" (Raj et al. 2015, p. 105)

According to Seyer and Müller (2011) and Iliev (2020), religious tourism has developed into a much larger and segmented market and niches such as pilgrimage, religious conferences and conventions, volunteer-oriented religious tourism,

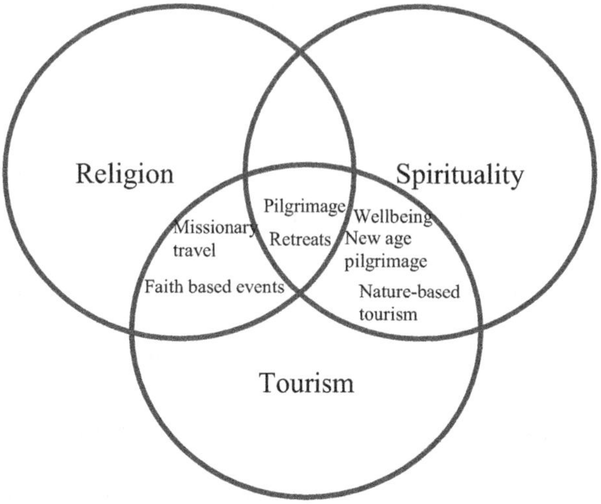

Fig. 1 Intersection between religiosity, spirituality, and tourism (elaborated by authors)

short-term mission travel, cruises, leisure/getaways, cyberpilgrimage (Digance 2006), retreats/guesthouses, Christian camps, adventure/active, student/youth, and family/intergenerational.

Schweinsberg (2023) highlighted the paradigm proposed by Barkathunnisha et al. (2019) on the notion of the seventh platform of tourism knowledge based on spirituality and the underlying four pillars—a spirit of knowing, a spirit of doing, a spirit of being, and a spirit of becoming (Fig. 2).

2.2 *Push-Drivers and (Intrinsic) Motivations and Benefits from Spiritual/Religious Tourism*

McKercher (2016) proposed a six-level taxonomy of tourism products, where religious travel is classified within the category 'Personal Quest', which refers to travel for more personal reasons associated with self-development and/or learning (Schutte and Dreyer 2006), personal transformation and self-actualization (Andriotis 2009; Heintzman 2013), and restoration (Ouellette et al. 2005).

Spiritual tourism as a retreat is defined as "one of escape from the everyday, or of sacred time or ritual renewal" (Norman 2012, p. 31). Moreover, Underwood (2011) proposed a 16-item self-report measure designed to assess ordinary experiences of connection with the transcendent in daily life—the Daily Spiritual Experience Scale (DSES). It includes constructs such as awe (Preston and Shin 2017), gratitude, mercy, and sense of connection with the transcendent and compassionate love. It also includes measures of awareness of discernment/inspiration and a sense of deep inner peace.

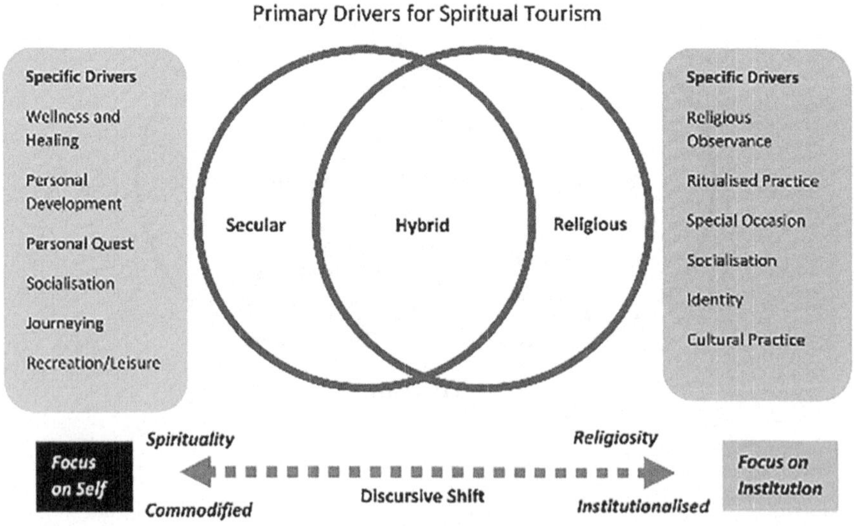

Fig. 2 Primary drivers of spiritual tourism (Cheer et al. 2017, p. 255)

Preston and Shin (2017) found that spiritual experiences seem to invoke:

a) A sense of small self—the infinite power of God, size of the Universe, and grandeur of creation can all make a person feel tiny in comparison (Meier et al. 2007).
b) Feelings of spiritual humility (Davis et al. 2010)—personal modesty and deference before more powerful forces—is a central value of all religions and correlated with spirituality (e.g. Powers et al. 2007).

According to Cheer et al. (2017), the current drivers of spiritual tourism oscillate between commodified or institutionalized aspects. Considering the secular motives often involved, it is evident that the fundamental motives of tourism, such as health, adventure, and relaxation, are deliberations based on the self.

2.2.1 Positive Emotions and the Upward Spiral Theory

Van Cappellen et al. (2021) claimed that positive emotions, such as everyday gratitude when praying or intense feelings of awe during mystical experiences, feel good and build psychological, social, and biological resources. Many of the positive emotions experienced in the context of religion practices, such as prayer, meditation, and collective worship, are considered self-transcendent, pulling people out of themselves. Based on the broaden-and-build theory of Fredrickson (1998, 2013), Van Cappellen et al. (2021) proposed the upward spiral theory of sustained religious practice and lifestyle change. The inner loop of the model (Fig. 1) suggests that sacralized positive emotions might be more potent than their secular version in driving engagement. Over time, the experience of positive emotions during religious/spiritual behaviours may well become intimately connected with this broader meaning system and any positive cues can become imbued with sacredness (i.e. sacralising or sanctifying the environment) (Kumar et al. 2022) (Fig. 3).

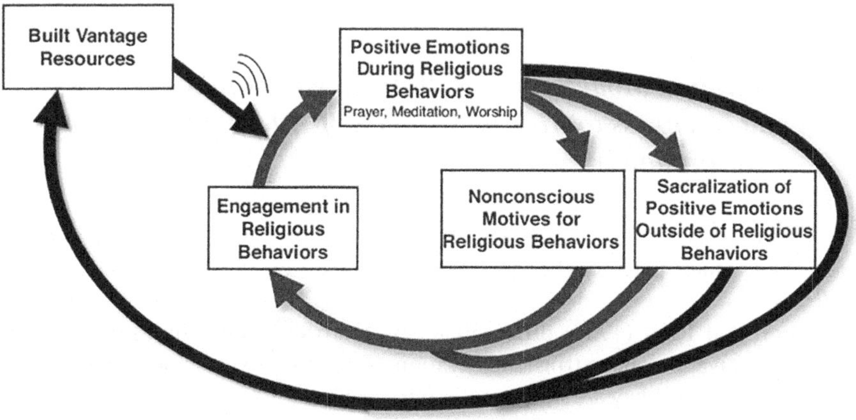

Fig. 3 Upward spiral theory of sustained religious practice (van Cappellen et al. 2021, p. 94)

2.2.2 Inner Peace and Life Satisfaction

For Aulet Serrallonga (2020), the experience of pilgrimage and the pilgrims' quest for inner peace can enhance a respectful attitude towards people and places peace. The peace-building process has three dimensions: (1) from a personal dimension and spiritual sphere, (2) peace from a social dimension, and (3) peace from an environmental dimension. The spiritual aspect of peace has received less attention in education and research related to peace than other dimensions like the political or the social ones. This is highly important in the current situation about, which we need to understand more deeply, the place where religion and spirituality play in individual and collective life (Danesh 2006).

Zullig et al. (2006) found an association between perceived spirituality, self-rated health, and life satisfaction. Furthermore, there is a relationship between memorable religious tourism experiences and place attachment (Albayrak et al. 2018; Silva et al. 2023; Vada et al. 2019; van Laer and Izberk-Bilgin 2019).

2.3 Potential Negative Outcomes: Self-Enhancement and Spiritual Narcissism

According to Vonk and Visser (2021), people may aim to become more successful, more respected, or loved because of their spiritual development. They may get a sense of excitement, or wisdom and serenity, and embrace the ideology that brought them this delight. The road to spiritual enlightenment may yield the exact same mundane distortions that are all too familiar in social psychology, such as self-enhancement, illusory superiority, closed-mindedness, and hedonism (clinging to positive experiences) under the guise of alleged 'higher' values. Vonk and Visser (2021) found that the self-enhancement function of spirituality was correlated with self-esteem and, more strongly, with communal narcissism, corroborating the notion of spiritual narcissism.

2.4 Individual Moderators: (Self-Ascribed) Religiosity and Openness to Absorbing and Self-Altering Experiences ("Absorption")

For Nowak et al. (2024), religiosity and spirituality offer a framework through which individuals perceive the world and direct their everyday decisions and lifestyle choices.

Allport and Ross (1967) differentiate the motivation for religiosity into two: intrinsic motivation and extrinsic motivation. Religion provides a way to lead their daily life, and thus, they are intrinsically motivated. For the disenfranchised people, religious congregations are sources of social support, seeking comfort, or even meeting their personal needs—extrinsic motivation. Several scholars (e.g. Patwardhan et al. 2012) even show a negative correlation between intrinsic and extrinsic religiosity.

DeJong et al. (1976) identified six dimensions of religiosity: belief, experience, religious practice, religious knowledge, individual moral consequences, and social consequences.

Furthermore, Allport and Ross (1967) developed the scale of intrinsic religiosity and extrinsic religiosity. More recently, other measurements were developed by several authors (Da Silva et al. 2019; Noronha et al. 2017; Sardana et al. 2018; Wilkes et al. 1986). For example, spirituality has been measured by several authors (e.g. Hatch et al. 1998; Sardana et al. 2018) by a scale primarily adapted from Delaney (2005).

According to Lifshitz et al. (2019), some people seem to have a 'talent' for spiritual experience: they readily sense the presence of supernatural beings, receive special messages from God, and report intense feelings of self-transcendence, awe, and wonder. Participants scoring highly on the Tellegen Absorption Scale (Tellegen and Atkinson 1974) report vivid experiences of hearing God's voice during prayer, intense mystical experiences in response to psychedelics or placebo brain stimulation, and strong feelings of presence and transcendence when confronted with natural beauty, virtual reality, or music.

2.5 *Pull/Extrinsic Drivers of Spiritual Tourism*

Nair and Dileep (2021) analysed the drivers of spiritual tourism. In the case of spirituality, there should be a differentiation between Eastern philosophies and Western materialism. The ontological standpoints (emic/etic understandings and cultural dimensions) of an individual further delineate these classifications.

Norman (2012) identified five distinct categories of spiritual tourism attractions: (1) spiritual tourism as 'healing', related to well-being tourism (Buzinde 2020); (2) spiritual tourism as 'experiment' (e.g. yoga meditation); (3) spiritual tourism as quest; (4) 'spiritual tourism as retreat' in which the traveller attempts an escape from every day and seeks a religious experience or a ritual renewal (the attractions for these tourists may include meditation retreats, health spas, and eco-tourism attractions); and (5) spiritual tourism as 'collective'.

2.5.1 Nature and Silence

Nature is a vital context for transcendent experiences fostering the spiritual dimensions of humankind (Wang et al. 2023). On the other hand, in pilgrimages, Vidal-González and Sánchez (2019) observed that participants seek to be part of these practices and their corresponding contact with nature, but inevitably in the company of others. These authors identified several secular drivers for hiking-oriented pilgrimages: (1) the search for original roots, (2) the experience of rural sociability, (3) the search for authenticity, (4) to know more about the tangible and intangible cultural heritage, (5) to travel back in time to the Middle Ages, and (6) to escape from the stress of urban life. Furthermore, Dowson (2020) developed a theoretical framework claiming that the shared experience of events, religious objects, and sacred spaces in the pilgrimage context builds the community cohesion, enhances the community identity, strengthens faith communities, and enlivens individual faith.

Moreover, Mourtazina (2020) suggested that silent landscapes are not only backdrops of spiritual retreat experiences but can be considered as inter-subjective performative and resourceful milieu of engagement that intertwine intimate embodied experiences with broader social and cultural values.

2.5.2 Sustainability and Materiality

According to Romanelli et al. (2021), religious tourism and pilgrimages help to drive the sustainable tourism development and social and economic growth of touristic destinations in sacred sites. Today, religious tourism is becoming an important source for social and public value creation processes within communities by enabling territories and local economies to strengthen cooperative processes and collaboration for sustainable development.

Terzidou (2020) made an ethnographic research highlighting the importance of materials' thingness, rituals, and staged performances in the religious experience. She argues that the 're-materialization' of religion has a decisive importance to religions and sacred sites, generating immanent and profound powers on people that enable (rather than overshadow) spiritual contact. Abstract symbols and metaphors (e.g. iconography) are important in communicating values and identities and ascribing meaning to events (Bowman 2016; Moufahim and Lichrou 2019). While there is a systemic function of objects, which is created through semiotic and social powers of religion that ascribe use—and sign—value to objects, there is also an improvisational function of objects, recognizing the unexpected power of the objects' materiality on people as a result of the notion of affect that determines their religious experiences and understandings. From the tourism perspective, religion materiality is an essential ingredient of the marketing mix (physical evidence) of the experience. In the post-experience stage of the journey, religious souvenirs also help to save the memories in tangible objects (Paraskevaidis and Andriotis 2015).

2.5.3 Inhibitors and Constraints

Dowson (2020) stresses some concerns with the lack of authenticity in some religious events. For example, Boorstin (2012) defines pseudo-events as having a public relations purpose, with an element of misleading or deception, some of which may be openly acknowledged while others may be hidden, intentionally or otherwise. Faith events are expected to retain their authenticity and overt sacred purpose.

3 Tourism Sector in Portugal

Bearing in mind that the empirical analysis that follows in this study is in Portugal, it is understood that the study of the tourism sector in its religious or spiritual component implies, albeit briefly, a reference to the composition of the religious identities of the Portuguese population. This is because religious identities, belief systems, and non-religious affiliations condition the use of spaces and places with very distinctive traits on the landscape and public space.

Portuguese society is characterized by a majority of Catholic population which, in the last population censuses, registered 80% of the resident population (INE 2021). This percentage has decreased slightly over the last few decades, but it still allows us to place the Catholic religion as the majority in our country, reflecting the Catholic matrix of the cultural identity of the Portuguese people. This matrix is reflected and implicit in the experiences, practices, and rituals of the population.

In the field of religion, the last decade has also seen an increase in the population with no religion (from 6.8% in 2011 to 14.1% in 2021), which is now the second most representative group, as well as an increase in Christian and non-Christian minority groups (3.9% in 2011 to 5.7% in 2021) (INE 2021).

Perhaps these identity traits end up leaving an indelible mark on the territory itself. The heritage component and the tourism sector are strongly marked by this history and by the use and consumption of religious places, spaces, and territories. The territoriality of the Catholic Church is evident, with the hierarchy of places of worship being very marked by parishes and places where tourists and pilgrims flock, as it is the case of the Sanctuary of Fátima, one of the largest Marian shrines in Europe.

Even in a context of progressive individualization and secularization of the Portuguese population, with a clear increase in the population with no religious belief or belonging, the population tends to seek out rituals or experiences with which they are more familiar and loyal, or which they recognize as effective in the search for inner well-being, encounter, and a sense of peace. Pilgrimages, spiritual retreats, the use of prayer and meditation apps, visits to places of worship, and religious trips continue to be popular among the Portuguese population and make it possible to create bonds, densify knowledge, and strengthen faith among members of the same community.

In the absence of exhaustive official data, estimates by Pereira (2008) suggest that between 10% and 14% of national tourism is religious tourism, involving around 7 million people a year, of which around 6 million visit Fátima, according to the official 2022 data from the Fátima Sanctuary itself. Also, this year, figures from the National Statistics Institute point to 211 million trips by Portuguese tourists for religious reasons. This figure has fluctuated over the years, and in the context of the COVID-19 pandemic, the figures were lower and are now tending to stabilize again.

4 Design/Methodology/Approach

The Catholic Church has also offered spiritual exercises as an immersive, transformative experience (Bento da Silva et al. 2022; Phillips 2021; Radvan 2021, 2024). These retreats were proposed by St Ignatius of Loyola and are now offered by Jesuit communities, among other communities, in a variety of time-based formats for those who wish to increase their self-awareness and develop their personal relationship with God under the supervision of a personal spiritual director.

Therefore, this study adopted the multiple case-study approach analysing the following case studies of spiritual/religious experiences in Portugal, and four out of six of them offer the St Ignatius Spiritual Exercises as part of their programme:

(a) 'Casa de Sicar', Guimarães (https://pontosj.pt/cvxp/iconografia-e-ecumenismo-cvx-de-braga-em-sicar/)
(b) 'Casa da Torre', Braga (https://pontosj.pt/casadatorre/)
(c) 'Seminário Maior', Coimbra (https://www.seminariomaiordecoimbra.com/en/visit-experiences/)
(d) The House ('Casa')—activity house of Saint Afonso, Cernache (https://pontosj.pt/acasa/)
(e) 'Casa de Exercícios de Santo Inácio', Sintra (https://pontosj.pt/casadesantoinacio/)
(f) 'Casa Madre Luisa Andaluz', Santarém (https://lacc.pt/casa-madre-2/)

The researchers conducted in-depth interviews with the main coordinators of these institutions and with some of the participants in the spiritual experiences in order to capture the conscious and unconscious motivations and benefits of engaging in these activities. As all authors played an active role in these experiences, either as participants or guides, the qualitative approach was complemented by a self-ethnographic approach.

4.1 Interview Script

Based on the literature review, the in-depth interview script included the following questions:

(a) What kind of activities are offered in this institution? (Non-) Religion-based versus spirituality-based? Individual versus group activities?
(b) What kind of experiences? Pilgrimage, retreats, culture and arts workshops, St Ignatius Spiritual Exercises, study of the Bible, or group meetings?
(c) What are the main motives of participants? Recommendation of friends or religious institutions or movements; economic, cultural, or security reasons; or cleanliness, local landscape, historical heritage, convenience, and easiness to access?

(d) What features do participants value more? Indoor versus outdoor facilities, silence level, tranquillity, simplicity, nature surroundings, artificial scenarios, etc.
(e) What benefits do participants aim to receive? Physiological/health needs, self-knowledge, self-esteem, self-development, restoration, inner peace, physical/psychological healing, faith/religion-based growth, escape to daily life, escape to routine, social relationships, contact with multicultural communities, or cultural and history cognitive knowledge.
(f) The spiritual experiences are not always religious. Do you agree?
(g) As activities coordinator, do you aim to intensify the religious purposes?
(h) How many persons are engaged in the activities?
(i) Is there always a guide or supervisor?
(j) How many days do the activities take?
(k) Do you think that the physical space where the activities take place is important? Why?
(l) During those experiences, what kind of emotions/sensations do participants aim to accomplish? (transformative, immersive, introspective, self-development experiences, connection to God)

4.2 Case Studies

4.2.1 'Casa de Sicar', Guimarães, Portugal

Casa de Sicar is a rural house located in a small village of Guimarães municipality, on the top of the hill 2 Km from the Penha Sanctuary. According to the New Testament of the Bible (John, 4:5–6), *Sicar (Sychar)* is the name of a city in Samaria (today Balata near Nablus in the West bank) where the "well of Jacob" is located and where Jesus had an encounter with a Samaritan woman.

Participants can use the lawn and a small (vegetable) garden to walk or meditate in the shadow of fruit trees. The house has three rooms with private bathroom and a common balcony for sightseeing on the surrounding valleys. The main building also has common spaces such as the dining room, the chapel, and the library with a wood burning stove. Through the glasses of the library the guests can see the garden, and on the winter this a is very cozy space.

The owner is the Jesuit sister Alzira Fernandes that very often supervises individual Saint Ignatius Spiritual Exercises customized from three to 8 days long. She is also a specialist in Christian Iconography. The house receives several non-Catholic/Catholic individuals (from all the country and from Spain and France) that want to make silence retreats or search spiritual guidance … "*I always try to satisfy the needs of each different person...they say if I am tired, and if I come to fountain, I get restoration, I heal, I gain more hope...*".

It hosts events and meetings of local groups (maximum 25 people) that want study the Bible, or to attend cultural and religion seminars (e.g. iconography). The word-mouth and recommendation of friends is the main information source for the choice of the 'Casa de Sicar'. Participants valuate the beauty of the landscape, the house, and the historical heritage. The atmosphere is homelike, "*people say that they feel like at home*".

Participants also use to hike to the hill of Penha Sanctuary… "*to see different landscapes is part of the spiritual retreat*"…"*the social contact is very important, for example, the Bible Studies group, they eat and drink, they like to chat…is part of the house dynamics*".

> Here we have the symbology of the water (Sychar well) …and participants usually leave written messages saying that they come here - to drink from the (pure/original) source…they quench the thirst, the thirst for more depth, for being loved, the thirst for life, for being listened to,…

.

> I speak of the well, the encounter, the water of life, and it translates experiences of satiety, they leave here satiated, they leave prepared to bring the water to those who were entrusted. It is a beautiful and enriching experience.

4.2.2 'Casa da Torre': The Spirituality and Culture Centre (CEC) of the Society of Jesus in Soutelo, Braga, Portugal

Donated by the second Viscountess of Torre, in 1951, 'Casa da Torre' housed the novitiate and juniorate—the stages of formation—of the Society of Jesus in Portugal. This spirituality and culture centre offers an annual programme of spiritual exercises, in a range of formats, marriage education and family life training, and a variety of meetings/courses.

In addition to the annual programme of activity, Casa da Torre aims to be a place open to all people who, in one way or another, seek to grow in their inner life. In this sense, Casa da Torre is open to other proposals that are in line with their identity and mission. The priest Luís Providencia, SJ, has been responsible for Casa da Torre, for 14 years, until 2023.

In Casa da Torre, people are looking for religious and spiritual experiences. The type of experience offered in this institution is a spiritual retreat. The choice of the place to carry out the spiritual experience is associated with an order/movement. Creating an environment of silence is the basic characteristic: search for knowledge/discovery and (self-knowledge/)introspection. Moreover, spiritual experiences are not always of a religious nature. The place is in demand because it is an opportunity for silence and contemplation.

4.2.3 'Seminário Maior': Major Seminary of Coimbra, Coimbra, Portugal

The Major Seminary of Coimbra is a place where culture and spirituality connect with the city, in the upper part of Coimbra, next to the Botanical Garden. It has an impressive presence of Italian art in Portugal, with more than 250 years. Within this historical religious place, people can make a visit and stay that combines art, history, and spirituality. Each person can add to their visit the experience of a meal or even a night's sleep in this eighteenth-century religious house.

The one responsible for this place is the priest Nuno Santos, who is the rector of the sanctuary of Coimbra, the key person interviewed. The institution offers a range of experiences related to religion and spirituality, as well as community and personal celebrations. The institution's typology of spiritual experiences encompasses retreats, meetings for inner reflection, and meetings for community reflection. These meetings may be conducted by individuals or groups and may be creative or artistic in nature. Additionally, the institution offers experiences related to community celebrations. The decision to select this destination was motivated by considerations pertaining to economic factors, cultural aspects, and the environmental context.

The suitability of a location is determined by its historical and cultural heritage and the accessibility of its facilities. The attributes that users seek to value with these experiences are the indoor environment (indoor/closed), the outdoor environment (outdoor/open) both differing from everyday life, the silent environment, the peaceful space, and the contact with natural environment.

When asked what people are looking for with these experiences, the answer is: The desired outcome may vary depending on the individual. In some cases, the aim is simply to recuperate and unwind, whereas in others, the motivation may extend to a deeper pursuit of well-being. This question is important, yet the answer is contingent upon the underlying reason for its inquiry. A considerable number of individuals seek to take time for rest. Such individuals seek and perceive these experiences for a variety of reasons, including religion, cultural knowledge, history, and biblical history. Father Nuno concurs that these types of spiritual experiences are not always religious in nature. Furthermore, he asserts that he is not preoccupied with the religious implications of these experiences.

The objective is to identify the spiritual issue more precisely. The mass is more religious on Sundays and Tuesdays, in small groups (up to 10 people) and in large groups (more than 20 people). Examples of large groups include youth meetings and preparation for confirmation.

In most cases, a guide or advisor is present in person. This is particularly the case with retreats, where the individual is the focus of attention. At this time, no script has been developed. It is not uncommon for individuals to reside at the site for extended periods, pursuing doctoral or master's degree programmes, among other academic endeavours. The demographic composition of the site's visitors varies considerably in terms of their length of stay.

The physical environment in which experiences occur is of paramount importance. About the physical configuration of the house, concerns encompass security, economic considerations, hygiene, and landscape. Currently, accessibility is ensured through the provision of an elevator. Visitors seek experiences that facilitate immersion, transformation, and collection. Furthermore, he commented that:

> The concept of spiritual tourism was being developed. This was distinct from mass tourism associated with religious tourism and sanctuary tourism. Instead, it was focused on spiritual experience. The objective was to reinforce the circuit of European monasteries.

In the webpage of this seminary, it can be acknowledged:

> Spiritual tourism is an industry that aims to respond to the needs expressed by a significant proportion of the population, who wish to take time out from their daily activities in order to rest, to reconnect with their spiritual selves, to experience nature, to reinforce the meaning of their life, to listen to silence, to experience simplicity, to savour the sacred. Spiritual tourism is distinct from mass tourism, which is more closely associated with religious tourism and pilgrimage to sacred sites. Spiritual tourism is characterized by an individual's pursuit of a spiritual experience, which may or may not be aligned with a specific religious tradition. Instead, it is often shaped by a fundamental aspect of human existence: the spiritual dimension. The creation of this novel space is intended to reinforce the European monastery, convent and seminar network, which serves to facilitate the pursuit of spiritual experiences by individuals seeking out these distinctive locations. Such pursuits may encompass a variety of activities, including retreat, meditation, and mindfulness.

4.2.4 The House ('Casa'): Activity House of Saint Afonso, Cernache, Portugal

'Casa' is situated within the grounds of the former Colégio da Imaculada Conceição Community, occupying a spacious, verdant farmstead. She was 'baptised' on 6 September, the day of the canonization of Saint Afonso Rodrigues, a Jesuit brother for more than 40 years at the College of Palma de Mallorca. The facility is intended for the conduct of spiritual exercises and activities of a formative and spiritual nature. The facility comprises 20 rooms, a large chapel, an oratory, a 24-h bar with a small bookstore, a library, and a kitchen. It is advisable to bring sheets, a sleeping bag, and towels. The Casa is available for use by members of the Jesuit order, as well as other orders, movements, and individuals. In September of 2023, the Casa welcomed the arrival of a new community comprising four individuals: three Jesuits and a lay couple.

The priest Marco Santos, SJ, who is the spiritual leader of the community and the head of the Jesuit community in Coimbra, answered the questions of our interview. The experiences offered are religious (connotation with a very specific movement/order), spiritual, celebrations of community, pilgrimage, organized trip of a spiritual nature (inside and outside the country), spiritual retreat, meeting for inner reflection or community reflection, creative/artistic meeting, or community celebrations.

The users do choose these experiences to have a spiritual experience, for religious reasons/movement, for environmental/landscape reasons. They are looking for a different environment from everyday life, an environment of silence, a peaceful space, a natural environment, or the creation of a specific built environment.

The personal reasons elicited by participants are search for knowledge/discovery, introspection, rest, peace of mind, religion/questions related to religion, seeing different landscapes, contact with other people, cultural knowledge, history, biblical story, etc. Spiritual experiences are always religious. Participants arrive in small groups (up to 10 people), in medium groups (from 11 to 20 people), or in large groups (more than 20 people). During the experience there is a personal guide. The duration is medium (2 to 8 days).

The landscape inspires the silence, reflection, prayer, and meeting. During the experience, users may experience sensations as immersion, transformation, recollection, prayer, or introspection.

4.2.5 'Casa de Exercícios de Santo Inácio': Santo Inácio Exercise House—Rodízio, Sintra, Portugal

The Santo Inácio Exercise House, located in Rodízio, next to Praia Grande (Colares), is a work of the Society of Jesus, whose objective is to provide a simple and pleasant environment, which encourages silence, meditation, and prayer, in a context of tranquillity and natural beauty.

Inserted in a vast wooded terrain and conducive to reflection, the surrounding area, with beach and pine forest, extends from Sintra to Colares, Praia das Maçãs, and Cabo da Roca. Lisbon is 45 min away and Estoril/Cascais is around 20 min away.

Priest Francisco Campos, SJ, is the director of the Exercise House and answered promptly to que questions. The experiences offered in the institution are religious (connotation of a very specific movement/order), religious and spiritual, community celebrations, spiritual, catechetical, and human formation; preparation for the sacraments, etc.; and catechesis around established communities, spiritual retreats, meetings to reflect on the community, creative/artistic meetings, meetings to experience community celebrations, meetings to experience religion, and also itineraries that are not pilgrimages but have a strong spiritual character.

Users choose these experiences on the basis of the place where the spiritual experience takes place and also on the basis of the choice of religious order/movement, choice for economic reasons, cultural reasons, hygiene reasons, environmental/landscape reasons, association with an order/movement, on the basis of travel time or on the basis of ease of access, the welcome, the food, the silence, rooms prepared for prayer, chapels, etc.; moreover, people on spiritual exercises want to have an encounter with God, the conditions offered by the house, the silence, and the conditions. Users do look for an interior environment different from the daily routine, or also for the exterior environment, for a quiet environment, simple yet passionate environment, natural but also a specific built environment.

"We offer a very sophisticated experience, even Buddhists come here to pray". People arrive alone, in small groups (up to 10 people), in medium groups (from 11 to 20 people), or in large groups (more than 20 people); "*we have everything, for all tastes. The proposals are very varied, for all tastes, for all needs*!"

During the experience, the user carries out his/her own experiment, which can last a single day, short (two days), medium (2 to 8 days), or long (more than a week). People do look for different landscapes, allowing silence/reflection, prayer, encounter, escape, or immersion, transformation, contemplation, prayer, escape, or introspection. Spiritual consolation and desolation are some of the feelings that users may experience.

4.2.6 'Casa Madre Luisa Andaluz': Santarém and 'Convento das Capuchas'—Santarém, Portugal

Sister Mafalda Leitão, a sister of the Congregation of the Handmaids of Our Lady of Fatima and coordinator of the *Luiza Andaluz Knowledge centre*, was interviewed. The centre is inspired by the life and work of Luiza Andaluz, who founded this religious congregation in 1923 and was recognized as Venerable by Pope Francis in 2017.

This Congregation has three houses in Portugal: the first, 'Casa Madre Luiza Andaluz', located in Santarém, which aims to be a place of welcome and spirituality; the second, 'Casa de São Mamede', a place of projects and achievements; and the third, 'Convento das Capuchas', a place of education and social transformation.

Sister Mafalda Leitão said that "*taking into account the characteristics of contemporary society (the increase in interest in the spiritual, non-ecclesial groups, mobility, the search for meaningful and uncompromising experiences, people on the move and looking for them, the 'wave' of tourism and the spiritual and cultural heritage of the SNSF, as houses with historical and cultural significance, the person and life of Luiza Andaluz, Identity - the search to redetermine/update the way we express our identity and Luiza in the process of being canonized, whose heroic virtues were proclaimed in December 2017), emerges as a possibility to provide a spiritual-cultural response in the area of "religious, cultural and nature tourism*".

Through a more up-to-date language, it seeks to help transform society with the values of the Gospel, creating moments of culture or welcoming pilgrims and tourists, but also providing space for conferences and other initiatives.

For our study, Sister Mafalda highlighted only the Casa Madre Luiza Andaluz, since the other two houses are more geared towards larger groups, holding conferences, exhibitions, or concerts, among other activities. The experience of Casa Madre Luiza Andaluz is more in line with the objectives of our research.

Those who come to Casa Madre Luiza Andaluz are not so much looking for self-knowledge or inner healing, but rather to increase their knowledge of this religious congregation, including its foundress, to learn about the historical and patrimonial legacy of this congregation, but also to seek personal transformation through silence, prayer, encounter, and recollection. Anyone who visits this house knows that they will find a different physical environment from their everyday life (space is considered to be very important); they can enter the house and go to the crypt to pray, or just be in silence. You can also go to the room where Luiza Andaluz was born, which is still a nineteenth-century palatial space, and read a book.

Those who go to this house are usually small groups of no more than 20 people, who visit for a short time (no more than a day) and who are looking for 'novelty' (these houses have only been open for a relatively short time) and a moment away from everyday life. They do not have to spend any money and have the opportunity to find a form of rest, introspection, and seclusion through a guide that accompanies their stay/visit, or even through digital support.

For Sister Mafalda Leitão, this is a "proposal for a new way of life, where people in our cities can stop, rest, find peace and silence, feel at home (in the library, in the living rooms, in the refectory, in the chapel, in the garden, on the balcony...) and try out proposals for new lifestyles, which can put pressure on those in political, economic and social power (Laudato Si', 206). Especially at times of greater tension, of economic and social crisis, they are disruptive oasis-places that lead us to think and plan with hope".

5 Conclusions

Religion, as the driving force behind human behaviour, has been confronted with the consequences of a globalized world. In contemporary times, the implicit social relationships between the members of each group or between communities and religious places, spaces, and territories are conditioned by space-time compression (Massey 1994) and the imminent instantaneity, greater speed, and scope. Linear boundaries, watertight categories, and all that is fixed have been transgressed.

Religion had to adapt to the processes of de-territorialization of people and communities, for economic, political, or cultural reasons, as well as the consequent sharing and diversity of religious beliefs and practices. The secularization of society has not been indifferent to this process. Nowadays, cultures are hybrid, and the identities and territories of belonging are multiple, implying the existence of a dense network of places of belonging and interactive nodes that structure and densify this network (multiple territorial organization that combines different dimensions, scales, and spatial and temporal interactions).

In a context of apparent uprooting, alienation, or psychological distance between individuals and groups, to which electronic proximity replaces the actual presence of people (Appadurai 2004), sacred places, religious territories, and spaces of spirituality allow human beings to overcome the adversities of a rhizomatic world and achieve and experience security, comfort, well-being, and emotional and relational stability. Post-modern societies are characterized by an environment of risk and vulnerability that arises from profound changes in the sources of trust and security, such as kinship relations, the local community, religious cosmologies, and tradition (Giddens 1991; Valeri 2022). However, religion, sacred places, the network of territories, and religious territoriality are now ways of dealing with the 'society at risk' and promoting 'ontological security, (Giddens 1991).

Associated with these issues concerning the geography of religious places and spiritual places, it is important to take into account the growing coexistence of secularization in Portuguese society and the emergence of other religions and forms of religiosity (pluralism and diversity as dimensions of secularization). The decline of Catholicism and the growing indifference to questions of faith and religious beliefs is key to this. The population has a belief but does not belong to any Church (belief without belonging) growing, giving consistency to Davie (2010) ideas on the concept of 'vicarious religion', or Hervieu-Léger's (2006) concept of 'belonging without belief'. However, despite the decline, this substantial change has not erased the Catholic matrix, with many rituals, rites, and ways of being that Catholicism imprints on Portuguese society remaining. The Sanctuary of Fátima continues to be the place that welcomes the most visitors and pilgrims in Portugal today, revealing the strength of religious ties, practices, and experiences.

We believe that these two issues, one geographical and the other a matter of identity, are not discordant and, in our opinion, may even coincide. Spiritual tourism and the promotion of spiritual retreats are a way of maintaining, in a way, a connection to religion or, if we like, to the Catholic matrix of the Portuguese population. They are a way of seeking, although more individually, or in small groups, the well-being, silence, and prayer that sacred places offer. These are spaces which, associated with a very specific atmosphere of seclusion and contact with nature, allow for prayer, reading, thought, and contemplation. In fact, it can be said that religious actors shape these spiritual, transformative, and religious experiences. And this could be confirmed from the supply side of this tourism product in the six case studies in Portugal.

Those in charge of spiritual tourism venues highlight these possibilities and describe very clearly that these are the characteristics that tourists/users are looking for. A search that is increasingly individualized and adapted to the changes in religion as a driving force behind human behaviour.

References

Albayrak T, Herstein R, Caber M, Drori N, Bideci M, Berger R (2018) Exploring religious tourist experiences in Jerusalem: the intersection of Abrahamic religions. Tour Manag 69:285–296

Allport GW, Ross JM (1967) Personal religious orientation and prejudice. J Pers Soc Psychol 5(4):432–443

Andriotis K (2009) Sacred site experience: a phenomenological study. Ann Tour Res 36(1):64–84. https://doi.org/10.1016/j.annals.2008.10.003

Appadurai A (2004) Dimensões Culturais da Globalização. A Modernidade sem Peias. Trad. Telma Costa. Ed. Teorema, Lisboa, 267 pp

Aulet Serrallonga S (2020) Pilgrim's motivations: a theoretical approach to pilgrimage as a peace-building tool. Int J Relig Tour Pilgr 8(6):7

Barkathunnisha A, Diane L, Price A, Wilson E (2019) Towards a spirituality-based platform in tourism higher education. Curr Issue Tour 22(17):2140–2156. https://doi.org/10.1080/13683500.2018.1424810

Bento da Silva J, Quattrone P, Llewellyn N (2022) Turning to mystery in institutional theory: the Jesuit spiritual exercises. Organ Stud 43(9):1379–1400

Blackwell R (2007) Motivations for religious tourism, pilgrimage, festivals and events. CABI Books, CABI International, Wallingford

Boorstin DJ (2012) The image: a guide to pseudo-events in America, 50th Anniversary edn. Vintage Books, New York

Booth L (2012) Spirituality and recovery: A classic introduction to the difference between spirituality and religion in the process of healing. Deerfield Beach, FL: Health Communications

Bowman M (2016) The contented collector: materiality, relationality and the power of things. Mater Relig 12(3):384–386

Buzinde CN (2020) Theoretical linkages between Well-being and tourism: the case of self-determination theory and spiritual tourism. Ann Tour Res 83:102920

Cheer JM, Belhassen Y, Kujawa J (2017) The search for spirituality in tourism: toward a conceptual framework for spiritual tourism. Tour Manag Perspect 24:252–256

Chhabra D, Kim EG (2024) Sustaining inner transformation through spiritual tourism. Int J Spa Wellness 7(1):43–64

Christou PA, Pericleous K, Singleton A (2023) Spiritual tourism: understandings, perspectives, discernment, and synthesis. Curr Issue Tour 26(20):3288–3305

Collins-Kreiner N (2020) Religion and tourism: a diverse and fragmented field in need of a holistic agenda. Ann Tour Res 82(102892):1–22

Da Silva JP, Pereira A, Monteiro S, Bártolo A (2019) Qualidade psicométrica de escalas portuguesas de espiritualidade: uma revisão de literatura. Psicologia, Saúde & Doenças 20(3):726–739

Danesh HB (2006) Towards an integrative theory of peace education. J Peace Educ 3(1):55–78

Das A, Kondasani RKR, Deb R (2024) Religious tourism: a bibliometric and network analysis. Tour Rev 79(3):622–634

Davis DE, Hook JN, Worthington EL Jr, Van Tongeren DR, Gartner AL, Jennings DJ (2010) Relational spirituality and forgiveness: development of the spiritual humility scale (SHS). J Psychol Theol 38:91

Delaney C (2005) Development and psychometric testing of a holistic instrument to assess the human spiritual dimension. J Holist Nurs 23(2):145–167

DeJong GE, Faulkner J, Warland R (1976) Dimensions of religiosity reconsidered; evidence from a cross-cultural study. Soc Forces 54(June):866–889

Digance J (2006) Religious and secular pilgrimages: journeys redolent with meaning. In: Timothy D, Olsen DH (eds) Tourism, religion and spiritual journeys. Routledge, London, pp 36–48

Dillette AK, Douglas AC, Andrzejewski C (2021) Dimensions of holistic wellness as a result of international wellness tourism experiences. Curr Issue Tour 24(6):794–810

Dowson R (2020) Religious pilgrimage: experiencing places, objects and events. Int J Relig Tour Pilgr 8(6):24–38. https://doi.org/10.21427/D7VC7D

Fredrickson BL (1998) What good are positive emotions? Rev Gen Psychol 2:300–319

Fredrickson BL (2013) Positive emotions broaden and build. Adv Exp Soc Psychol 47:1–53

Giddens A (1991) As Consequências da Modernidade. Trad. Raul Fiker, ed. UNESP. 5.ª reimpressão. São Paulo, 156 pp.

Davie G (2010) Vicarious religion: a response. J Contemp Relig 25:261–267

Hatch RL, Burg MA, Naberhaus DS, Hellmich LK (1998) The spiritual involvement and beliefs scale: development and testing of a new instrument. J Fam Pract 46(6):476–487

Heintzman P (2013) Retreat tourism as a form of transformational tourism. In: Reisinger Y (ed) Transformational tourism: tourist perspectives. CAB International, Wallingford, pp 68–81

Heydari Chianeh R, Del Chiappa G, Ghasemi V (2018) Cultural and religious tourism development in Iran: prospects and challenges. Anatolia 29(2):204–214

Hervieu-Léger D (2006) The role of religion in establishing social cohesion. In: Michalski K (ed) Religion in the new Europe. Central European University Press, Budapest, Hungary, pp 45–63. https://doi.org/10.1515/9786155053900-003

Huss B (2014) Spirituality: the emergence of a new cultural category and its challenge to the religious and the secular. J Contemp Relig 29(1):47–60

Iliev D (2020) The evolution of religious tourism: concept, segmentation and development of new identities. J Hosp Tour Manag 45:131–140

Instituto Nacional de Estatística (2021) Recenseamento Geral da População, Lisboa

Kim B, Kim S, King B (2020) Religious tourism studies: evolution, progress, and future prospects. Tour Recreat Res 45(2):185–203

Kirillova K (2019) Existentialism and tourism: new research avenues. Int J Tour Cities 5(3):429–442

Kujawa J (2017) Spiritual tourism as a quest. Tour Manag Perspect 24:193–200

Kumar S, Shekhar V, M., Sageena, G. (2022) Spiritual tourism transition: an ethos of Hope during COVID-19 crisis from Indian perspectives. In: Valeri M (ed) Tourism risk. Crisis and recovery management. Emerald Publishing

Lepherd L (2014) Spirituality in men with advanced prostate cancer. J Holist Nurs 32:89–101. https://doi.org/10.1177/0898010113504492

Lifshitz M, Van Elk M, Luhrmann TM (2019) Absorption and spiritual experience: a review of evidence and potential mechanisms. Conscious Cogn 73:102760
Lin C, Fu C (2019) Cognitive implications of experiencing religious tourism: an integrated approach of means-end chain and social network theories. Int J Tour Res 22(1):71–80
Massey D (1994) A global sense of place from space. Place and Gender
McKercher B (2016) Towards a taxonomy of tourism products. Tour Manag 54:196–208. https://doi.org/10.1016/j.tourman.2015.11.008
Meier BP, Hauser DJ, Robinson MD, Friesen CK, Schjeldahl K (2007) What's "up" with god? Vertical space as a representation of the divine. J Pers Soc Psychol 93:699
Moufahim M, Lichrou M (2019) Pilgrimage, consumption and rituals: spiritual authenticity in a Shia Muslim pilgrimage. Tour Manag 70:322–332
Mourtazina E (2020) Beyond the horizon of words: silent landscape experience within spiritual retreat tourism. Int J Cult Tour Hosp Res 14(3):349–360
Nair B, Dileep MR (2021) Drivers of spiritual tourism: a destination-specific Approach. Int J Relig Tour Pilgr 9(5):2. https://doi.org/10.21427/8VK2-JB56
Norman A (2012) The varieties of the spiritual tourist experience. Literature Aesthetics 22(1):20–37
Noronha APP, Rosa PAC, Bernardes LFA (2017) Estudos psicométricos da Escala de Avaliação da Percepção da Religiosidade. Avaliação Psicológica 16(2):215–224
Nowak B, Myers LA, Brzóska PM, Duffy N, Good N, Piotrowski JP (2024) Looks like it is not causal: effects of activation of religiosity and spirituality on the contaminated mindware. Psychol Relig Spiritual
Ouellette P, Kaplan R, Kaplan S (2005) The monastery as a restorative environment. J Environ Psychol 25(2):175–188. https://doi.org/10.1016/j.jenvp.2005.06.001
Paraskevaidis P, Andriotis K (2015) Values of souvenirs as commodities. Tour Manag 48:1–10
Pargament KI (1999) The psychology of religion and spirituality? Yes and no. Int J Psychol Relig 9(1):3–16
Patwardhan AM, Keith ME, Vitell SJ (2012) Religiosity, attitude toward business, and ethical beliefs: Hispanic consumers in the United States. J Bus Ethics 110(1):61–70
Paul Victor CG, Treschuk JV (2020) Critical literature review on the definition clarity of the concept of faith, religion, and spirituality. J Holist Nurs 38(1):107–113
Pereira V (2008) Turismo Cultural e Religioso em Braga e Santiago de Compostela: Proposta de criação de um produto conjunto. (Ed. 1) (pp. 333-345). Santiago de Compostela: Xunta de Galicia
Phillips SS (2021) Walking the spiritual exercises: praying with scripture on the pilgrim trail. The Way 60(1):100–110
Powers C, Nam RK, Rowatt WC, Hill PC (2007) Associations between humility, spiritual transcendence, and forgiveness. Res Soc Sci Study Relig 18:75–94
Preston JL, Shin F (2017) Spiritual experiences evoke awe through the small self in both religious and non-religious individuals. J Exp Soc Psychol 70:212–221
Radvan I (2021) The transformative nature of the spiritual exercises: a radical change of meaning-system. PhD Thesis. Sydney College of Divinity
Radvan I (2024) Conversion through the spiritual exercises of St Ignatius of Loyola. J Psychol Theol 52(1):91–114
Raj R, Griffin K, Blackwell R (2015) Motivations for religious tourism, pilgrimage, festivals and events. In: Religious tourism and pilgrimage management: an international perspective. Cabi, Wallingford
Romanelli M, Gazzola P, Grechi D, Pollice F (2021) Towards a sustainability-oriented religious tourism. Syst Res Behav Sci 38(3):386–396
Sardana D, Gupta N, Sharma P (2018) Spirituality and religiosity at the junction of consumerism: exploring consumer preference for spiritual brands. Int J Consum Stud 42(6):724–735
Schutte CHK, Dreyer Y (2006) Monastic retreat and pastoral care in the Dutch Reformed tradition. HTS Teologiese Studies/Theological Studies 62(4):1453–1468
Schweinsberg S (2023) Religion, spirituality, and the formation of tourism knowledge. Tour Recreat Res 48(4):593–604

Seyer F, Müller D (2011) Religious tourism: niche or mainstream? In: Papathanassis A (ed) The long tail of tourism: holiday niches and their impact on mainstream tourism. Gabler Verlag, Wiesbaden, pp 45–56
Sharpley R, Jepson D (2011) Rural tourism: A spiritual experience? Ann Tour Res 38(1):52–71
Shekhar VM (2023) Modelling the effect of spiritual tourism motivators on spiritual tourism. Tour: Int Interdiscip J. https://doi.org/10.37741/t.71.3.13
Sheldon PJ (2020) Designing tourism experiences for inner transformation. Ann Tour Res 83:102935
Silva C, Abrantes JL, Reis M, Seabra C (2023) Exploring memorable sacred tourism experiences and place attachment. Int J Relig Tour Pilgr 11(5):9
Taylor EJ (2007) Client perspectives about nurse requisites for spiritual caregiving. Appl Nurs Res 20:44–46.
Tellegen A, Atkinson G (1974) Openness to absorbing and self-altering experiences ("absorption"), a trait related to hypnotic susceptibility. J Abnorm Psychol 83(3):268–277
Terzidou M (2020) Re-materialising the religious tourism experience: a post-human perspective. Ann Tour Res 83:102924
Tomass M. K (2012) Religious identity, informal institutions, and the nation-states of the near east. Journal of Economic Issues, 46:705–728.
Tzeng H-M, Yin C-Y (2008) Religious activities of inpatients and their family visitors in Taiwan. J Holist Nurs 26:98–106.
Underwood LG (2011) The daily spiritual experience scale: overview and results. Religions 2(1):29–50
Vada S, Prentice C, Hsiao A (2019) The influence of tourism experience and Well-being on place attachment. J Retail Consum Serv 47:322–330
Valeri M (2022) Tourism risk. Crisis and recovery management. Emerald Publishing, UK
van Cappellen P, Edwards ME, Fredrickson BL (2021) Upward spirals of positive emotions and religious behaviors. Curr Opin Psychol 40:92–98
van Laer T, Izberk-Bilgin E (2019) A discourse analysis of pilgrimage reviews. J Mark Manag 35(5–6):586–604
Vidal-González P, Sánchez V (2019) Hiking paths and intangible heritage: a quest for cultural roots. Cases in the province of Castellón, Spain. Sport Soc 22(12):2065–2076
Victor CG, Paul, Treschuk JV (2020) Critical literature review on the definition clarity of the concept of faith, religion, and spirituality. J Holist Nurs 38(1):107–113
Vonk R, Visser A (2021) An exploration of spiritual superiority: the paradox of self-enhancement. Eur J Soc Psychol 51(1):152–165
Wachholtz AB, Pearce MJ, Koenig H (2007) Exploring the relationship between spirituality, coping, and pain. J Behav Med 30:311–318.
Wang S, Blasco D, Hamzah A, Verschuuren B (2023) Tourists and 'philosophers': nature as a medium to consciousness and transcendence in spiritual tourism. Ann Tour Res 99:103543
Wilkes RE, Burnett JJ, Howell RD (1986) On the meaning and measurement of religiosity in consumer research. J Acad Mark Sci 14:47–56
Willson GB, McIntosh AJ, Zahra AL (2013) Tourism and spirituality: a phenomenological analysis. Ann Tour Res 42:150–168
White ML, Peters R, Schim SM (2011) Spirituality and spiritual self-care: expanding self-care deficit nursing theory. Nurs Sci Q 24:48–56
Wixwat M, Saucier G (2021) Being spiritual but not religious. Curr Opin Psychol 40:121–125
Zinnbauer BJ, Pargament KI, Cole B, Rye MS, Butter EM, Belavich TG et al (1997) Religious and spirituality: Unfuzzying the fuzzy. J Sci Study Relig 36(4):549–564
Zrinščak S (2004) Why, at all, do we need religion? In: Religion and morality in post-communist Europe. A paper was presented at the annual meeting of the Association for the Sociology of religion, San Francisco, California
Zullig KJ, Ward RM, Horn T (2006) The association between perceived spirituality, religiosity, and life satisfaction: the mediating role of self-rated health. Soc Indic Res 79:255–274

Religious Tourism as an Educational Experience

Daliborka Puric, Ljiljana Kostic, and Elena Mirela Samfira

Abstract The purpose of this study was to examine the possibilities of gaining educational experience through the developed model of religious tourism that includes historical, artistic, and cultural elements.

The study design included theoretical research, focus groups, and survey. The sample in the empirical part consisted of in-service teachers and students from Serbia.

Research findings show that organized religious tourism can be used for educational purposes. The developed model of educational-religious tourism can contribute not only to the achievement of teaching goals and subject outcomes, but also to the implementation of educational intentions of a wider scope, which go beyond subject, class, and even institutional frameworks. Content and implementation possibilities of this model, which represents only an initial step in achieving the indicated goals, can be further improved.

1 Introduction

Tourism represents an important element of the global social system and a type of social process, the influence of which is particularly visible in a cultural and economic sense. The constant development of tourism in the last few decades has resulted in great mobility of planet inhabitants and their engagement in discovering other cultures (Isakov 2016). The contact of individuals and human communities caused mutual touches of different cultures and civilizations, which is a key prerequisite for establishing intercultural dialogue. Since tourism is the activity that "most

D. Puric (✉) · L. Kostic
Faculty of Education in Uzice, University of Kragujevac, Uzice, Serbia
e-mail: puric@pfu.kg.ac.rs; kostic@pfu.kg.ac.rs

E. M. Samfira
Teacher Training Department, University of Life Sciences from Timisoara, Timisoara, Romania
e-mail: mirelasamfira@usvt.ro

M. Valeri (ed.), *Contemporary Religious Tourism*, Tourism, Hospitality & Event Management, https://doi.org/10.1007/978-3-031-79074-4_5

comprehensively encourages and promotes the cultural heritage" of a nation, mutual relations between tourism and culture have become one of the basic points for cultural policy at the national and international level (Vidić 2002). In this regard, one of the most significant propulsive forms of tourism at the global level is religious tourism, a kind of connection to the human need to move and learn new things and to satisfy spiritual, cultural, and social needs.

2 Religious Tourism

Religious tourism represents one of the selective forms of tourism, which can be motivated by confessional affiliation to a religion, religious feeling, and the desire to strengthen faith (pilgrimages and devotional trips), as well as the need to get to know cultural (historical, ethnographic, artistic, and other) aspects of a religious object or destination (Micunovic et al. 2015). Originated in the earliest periods of human civilization, as an expression of the desire and obligation to participate in a religious ceremony, this form of tourism has become one of the most important sectors of global tourism, as over 20% of tourists on the entire planet travel within it today (Isakov 2016). In addition to the devotional element, religious tourism has acquired a cultural and educational component over time, that is, it does not necessarily mean religious tourists, since the majority of religious tourists visited holy places and sacred objects on the way to their final destinations, even though they are not traditional believers, but, above all, art and culture devotees (churches, monasteries, monastery libraries, icons, frescoes, calligraphy, etc.) (Radović and Radović 2017). In this sense, religious tourists (homo turisticus religiosus) are categorized on the basis of their motivation into religious and secular (Adler 1985; Andriotis 2009; Smith 1992). Blackwell (2007) distinguishes three types of travelers: travelers who are more believers than tourists, travelers who are equally believers and tourists, and travelers who are more tourists than believers. They can be members of the same or different religions or atheists or visit a religious place for some other, nonreligious reason. According to the degree and intensity of beliefs and feelings, there are differences between religious tourists who are convinced of their religious attitudes and religious tourists who have a certain reserve towards religious principles, are less familiar with theological dogmas, and do not visit holy places exclusively for religious reasons (Stamenković 2005). Thus, the example of a religious tourist reflects Eliade's position that "the sacred and the profane represent two ways of being in the world, two existential situations that man has taken on during his history" (Elijade 2003).

Religious tourism is multidisciplinary by nature, since it appears in various forms—from pilgrimage to tourism that is not exclusively religiously motivated (Bulatović and Najdić 2017). More precisely, it can be motivated by religious, health, cultural, and educational reasons, while for newer generations (Generation G, Generation Z) the sources of motivation can be curiosity, research, socializing, getting to know each other, entertainment, etc. (Zhou 2013).

3 Religious Tourism as a Cultural and Educational Experience

At the first congress of religious tourism in Ravenna in 1990, several important features of this form of tourism were highlighted—"humane promotion, form of solidarity and friendship, meeting of peoples, their cultures and religions, teaching peace among people, respect for nature and valorization of natural and cultural assets" (Plavša 2000).

Every journey represents "questioning and pushing the limits of one's own being" (Stamenković 2005). Elijade (2003) viewed the journey as a re-creation or reorientation of the spiritual cosmos. It broadens horizons, encourages thinking and rethinking attitudes, frees prejudices, and contributes to getting to know others and greater openness in communication and cooperation with others (members of other nations, other religions, etc.) (Valeri 2023). Religious tourism contributes to the strengthening of religious and ethnic tolerance and reduces ethnocentrism and bigotry. It also makes it possible to perceive one's own history, culture, beliefs, attitudes, and lifestyle from a new perspective. Therefore, religious tourism contributes to developing cultural competencies of a person and building tolerance, as an important factor in reducing social tensions, preserving peace, and solidarity.

Religious tourism has an "identity, integrative and intercultural function in modern society" (Radisavljević Ćiparizović 2011). In conditions of the modern era, the time of migration, globalization, and transition, but also a kind of desecularization, religious tourism has a role in the process of redefining the identity of an individual and part of a nation (Mitrović 2011). In this regard, religious tourism should be associated with a rational view of the world—the ultimate goal is not an encounter with the deity, but an intellectual and spiritual exploration of one's own cultural and religious roots. In this way, the individual will establish his/her religious identity in the context of wider society and the community of which he/she is an integral part (Cohen 2006). The religious tourist will thus become an active participant of the process of developing interculturalism in the modern world, which promotes dialogue between different cultures and peoples, spreads the values of the culture of peace, and understanding among peoples (Radisavljević Ćiparizović 2011).

Religious tourism has an important role in learning national history and culture, in preserving and promoting folk customs and traditions, since within this form of tourism, objects that date back to earlier centuries are visited and which, in addition to religious and historical, have cultural significance (in terms of architecture, art, literature, language, etc.). Getting to know and preserving cultural heritage as a part of national identity is an important educational task, but also a social obligation, and, depending on the way it is implemented, it can have two effects: a direct effect, through educational motivation and educational outcomes, and an indirect effect, through other forms of tourist motivation, such as religious, environmental, rural, and cultural. Therefore, the educational motivation of tourists can be effectively considered a factor through which to expand, not only their knowledge, but also their religious, cultural, and environmental experiences (Ramírez and Portillo 2018).

The need to change places is immanent to man—he travels, first of all, out of the need for knowledge, experience, and discovery of the unknown (Živković et al. 2017). In the basis of religious tourism is the question of spirituality and faith, and the answers are found in journeys that, in addition to religion, include culture, history, and education, since the places that are visited are also viewed in the function of enriching knowledge about the world and all aspects of life. In contact with the new, knowledge is enriched, experiences are expanded, and creative thinking is stimulated. In this context, religious tourism represents a kind of cognitive tourism since it satisfies the tourist's need for knowledge (Kurmanaliyeva et al. 2014; Baydeniz et al. 2024). The cognitive aspect of tourism in general, including religious tourism, is based on experiential learning, i.e., on "co-creation of desired experiences" (Richards 2011), which includes things and people outside their immediate environment, thus representing a prerequisite for effective global learning (Deloach et al. 2015; Lilley 2014).

Medieval pilgrimages were completely devoid of the desire for education (primary was the quest for redemption or healing). In the new century, travel acquired an educational dimension—in the Age of Enlightenment it became an "act of culture," but also a form of self-discovery (Konstantinović 1966), since a person can judge himself and his wider and narrower environment only when he meets others and foreign culture (Stefanović 2009). The popularization of institutionalized religious education, the development of heritage tourism, and the importance of informal education in identity formation have all influenced the emergence of educational-religious tourism for religious reasons but also for educational purposes (Cohen 2006). Visits to religious objects became part of cultural education, anthropological interest in the Other, and in one's own religious roots (Cohen 2006).

According to the research findings, the educational perspective has a secondary role in the context of religious tourism (Ramírez and Portillo 2018) and in some studies it is mentioned in the theoretical part, but not in the empirical part (Amaro et al. 2018; Terzidou et al. 2018). The authors point out that "educational reasons" belong to the category of less important elements of tourists' motivation to visit religious sites (Kamenidou and Vourou 2015). Research attention is paid to educational motivation in the context of religious tourism, first of all from the psychological aspect—to acquire and/or expand knowledge about their immediate environment, individuals are encouraged by feelings of enjoyment stimulated by a tourist visit to a religious object (Ramírez and Portillo 2018). The educational implications of religious tourism are, furthermore, related to the specificity of attitudes towards religion, expectations from journey, and priorities in education (Cohen 2006).

4 Model of Educational-Religious Tourism

Starting from the potential of religious tourism to satisfy the educational motives of tourists, who learn from the historical, cultural, and artistic heritage of the area and/or object they visit (Olsen and Timothy 2006), and bearing in mind the goal and outcomes of the education system in Serbia, as well as respecting the developmental characteristics of younger school-age pupils, a model of educational-religious tourism (ERT) was developed, with the aim of examining the possibilities of gaining educational experience through systematically organized tourist visits to religious facilities. The bearers of model creation and implementation are primary schools (teachers and school management) and the local community (local self-government, local/regional tourism organization, local/regional religious communities, cultural institutions, experts from the fields covered by the model, pupils' parents). The model is intended for pupils from the first to the fourth grade of elementary school, and the indirect users are also their families.

Considering that the *Teaching and Learning Program* (2019) highlights as desirable the organization of activities outside the classroom, i.e., in the natural and social environment, the implementation of the ERT model is foreseen through outdoor education, organized visits, and excursions. The model includes three basic stages: (1) preparation for the visit, (2) visiting a religious facility, and (3) curricular and extracurricular activities motivated by the tourist visit. The ERT model envisages one educational-religious tourist visit in each of the younger grades, namely, in the first grade, to a Christian religious facility in the wider area of the local community; in the second grade, a Christian religious facility in the area of the district; in the third grade, a selected religious facility in the state; and in the fourth grade, an Islamic or Judaic religious facility in the state.

The educational content that is an integral part of the model consists of three basic elements:

- Historical: the origin of religious object in the historical context; role and significance in other historical events; famous personalities—interesting stories based on historical sources, suitable for younger school age
- Artistic: literature, fine arts, painting, sculpture, architecture, music—written, material and oral historical sources (epic poems, legends, fairy tales) considered in the context of certain types of art, as well as in the context of historical chronology (artistic works of contemporary authors inspired by certain religious facilities), with adaptation to younger school age
- Cultural: written, material, and oral historical sources within the religious facility considered in the context of national (development of language, literature, art) and international cultural heritage with adaptation to younger school age

Creation and implementation of the ERT model includes the following activities:

- Organization of religious object visit—by the local government, local/regional tourist organization, local/regional religious community, and school management.
- Creation of educational content intended for pupils, related to a religious object—by experts of various profiles (historians, art historians, architects, curators, literature or music teachers, priests, religious teachers) from the local community, hired by the local self-government and schools. Also, interested parents can participate in the creation of educational content.

- Introducing educational content to pupils—this segment can be implemented by experts who participated in the creation of educational content within the model, but also by teachers or others (curators, actors, etc.), competent to adequately present the created content to pupils.
- Activities after the tourist visit, motivated by the educational experience—teaching and/or extracurricular activities are carried out by teachers, as well as interested parents.

In order to make the model sustainable, schools can be "networked" through its implementation, in cooperation with other participants from the local community. Also, in the later stages of model development and implementation, interested pupils who were previously included in the model can take over certain elements of creating educational content and introducing students to them. Peer education can represent a significant element of learning support, taking into account cognitive, emotional, and social interaction between pupils.

Implementation of the developed ERT model is justified in multiple ways—from the educational, cultural, and economic aspects. Connecting religious and educational elements contributes to the achievement of educational goals and prescribed subject outcomes. Also, in a broader sense, created educational content—historical, artistic, and cultural—can become an element of the local and/or regional tourist offer, which enables the acquisition of educational and cultural experience not only for pupils involved in the model implementation, but also for other categories of tourists. Getting to know cultural heritage contributes to developing awareness of the need to preserve it. The model implementation contributes directly and indirectly to the economic empowerment of the local community. Direct impact is achieved by pupils' visits, which increases the number of tourists, and indirect impact—by enriching the tourist offer with educational content created within the model, but also by motivating experts of various profiles to think about the content related to the religious place, which can result in further tourist offer development and enrichment.

5 Possibilities of Gaining Educational Experience by Implementing the ERT Model

The objective of the empirical research was to examine the possibilities of gaining educational experience by implementing the developed ERT model. Research tasks were related to (1) defining the success factors of the developed ERT model and (2) examining teachers' and future teachers' attitudes about (a) success factors of the developed ERT model, (b) the possibilities of achieving prescribed educational goals and outcomes by implementing the ERT model, and (c) teacher competencies for implementing the ERT model.

Techniques and instruments Success factors of the developed ERT model were defined in focus groups that consisted of teachers and students. Teachers' and future teachers' attitudes were examined using a survey questionnaire created for the purposes of this study. The instrument is a 35-item measure, using a five-point Likert scale, from (1) strongly disagree to (5) strongly agree. Alpha Cronbach for the scale was 0.87 for the present research.

Participants A total of 15 teachers and 11 students from Serbia participated in focus groups. The online platform Zoom was used to conduct group interviews. A total of 237 in-service teachers from Serbian public schools and 171 students from Serbian universities participated in the survey, while members of the focus groups did not participate in the survey. Attitudes of teachers and future teachers were examined as two separate sample groups because the perspective of teachers, who have different experiences in direct practice, as well as students, whose attitudes, among others, reflect the modern approach to the formal education of experts in this field, is important for evaluating the model elements.

Procedure Two separate focus groups were held—teachers participated in one, and students in the other. The group interview was focused on the success factors of the developed ERT model—strengths, weaknesses, opportunities, and threats (SWOT).

Teachers and students, potential participants in the survey, were first presented with the objective of the study and the research procedure. Then they were introduced to the ERT model in detail, in written form. The measurement was carried out using an online survey system (Google Forms). Respondents who agreed to participate in the research were sent a link to the web questionnaire, which contained the agreement, instructions for filling it in, as well as ethical issues (respecting the privacy and the right to withdraw, without any negative consequences). Filling out the questionnaire takes about 15 min.

Data analysis Survey results were analyzed using statistical software package SPSS 20.0. Descriptive statistics and the chi-square test were used to examine the differences in teachers' and students' attitudes about the success factors of the ERT model and the possibilities of its implementation, depending on the professional qualification and work experience of teachers, and the year of study and academic success of students.

Results and discussion Discussions in focus groups resulted in defining the success factors of the developed ERT model (Table 1).

Examining teachers' and students' attitudes regarding the defined success factors of the ERT model shows a high degree of agreement among respondents. When it comes to the model advantages, teachers (SD = 0.662) and students (SD = 0.505) agree to the greatest extent with the statement that *the model implementation enables experiential learning*. This type of learning, in the context of the tourist perspective,

Table 1 *SWOT analysis of the ERT model*

Strengths	*Weaknesses*
– Achieving the educational outcomes of different teaching subjects by implementing the model – Systematic linking of knowledge from different subject areas through the model – Developing pupils' cross-curricular competencies – Direct introduction to cultural heritage through the implementation of the model – Enabling experiential learning within the model	– Pupils' motivation to acquire knowledge within the model – The impossibility of simply differentiating educational and religious content when implementing the model – External experts' motivation for involvement in the model implementation – Complexity of organizing external experts' activities during the model implementation
Opportunities	*Threats*
– Developing interculturality by implementing the model – Encouraging intrinsic motivation for learning through the model – The model's contribution to durability of experientially acquired knowledge – Educational contents thematic planning within the model	– Teachers' motivation for the model implementation – Parents supporting children to participate in the model implementation – External experts' competences for working with younger school-age children

enables the acquisition, transformation, application, and testing of what has been experienced outside the borders of everyday life (Deloach et al. 2015), which turns out to be a prerequisite for effective global learning (Lilley 2014). On the other hand, the greatest disagreement occurs when assessing the model contribution to *the achievement of educational outcomes of different teaching subjects*, both by teachers (SD = 0.961) and by students (SD = 0.748). Therefore, respondents differently evaluate the potential of the model for achieving prescribed educational outcomes, which may be caused by differences in the perception of the created model, but also by ambiguities regarding the formulation of learning outcomes and their connection with educational goals and outcomes (Mikanović 2014).

Teachers with a higher level of education recognize to a greater extent the importance of the ERT model for cross-curricular competencies development ($\chi^2 = 5.461$; df = 3; $p = 0.141$), which can be explained by the contribution of teaching master's program to the improvement of their professional competencies, especially those related to changes in the educational system.[1]

When assessing the weaknesses of the model, teachers (SD = 0.836) and students (SD = 1.041) agree about the complexity of *organizing the external experts activities in the model implementation*, and the most disagreements in teachers' attitudes occur when assessing *pupils' motivation to learn within the model* (SD = 1.124), while students differ to the greatest extent in assessing *the external experts' motivation for involvement in the model implementation* (SD = 1.105). Therefore, the key

[1] *The Law on the Basics of Education and Training*, which prescribes general cross-curricular competencies, was adopted in 2017.

weakness of the model is the motivation of the participants in its implementation. Considering that educational motivation is gradually becoming an element of modern religious journeys (Ramírez and Portillo 2018) and that "there is no apparent dichotomy between pilgrims and tourists, who are only differentiated by their travelling purposes" (Weidenfeld 2006), it is possible to motivate younger school-age pupils to gain educational experience within the model, if the contents are designed in a way to engage their curiosity and research spirit, which raises the question of the external experts' motivation to design an adequate approach to pupils of this age. Reasons for the disagreement of respondents from both subgroups can be explained precisely by their awareness of the complexity of such an approach.

Examining the attitudes regarding the ERT model possibilities shows the highest agreement of teachers (SD = 0.879) and students (SD = 0.770) when assessing the contribution of the model implementation *development of intrinsic motivation for learning*, which corresponds to research according to which feelings of enjoyment stimulated by a tourist visit to a religious object motivate tourists to new research and, therefore, for further learning (Ramírez and Portillo 2018). Teachers (SD = 1.021) and students (SD = 1.076) mostly differ in the attitude that this model *enables thematic planning*. The ERT model is harmonized with the model of thematic-integrative approach to contents of subjects of importance for nurturing the Serbian culture and developing the national identity, defined at the national level (*Priručnik za nastavnike* 2024).

When assessing potential threats to the model implementation, teachers (SD = 1.076) and students (SD = 1.003) mostly agree that *parents' support for their children participation in the model implementation is unquestionable*. The lowest agreement of teachers (SD = 1.098) is observed when assessing *external experts' competences for working with younger school-age children* and the lowest agreement of students (SD = 1.243) when assessing *teachers' motivation for implementing the model*. Teacher motivation is an important prerequisite for both the creation and implementation of the ERT model, and it can be conditioned by the motivation of students and other participants, effective working conditions, autonomy in the implementation of responsibilities related to the model, and the perception of the model implementation success (Börü 2018). Potential negative effects of inadequate adaptation of educational content to younger school-age pupils can be prevented by teamwork, in which teacher should play a key role.

Evaluating the educational goal achievement through the model implementation, teachers (SD = 0.548) and students (SD = 0.495) mostly agree that *the model develops cross-curricular competencies*, while the greatest disagreement between teachers (SD = 0.860) and students (SD =0.753) appears regarding *the contribution of the model implementation to personal and national identity development*. The inconsistency of teachers' and students' attitudes may indicate that they do not perceive development of personal and national identity as a primary element in the created model, but as an added value, bearing in mind the age of pupils included in the model.

Teachers with different work experience expressed significantly different views regarding the claims that the model implementation *contributes to the respect and preservation of national and world cultural heritage* ($\chi^2 = 19.948$; df = 9; $p = 0.018$)

and to *the development of positive human values* ($\chi^2 = 13.568$; df = 6; $p = 0.035$). The obtained results corresponds with research findings, according to which the contribution of educational experience in the context of religious tourism is reflected in the formation of identity, values, and beliefs, building positive interpersonal relationships and group dynamics (Jeffs and Smith 1990; Cohen 2006).

Teachers and students mostly do not agree about the impact of the model on the outcomes of certain teaching subjects. Teachers to the greatest extent recognize suitability of the model for achieving the outcomes of *Nature and Society* (SD = 0.635) and students, the outcomes of *Fine Arts* (SD = 0.607). The assessments of teachers (SD = 0.966) and students (SD = 0.950) differ the most regarding the outcomes of *Music Education*.

Teachers' professional training affects the outcome achievement evaluation of *Music Education* ($\chi^2 = 14.455$; df = 3; $p = 0.002$), *Religious Education* ($\chi^2 = 10.703$; df = 3; $p = 0.013$), and *Civic Education* ($\chi^2 = 14.887$; df = 3; $p = 0.002$), and their work experience affects the assessment of outcomes of *Fine Arts* ($\chi^2 = 13.443$; df = 6; $p = 0.037$) and *Civic Education* ($\chi^2 = 18.596$; df = 9; $p = 0.029$). Significant differences in students' attitudes are noted regarding the model influence on outcomes of *Nature and Society* ($\chi^2 = 9.855$; df = 4; $p = 0.043$) and *Fine Arts* ($\chi^2 = 13.329$; df = 4; $p = 0.010$). Differences in teachers' attitudes regarding the teaching subject outcomes can be interpreted by their assessing the representation of subject content in the ERT model. The difference in attitudes of both teachers and students indicates that in further model development, greater attention should be paid to connecting the educational content within the model with the prescribed teaching subjects educational outcomes, which will enable adequate teachers respond to the complex requirements within the model and will also contribute to improving the quality of its implementation.

The self-assessment of teachers competence for the model implementation shows the highest agreement regarding *teachers qualification to implement the model* (SD = 0.637), while the students agree to the greatest extent that *teacher has communication competences for implementing the model* (SD = 0.542). The most different attitudes of teachers (SD = 0.940) and students (SD = 0.980) appear in connection with *the necessity of nonformal and informal, in addition to formal education required for the model implementation*. Therefore, teachers are focused on competencies for implementing the model, while students are focused on the possibilities of developing those competencies.

Teachers with different professional qualifications express significantly different attitudes regarding their *professional* ($\chi^2 = 11.297$; df = 3; $p = 0.010$), *organizational* ($\chi^2 = 16.036$; df = 3; $p = 0.001$), and *motivational competencies for the model implementation* ($\chi^2 = 12.300$; df = 3; $p = 0.006$), while teachers' work experience affects the assessment of indicators related to *professional* ($\chi^2 = 19.773$; df = 9; $p = 0.019$) and *motivational competencies for the model implementation* ($\chi^2 = 20.901$; df = 9; $p = 0.013$). Year of study has a significant effect on students' attitudes regarding *teachers' motivational competencies* ($\chi^2 = 16.976$; df = 6; $p = 0.009$). The attitudes of respondents from both subgroups differ to the greatest extent in relation to motivational competences of teachers, which can be explained by the complexity of this problem, considering that motivational competences can refer to different forms of teacher activity—motivating pupils to learn educational content, as well as to developing a positive school climate, motivating parents and other interested parties to support pupils and support the educational process, motivating decision-makers for systemic support during necessary changes of the educational system elements, which as a "living system" modificates and adapts to broader social processes.

6 Theoretical and Practical Implications

Although there is extensive literature on the various benefits of religious tourism, both theoretical and empirical findings on religious tourism in an educational context are quite scarce. In addition, the concept of religious tourism in the educational system is not comprehensively defined, which is one of the reasons why it is not seen as a relevant source for knowledge and a broader understanding of the history, culture, and art of a nation or an era. This study draws researchers' attention to the importance and educational benefits of religious tourism and creates an original model of educational-religious tourism based on relevant studies.

From a practical aspect, we hope that the developed ERT model will be implemented in schools and that other researchers and practitioners will continue to develop it and adapt it to educational practice, not only in Serbia. Also, the model can represent a starting point for teachers and textbook authors to create an environment for learning and personal development of students, in the domain of developing skills and abilities, meeting new and varied people, and gaining a new perspective in life (Ramírez and Portillo 2018).

7 Limits of the Study and Future Directions

The model of educational-religious tourism in this study was created based on the educational system in Serbia, and the obtained results should be interpreted in that context. Bearing in mind cultural differences and different models of national education systems, it is recommended to conduct similar research in other countries and in other geographical and cultural regions, which would contribute to clarifying this problem and generalizing the conclusions. Since this study deals with the possibilities of achieving teaching goals, subject outcomes, and teacher competencies in the function of implementing the ERT model, it is necessary to examine how the created model reflects on other elements of teaching, primarily on the pupil as a participant in teaching process. However, certain limitations of this study in terms of survey bias can be avoided by designing future research procedures using more objective measurement techniques.

Although religious tourism is the focus of numerous studies, its educational potential has not yet been recognized to a significant extent. The model of educational-religious tourism created in this study contains three key elements—historical, artistic, and cultural—which appear in the cross section of most studies dealing with the educational aspect of religious tourism. Considering the fact that the model was designed in a theoretical framework, it would be important to examine the possibilities and limitations of its practical application.

At the same time, we should take into account research that states the destructive effects on cultural unity and natural environment (ecology) of the area where the religious object is located (Olsen and Timothy 2006), despite the fact that tourism represents one of the ways of preserving cultural heritage and religious facilities, as well as a form of stimulating the economy of a region or country. Religious tourism has also had negative effects on some of the most visited touristic places—for example, some monasteries in Meteora in Greece have lost their primary function since they are no longer inhabited by monks. Therefore, during the further development and implementation of the ERT model, it is necessary to take into account the need to prevent its negative effects (Micunovic et al. 2015), in the sense of disrupting the existence and operation of religious objects.

8 Conclusions

The results of this study show that organized religious tourism can be applied for educational purposes. The developed model of religious tourism, according to the assessment of teachers and students—future teachers—can contribute to the acquisition of educational experience for younger school-age pupils. The ERT model implementation enables direct acquaintance with the elements of national cultural heritage, as well as the co-creation of experiential knowledge, which contributes to the development and preservation of national and cultural identity, the development of tolerance, and appreciation for diversity. The ERT model contributes to the realization of educational intentions of a wider range, which go beyond subject, class, and institutional frameworks, such as developing a sense of authentic aesthetic values or educating young people in the spirit of the highest moral values (humanism, solidarity, patriotism, altruism). At the same time, the ERT model represents only the initial step in achieving the indicated goals, and its content and implementation possibilities could be further improved.

References

Adler J (1985) Youth on the road: reflection on the history of tramping. Ann Tour Res 12(3):335–354

Amaro S, Antunes A, Henriques C (2018) A closer look at Santiago de Compostela's pilgrims through the lens of motivations. Tour Manag 64:271–280. https://doi.org/10.1016/j.tourman.2017.09.007

Andriotis K (2009) Sacred site experience. Ann Tour Res 36:64–84. https://doi.org/10.1016/j.annals.2008.10.003
Baydeniz E, Çilginoglu H, Valeri M (2024) Impact of health tourism visitor experiences on behavioral intention. J Organ Chang Manag 37(5):1133–1153. https://doi.org/10.1108/JOCM-09-2023-0381
Blackwell R (2007) Motivations for religious Tourism, pilgrimage, festivals and events. In: Raj R, Morpeth ND (eds) Religious tourism and pilgrimage festivals management: an international perspective. Cab International, Oxfordshire, pp 35–47
Börü N (2018) The factors affecting teacher-motivation. Int J Instr 11(4):761–776
Bulatović V, Najdić M (2017) Izazovi i mogućnosti razvoja verskog turizma u multikulturalnoj i multireligijskoj sredini. In: Stanišić M (ed) Religious tourism and the contemporary tourism market. Singidunum University, Belgrade, pp 75–80. https://doi.org/10.15308/Sitcon-2017-75-80
Cohen E (2006) Religious tourism as an educational experience. In: Timothy D, Olsen D (eds) Tourism, religion and spiritual journeys. Routledge, New York, pp 78–93
Deloach SB, Kurt M, Olitsky NH (2015) Does content matter? Analyzing the change in global awareness between business- and nonbusiness-focused short-term study abroad courses. J Teach Int Bus 26(1):4–31. https://doi.org/10.1080/08975930.2014.929512
Elijade M (2003) *Sveto i profano*. Sremski Karlovci–Novi. Izdavačka knjižarnica Zorana Stojanovića, Sad
Isakov S (2016) Umetničke prakse u kontekstu kulturnog turizma. Fakultet za hotelijerstvo i turizam, Vrnjačka Banja
Jeffs T, Smith M (eds) (1990) Using informal education. Open University Press, Buckingham
Kamenidou I, Vourou R (2015) Motivation factors for visiting religious sites: the case of Lesvos Island. Euro J Tour Res 9:78–91
Konstantinović Z (1966) Putovanje kao čin kulture (Dositej i naše veze sa svetom). In: Pavić M (ed) Od baroka do klasicizma. Nolit, Beograd, pp 453–456
Kurmanaliyeva A, Rysbekova S, Duissenbayeva A, Izmailov I (2014) Religious tourism as a sociocultural phenomenon of the present "the unique sense today is a universal value tomorrow. This is the way religions are created and values are made". Procedia–Soc Behav Sci 143:958–963. https://doi.org/10.1016/j.sbspro.2014.07.534
Lilley K (2014) Educating global citizens: translating the 'idea' into university Organisational practice. Discussion. International Education Association of Australia, Melbourne
Micunovic G, Novakovic N, Stefanovic G (2015) Religious tourism and possible negative effects in relation to the monasteries. BizInfo J 6(1):43–52. https://doi.org/10.5937/BIZINFO1501043M
Mikanović B (2014) Ishodi učenja i standardi znanja u osnovnom obrazovanju. Inovacije u nastavi XXVII(1):84–93
Mitrović L (2011) Sociološko traganje za dubinskim matricama "nevidljivog društva". In: Radisavljević Ćiparizović D, Todorović D (eds) Hodočašća—između svetog i svetovnog. JUNIR, Niš, pp 17–19
Negovanje kulture srpskog naroda i razvijanje nacionalnog identiteta (2024) Priručnik za nastavnike. ZUOV, Beograd
Olsen D, Timothy D (2006) Tourism and religious journeys. In: Timothy D, Olsen D (eds) Tourism, religion and spiritual journeys. Routledge, New York, pp 1–21
Plavša J (2000) Turističke regije sveta—skripta. PMF, Novi Sad
Radisavljević Ćiparizović D (2011) Sociologijski zapis o verskom turizmu. In: Radisavljević Ćiparizović D, Todorović D (eds) Hodočašća—između svetog i svetovnog. JUNIR, Niš, pp 9–16
Radović N, Radović M (2017) Verski turizam u funkciji razvoja turističkih destinacija Srbije. In: Stanišić M (ed) Religious tourism and the contemporary tourism market. Singidunum University, Belgrade, pp 51–54. https://doi.org/10.15308/Sitcon-2017-51-54
Ramírez RR, Portillo AF (2018) What role does touristś educational motivation play in promoting religious tourism among travellers? Ann Leisure Res 23(4):1–22. https://doi.org/10.1080/11745398.2018.1561309

Richards G (2011) Cultural tourism trends in Europe: a context for the development of cultural routes. In: Khovanova-Rubicondo K (ed) Impact of European cultural routes on SMEs' innovation and competitiveness. Council of Europe Publishing, Strasbourg, pp 21–39

Shekhar, Valeri M (2023) Modelling the effect of spiritual tourism motivators on spiritual tourism. Tour: Int Interdiscip J 71(3):634–646. https://doi.org/10.37741/t.71.3.13

Smith VL (1992) Introduction: the quest in guest. Ann Tour Res 19(1):1–17. https://doi.org/10.1016/0160-7383(92)90103-V

Stamenković I (2005) Religious tourism, types and reasons regarding the ortodox. Zbornik radova Departmana za geografiju, turizam i hotelijerstvo 33–34:173–188

Stefanović M (2009) Leksikon srpskog prosvetiteljstva. Službeni glasnik, Beograd

Terzidou M, Scarles C, Saunders MN (2018) The complexities of religious tourism motivations: sacred places, vows and visions. Ann Tour Res 70:54–65. https://doi.org/10.1016/j.annals.2018.02.011

Vidić N (2002) Revitalizacija fruškogorskih manastira u funkciji kulturnog i verskog turizma. Turizam 6:78–79

Weidenfeld A (2006) Religious needs in the hospitality industry. Tour Hosp Res 6:143–159

Zhou T (2013) Understanding the effect of flow on user adoption of mobile games. Personal and Ubiquitous Computing 17(4):741–748

Živković R, Brdar I, Gajić J, Stanković J (2017) Digitalna promocija verskog turizma. In: Stanišić M (ed) Religious tourism and the contemporary tourism market. Singidunum University, Belgrade, pp 132–139. https://doi.org/10.15308/Sitcon-2017-132-139

Exploring the Multifaceted Impact of Religious Tourism: A Multidisciplinary Perspective

Bapon Chandra Kuri, Ayan Banik, and Marco Valeri

Abstract Religious tourism impacts both destinations and tourists alike by embracing elements from marketing, economics, spirituality, and sustainability. The aim of this chapter is to review literature on religious emotions and their association with travel and, through case studies and data analysis, explaining why people travel to sacred places. The result reveals that religious destinations come with issues such as overcrowding and unsustainable practices but at the same time offer opportunities for community engagement and cultural exchange. The chapter states a multidisciplinary approach in dealing with the religious sites of tourism and acknowledges limitations related to qualitative methods and the need for future research. Moreover, in the near future, sacred destinations will take the form of a virtual platform; therefore, this study highlights a model that describes how the digital transformation of religious tourism can augment tourist experiences through AI, VR, and AR.

1 Introduction

The size of the worldwide market for religious tourism is predicted CAGR (compound annual growth rate) of 10.2% between 2024 and 2032, and the global religious tourism market, estimated to be worth USD 15.5 billion in 2023, is expected to grow to USD 37.2 billion by 2032 (Vantage 2023). This data triggers the paramount significance of religious tourism which introduces pilgrimage journeys, visiting sacred places, and spiritual quests. Nowadays, such a course of travel is

B. C. Kuri (✉) · A. Banik
Department of Tourism and Hospitality Management, Gopalganj Science and Technology University, Gopalganj, Bangladesh

M. Valeri
Faculty of Economics, Niccolò Cusano University, Rome, Italy

Faculty of Social Sciences and Leisure Management, School of Hospitality, Tourism and Events, Taylor's University, Subang Jaya, Selangor, Malaysia
e-mail: marco.valeri@unicusano.it

M. Valeri (ed.), *Contemporary Religious Tourism*, Tourism, Hospitality & Event Management, https://doi.org/10.1007/978-3-031-79074-4_6

important, being channeled through faith, culture, and travel itself. It fuses the dimensions of faith, culture, and traveling, crafting experiences in different ways for destinations and pilgrims. Therefore, this chapter brings us closer to the complicated world of religious tourism.

However, religious tourism touches on a soul, seeking spiritual fulfillment or personal transformation (Kim et al. 2020). When people travel to famous pilgrimage sites, they not only take part in religious festivals or retreats but also attach themselves to their own beliefs and heritage (Rashid 2018). In this regard, the current chapter delves into some of the motivations that are behind religious tourism. Moreover, we take into consideration the quests behind those journeys: whether it is for divine blessings or cultural experiences.

From the economic perspective, religious tourism plays a big part in helping local economies to grow and developing infrastructure in visited areas (Rinschede 1992). Moreover, pilgrimage sites and festivals attract millions of people per year, in turn raising the revenue of businesses from hotels to shops selling artifacts. These pose some other setbacks such as overtourism that might end up losing allurement and sense of religious place. In addition, exploring the economic dimension of religious tourism reflects ways that improve its benefits and at the same time avoid the negative impacts on communities and the environment (Jongmeewasin 2016). Cultural importance is yet another driving force behind religious tourism which carries enormous cultural heritage and spiritual traditions (Kim and Kim 2019). For example, Angkor Wat in Cambodia and the holy Ganges River in India are places full of rich cultural and spiritual meaning. However, a flow of tourists can sometimes threaten cultural identity and preservation, which is challengingly related to their cultural identity (Durán-Sánchez et al. 2018). In this context, we explore really how cultural components can uplift spiritual tourism through religion, culture, and heritage, without the temptations to succumb to growing tourism. More often, religious tourism ignites social development in that it creates employment and is supportive of education and local infrastructure (Olsen and Timothy 2022). However, caution must be observed about issues that concern social justice and the preservation of indigenous cultures from being swamped out. On the other hand, in psychological aspects, most travelers go through moments where they are reflecting or renewing their spirits. Furthermore, feelings experienced at sacred points can be a source of personal growth, springing from common rituals and experiences. By looking at the psychological side of it, we are going to learn how it enhances personal well-being and boosts spiritual growth.

In summary, religious tourism, through the interplay of faith, culture, and community, features promiscuously in the life of the conscientious traveler, seeking roots and meaning in a sea of affinity and choice. In this chapter, we want to open up, in a friendly way, its many dimensions: economic, cultural, social, and psychological dimensions that make religious tourism special, both for destinations and for visitors.

2 Literature Review

Religious tourism has attracted scholars' interest across a wide array of subjects, including marketing (Kartal et al. 2015), economics (Vukonić 1998), psychology (Lochrie et al. 2019), sociology (Firdaus and Rahmat 2019), anthropology (Badone and Roseman 2004), and tourism studies. It helps to see the vibrancy and effectiveness of religious tourism which brings the world's people on one platform to explore the religious authenticity of the local communities. However, some view these trips as a way by which one could find spiritual enlightenment, create a link with the divine, or find meaning in life (Hassan et al. 2022). Others may take such journeys because they need to fulfill religious obligations or wish to experience the blessings of God or to be a part of sacred ceremonies (Olsen and Timothy 2006). However, religious tourism enables tourists to get immersed in various cultures and get an on-site experience of different religious customs and practices (Cànoves et al. 2012). Moreover, pilgrimage spots and festivals create job opportunities for the community people by which they can develop their standard of living (Luz 2020). However, sustainability challenges and issues like over-tourism and terrorism can exasperate the overall authenticity of religious destinations (Chowdhury et al. 2017; Valeri 2022a, b). According to Heidari et al. (2018), these special places symbolically mean so much as people get around them for devotion and ritual practices. For example, Jerusalem is a holy site for Christians, Jews, and Muslims, and it draws worshippers and tourists to see its stunning beauty and powerful meaning (Reiter et al. 2000). Yet there is this worry about whether commercialization ruins their essence or authenticity gets lost along the way. However, when people go on religious journeys together, it promotes social connection—talk about bringing communities closer (Collins-Kreiner 2018). Vijayanand (2012) stated that these pilgrimages helped different faith groups understand each other better. For instance, India's Kumbh Mela festival is amazing since millions join in celebrating diverse beliefs together which pushes community development projects along—creating jobs or improving education for local folks (Dubey 2003). Still, this does not mean everything is perfect; concerns raised about equity and cultural appropriation needs are the major challenges in religious destinations.

3 Methodology

This section illustrates the methodology used to examine the multidimensional impact of religious tourism more thoroughly from different sectors. While drawing on insights from marketing, economics, psychosocial development, social development, sustainable development, and management in combination, this research offers nuanced understandings of the complexities of religious tourism and its implications for destination comparison.

In this qualitative study, a literature review and case study methods were adopted to highlight various dimensions of religious tourism. These qualitative methods, literature review, and case study permit the exploration of complex issues to gain in-depth and detailed insight into participants' psychological experiences and insights into the religious tourism phenomenon. The integration of findings from the different disciplines gives an approach that facilitates an inclusive understanding of the multidimensional impact of religious tourism.

4 Historical Development of Religious Tourism: Cultural, Mystical, and Social Value

Religious tourism can be traced back to the time of ancient civilizations when trips for pilgrimage purposes and execution of sacred rituals were core in religious practice and part of cultural identity. The ancient origin of religious tourism spans thousands of years and encompasses different spiritual traditions, routes of pilgrimages, and sacred places that trace human history and spirituality. One of the very ancient forms of non-secular tourism dates back to Mesopotamia, whereby pilgrims traveled to Ur to admire the Sumerian moon god Nanna. Similarly, in historic Egypt, pilgrims traveled to Karnak and Luxor to pay homage to gods like Amun-Ra and Isis for divine benefits and non-secular enlightenments in sacred temples. The growth of Christianity in the Roman Empire marked a significant change in travel history. This made pilgrimage quite a widespread phenomenon, with every believer from any walk of life flocking to holy sites related to the life and teachings of Jesus Christ. Jerusalem, Bethlehem, and Nazareth became important places of pilgrimage that attracted tourists from Europe, Asia, and Africa. Indeed, the Christian impact is devastating on travel, representing the deep influence religion has on human behavior and culture. However, during the Middle Ages, the Camino de Santiago in Spain, Via Francigena in Italy, and Hajj to Mecca in Islam emerged. Yet it was also invigorating, complete with spiritual contemplation, communal liturgy, and personal meetings with co-pilgrims called from all walks of life. The development of Christianity in the Roman Empire was one of the most outstanding moments in the history of pilgrimage which eventually became the revered pilgrimage centers that attracted pilgrims from Europe, Asia, and Africa. These pilgrims visited to show reverence to the Church of the Holy Sepulcher and the Church of Nativity to find solace in their faith.

5 Primary Motivations of Religious Tourists

Primary motivations of religious tourists are classified into the following types: spiritual quests, cultural exploration, personal transformation, ritualistic practice, social interaction, and adventure.

(a) *Spiritual quest:* These tourists make efforts to experience the presence of God and to participate in spiritual activities at their sacred attractions and places of pilgrimage.
(b) *Cultural exploration:* Cultural exploration motivates visitors to explore the religious roots of a particular community and their religious evolvement through different periods. This kind of visitor will be attracted by the stories, traditions, and cultural values within religious destinations that help increase their awareness of other cultures and religions.
(c) *Personal transformation:* This type of tourist desires to experience events during travel that may result in self-realization and a personality change. Their travels mostly comprise activities that bring them closer to introspection or spiritual breakthroughs that challenge or transform their perspectives.
(d) *Ritualistic practice:* Ritualistic practice pertains travelers to attending some religious ceremonies and rites connected with their religion. These practices are very important in their spiritual life and bring about a sense of trust and belief.
(e) *Social interaction:* Travelers who wish to relate with others and have similar interests go for activities that include religious meetings, pilgrimages, and events.
(f) *Adventure-seeking visitors:* They are the ones who combine their spiritual trips with sports activities like trekking, mountaineering, and visits to certain far-flung holy places where they get a lot of unusual and challenging experiences.

The understanding of these diverse motivations will ultimately provide a fulfilling and respectful experience.

6 Impact on Religious Tourism

Economic, environmental, social, and spiritual dimensions make religious tourism special, both for destinations and visitors. First, the economic impact is the single largest area, indicating how religious tourism can contribute toward stimulating local economies concerning jobs, small businesses, and other incomes that may be derived from activities related to tourism. In this sense, religious tourism has emerged as an economic growth and expansion of destinations by way of generating employment and income for the community people. Similarly, the financial impacts of religious tourism extend beyond the tourism sector, such as hospitality, transportation, retailing, and cultural heritage preservation. Moreover, job creation and employment generation have emerged as major economic benefits of religious tourism. Additionally, this sector attracts millions of visitors to explore pilgrimage sites, religious festivals, and cultural attractions, and the requirement immediately goes to services like accommodation, transport, food, and buying souvenirs. Therefore, local businesses, hotels, restaurants, and tour operators are the main beneficiaries of religious tourism where tourists get unique experiences in return. Apart from that, it offers opportunities for entrepreneurship and small business development, thus empowering the local community people. For example, religious tourism is the economic propellent in generating revenue, by which pilgrims and tourists contribute to destination economies through expenditures on accommodations, transport, food, shopping, and cultural activities. This is a billion-dollar industry that generates huge GDP across destinations annually. Furthermore, revenue generated from religious tourism is used to finance government budgets. The process of infrastructure development projects and social welfare programs are essential initiatives to reduce poverty and ensure sustainable development in destination regions. On the other hand, it plays an indispensable role in protecting and branding cultural heritage. This interlinks cultures through intercultural dialogue, thus realizing the possibility of maintaining and celebrating traditions, customs, and religious-historical values that already exist at the location. Similarly, transformational spiritual experiences are made available to tourists to deepen their faith and provide opportunities for personal growth and fulfillment in faith. In this context, religious tourism promotes community cohesion and increased social interaction, as it avails an opportunity for the locals to interact with the visitors that builds mutual respect and understanding. Moreover, religious tourism is at the cost of over-tourism and habitat degradation. Adopting sustainable tourism practices and environmental protection measures will reduce the detrimental impacts of over-tourism.

However, the religious impacts of tourism have multiple dimensions, indicating a balanced approach toward its management and promotion in creating sustainable tourism practices.

7 Challenges and Opportunities for Destination Management in Religious Tourism

While challenges are to be mastered, destination managers of religious tourism also find opportunities for sustainable development. This finding may include a preservation of cultural heritage while balancing modern demands for tourism. The environment, culture, and infrastructural development present major challenges but offer numerous opportunities for innovation, collaboration, and responsible stewardship. However, it is against this dynamic backdrop that destination managers have to proactively face the challenges of religious tourism and seize the potential benefits of all concerned.

(a) *Over-tourism:* On this note, extremely famous pilgrimage sites have to do with issues such as environmental degradation, cultural erosion, and social tension related to overcrowding. Consequently, management of visitor flows and carrying capacity measures reduce pressure from overcrowded sites and share equitably gains from all religious destinations.

(b) *Infrastructure development:* The infrastructural facilities need to be developed and maintained from the point of view of enhancing the experience for visitors, as a prerequisite for fostering the growth of religious destinations. However, the development of infrastructures needs to be offset to mitigate the harmful influence on natural and cultural resources. This would also ensure that the infrastructure being developed remains responsive to destination needs and priorities through public-private participation and community dialogue.

(c) *Cultural sensitivity and respect:* Showing respect for the community culture, heritage, and tradition at religious destinations is the key to positive tourist-host community interactions. This should also involve cultural training for tourism stakeholders, codes of conduct for visitor management, and intercultural cooperation for mutual understanding and respect. Moreover, the involvement of the host community at the levels of policy-making will also ensure the sensitivity of the cultural values and priorities relevant to the concerned region in tourism development.

(d) *Environment conversation:* Preserving the natural environment and biodiversity at religious tourist destinations is important to retain their ecological integrity and cultural sustainability. In this context, the destination manager must ensure provisions for environmental conversation, habitat restoration, waste management, and eco-friendly practices that lessen the impact of the tourists' activities on the environment. Moreover, it is crucial to build awareness among tourists concerning environmental protection and responsible tourist practices that place efforts toward preserving the environment by reducing the imbalance-ecological footprint of religious tourism.

(e) *Community empowerment:* The ability of local communities to get involved in tourism initiatives, as well as building the capacity of local economic opportunities, is fundamental to the sharing of benefits in tourism which will enhance socio-economic development at the religious sites. For example, destination managers can facilitate community-based tourism enterprises and cultural heritage preservation by providing access to training and resources for local entrepreneurs. This could further raise the level of involvement of local communities by developing partnerships between the stakeholders in tourism.

(f) *Interfaith dialogue and cooperation:* Destinations in religious tourism ought to be places where peace, harmony, and understanding reign. More particularly, this would help in having regional stability and reconciliation processes by participating in religious events. Additionally, this process will accelerate peacebuilding, conflict resolution, and social cohesion.

Despite raising many challenges in destination management, religious tourism opens up huge opportunities for sustainable development, cultural exchange, and community empowerment. By seizing such opportunities, destination managers would encourage responsible tourism and improve the quality of visitor experience.

8 Marketing and Promotion Strategies in Religious Tourism

Effective marketing and promotion activities are done with enhanced competitiveness in the tourism industry in order to maximize the associated economic and sociocultural benefits related to religious tourism. Moreover, destination managers in this regard can avail the aid of digital technologies and storytelling to come up with focused marketing campaigns that engage audiences of great diversity, foster brand loyalty, and create truly memorable experiences.

(a) *Digital presence and destination marketing:* The tourism stakeholders would highlight some of the unique attractions of these religious places through social media platforms, destination websites, or mobile applications by sharing compelling stories with these travelers and engaging them in real-time experiences.
(b) *Storytelling through content marketing:* Storytelling can engage audiences, elicit emotion, and engender an authentic connection with the visitor. Destination managers can employ storytelling while emphasizing the cultural heritage, spiritual values, and transformational experiences that characterize religious tourism destinations. Furthermore, they can be better positioned to hold audiences, ignite wanderlust, and distinguish their destination in a very competitive market by developing multimedia production in cooperation with local storytellers, artists, and influencers.
(c) *Targeted marketing campaigns:* Segmentation of the target markets and then designing campaigns for each segment will make promotional tools more effective and ensure maximum ROI (return on investment). In this sense, destination managers can identify religious pilgrims, heritage enthusiasts, or wellness travelers as niche markets and develop marketing campaigns that take into account special interests, preferences, and motivations. Moreover, they can sharpen marketing messages and resource efficiency with the help of data analytics, market research, and consumer insights.
(d) *Partnerships and collaborations:* Partnering with a travel agent, tour operator, religious body, or local community for far-reaching coverage is essential for building a communication hub to serve religious tourists effectively. It can engage in strategic partnering, joint marketing efforts, and cross-promotional activities with the major stakeholders to help and utilize their network, resources, and competencies in understanding and accessing the target customers.
(e) *Cultural events and festivals:* Cultural events, festivals, and religious celebrations can easily capture visiting tourists. In this context, destination managers can use the different religious holidays, anniversaries, or special occasions to stage cultural performances, religious rites, or even community events that will showcase the local traditions, rites, and hospitality. The development of memorable experiences would build a feeling of belonging and loyalty among visitors.

However, marketing and promotion can influence the degree of success and sustainability at a religious tourism destination. In addition, destination managers should change with the unending changing tourism industry to unlock the full potential of religious tourism as an accelerator of economic growth and cultural exchange by becoming competitive.

9 Digital Transformation of Religious Tourism

International tourism has been staggering for a few years due to COVID-19 by which people are deprived of tourism facilities. The advent of new technologies including AI (artificial intelligence), robotics, augmented reality (AR), and virtual reality makes this possible to travel to any religious destination from any location in real-time concept. However, digital marketing and branding strategy through social media (Facebook, Twitter, YouTube) can be a feasible tool to provide necessary information to the targeted customer who can get access to visit holy destinations through this technological domain. Therefore, it is peremptory for the stakeholders who are technologically updated. For example, "SMART tourists" utilize digital tools for online booking, getting credible information through destination webpages, and paying the expenses by credit card (Femenia-Serra et al. 2019). SMART government, on the other hand, facilitated its religious tourists to get one-stop online services including online visa processing, online ticketing, and online payment system (Kankanhalli et al. 2019). Similarly, the SMART community creates video documentation, blogs, and digital promotions about religious places where tourists can find religious serenity and cultural authenticity (Iqbal and Olariu 2020). Even, other SMART stakeholders (travel agencies, tour operators, hotels, food and beverage, transportation, and events) arrange virtual religious tourism for all kinds of tourists who can see the panoramic view of sacred places through technological sophistication (Ivars-Baidal et al. 2024. The outcomes of virtual religious tourism have a social, cultural, economic, and environmental impact on the religious destinations where peaceful bonding among the religious group and community is evident (Fig. 1).

10 Case Study: 01

10.1 Contrasting the Economic Impact, Religious Harmony, and Tourism Development in Bangladesh

Religious harmony is a possible state that denotes the harmonious coexistence of adherents of different religions to remove the potential for conflict. To address such a situation, there must be a universal recognition that all religions, in terms of their core beliefs and how they emerge in human behavior, are mutually acceptable. This understanding must also extend to all traditions and their development into modern

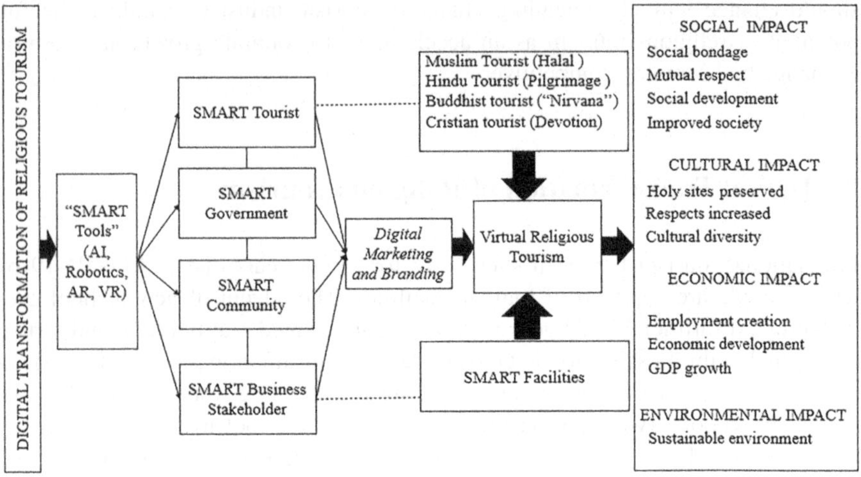

Fig. 1 Digital transformation of religious tourism. Source: Authors' compilation

life. In this context, Bangladesh is a Muslim-majority nation and a religious tourism landscape where vibrant holidays are celebrated all year round with great fervor and a proper approach. This country has a long history within the social organism and several religious festivals. For example, different communities include the Muslim's Eid-ul-Fitr and Eid-ul-Ajha, Buddhists' Buddha Purnima, the Christians' Christmas, and the Hindus' Durga Puja which reflect the emblems of the nation and the society. Although Islam is the predominant religion, rich cultural heritage does not seem like a single religious entity but an aggregate of various religious traditions and sites of sacredness. This religious diversity gives birth to a multifaceted panorama of religious tourism in terms of pilgrimages and festivals of different faiths, which attract domestic and international tourists. This country has always served as one of the departure points for large numbers of Muslims to make their way to Mecca for Hajj and Medina for Umrah. In the context of economic value, these journeys see thousands of Bangladeshi pilgrim's travels to Saudi Arabia annually. More specifically, spending on the Hajj pilgrimage toward travel, accommodation, and religious services generates revenue for travel agents, airlines, and the hotel industry. For instance, Bangladesh is the only host country that organizes "World Ishtema" every year and invites numerous foreign tourists to participate in this holy occasion. On the other hand, a few of the key Hindu pilgrimage sites are found in Bangladesh: the Dhakeshwari Temple in Dhaka, the Kantaji Temple in Dinajpur, and the archeological site of Mahasthangarh in Bogura. Thousands of pilgrims and tourists visit these sites from every part of Bangladesh and neighboring countries who spend money in the local economy on accommodations, transportation, and religious offerings. Most of these sites still have poorly developed infrastructure and facilities compared to other religious tourism destinations around the world. Most religious sites in Bangladesh bear cultural and religious importance; however, the development of tourism infrastructures and accessibility to religious spots have become a challenge for this country. For example, limited transportation, poor marketing, ineffective branding tactics, bad road conditions, and inadequate accommodation facilities are some of the major problems in promoting religious tourism in remote and rural areas. Moreover, the country has also been suffering from other problems that have turned out to be impediments to the development of tourism. Issues like political imbalance, terrorism, the Rohingya refugee problem, poor service quality, and frequent road accidents are the major cause that demotivates international tourists from seeing the religious beauty of Bangladesh.

On the other hand, historic mosques, temples, and archaeological sites offer prospects for developing religious tourism. However, the preservation and maintenance of such cultural assets call for huge heritage conservation and restoration investments. The lack of funds and institutional capacity related to cultural heritage preservation imposes a challenge to the upliftment of religious tourism in Bangladesh. In addition, community engagement is required to promote culture, heritage, and tradition on the international platform. Therefore, empathy toward the local communities through initiatives, training programs, and capacity building will ensure an equitable distribution of economic returns at the local level among the people. The adoption of promotional materials for sustainable tourism practices, such as responsible tourism, community-based tourism, and eco-friendly initiatives, is likely to diminish the detrimental effects of mass tourism in religious sites. However, in matters of infrastructural development, cultural heritage preservation, and community involvement, potential religious tourism in Bangladesh remains grossly untapped. Consequently, Bangladesh shall be in a position to offer much value addition by showcasing its religious diversity, developing tourism infrastructure, and pursuing sustainable tourism practices while preserving its rich cultural heritage and natural resources for posterity.

10.2 Queries That Might Justify This Following Case Study

Q1: How can religious tourism capitalize on tourist experiences? Draw a SWOT analysis and provide recommendations on how we can mitigate the challenges and restore cultural authenticity.

Q2: Do you think that Bangladesh is a paradigm example for synthesizing religious harmony among religious groups including Muslims, Hindus, Buddhists, and Christians? Give your judgment by using your logical thought process.

11 Case Study: 02

11.1 Illuminating Insights: Diverse Religious Traditions (International Perspectives)

Religious pilgrimage places have served over the ages as cultural heritage and community worship which focuses on spiritual devotion and a spectrum of diverse religious traditions. The paramount importance of religious tourism opens new prospects for exchanging cultural authenticity among devotees, tourists, and the community. For instance, from Spain's Camino de Santiago to India's Kumbh Mela, each pilgrimage site speaks of different traditions, rituals, and experiences that shape the pilgrimage journey for millions of devotees.

The Camino de Santiago is a sacred pilgrimage of Spain that thousands of Christian pilgrims from different parts of the world visit every year. The pilgrimage concludes in the Cathedral of Santiago de Compostela, under which St. James the Apostle lies buried. However, the Camino de Santiago means a changing journey to spiritual growth, self-realization, and culture. Walking through these beautiful landscapes with their picturesque villages, medieval towns, and historic places, one feels submerged in the sacred. For instance, through the pilgrimage experience, they will feel brotherhood and solidarity since they have to do their physical and spiritual challenges all together. On the other hand, Kumbh Mela, in India, is the largest religious festival globally. Thereby, it becomes a kaleidoscope of religious rites, cultural jubilation, and spiritual discourses in which pilgrims come together to be cleansed, blessed, and enlightened. It evidences the diversity and dynamism of Hinduism which brings the devotees together from all walks of life and traditions in a joyous celebration of shared faith and heritage. Similarly, Varanasi also called Kashi, or Light City in India, is regarded as one of Hinduism's most holy pilgrimage sites. Moreover, it is the holiest of the seven sacred cities in Hinduism visualizing a way toward spiritual liberation, which lies alongside the banks of the Ganges River. Every year, Varanasi sees a myriad of pilgrims who come to bathe in the purifying waters of the Ganges River and perform rituals for their forefathers at cremation ghats. The city has evolved into a melting pot of religious diversity, cultural heritage, and spiritual devotion the timeless mystique and sacred aura of which have been drawing pilgrims and tourists alike. On the other hand, Saudi Arabia invites Islamic pilgrimage (Haji) to gain spiritual serenity in the body, mind, and soul. For example, Macca is the most jaw-dropping illustration of dynamic worship executed by a group of people called Haji who can perform religious obligations by visiting the place physically at least once during their lives. This is how Muslims can spiritualize and transform them with activities like Tawaf or circumambulation around the House of Allah or Holy Kaaba.

Considered the holiest place of Buddhism, Bodh Gaya represents places that can be associated with personal transformation and spiritual growth. It is within the UNESCO Mahabodhi Temple Complex that a holy Bodhi Tree and several other Buddhist shrines are situated. It is among one of the biggest centers of pilgrimage for Buddhists all over the world. Buddhists visit Bodh Gaya to pay homage to the Buddha, meditate under the Bodhi Tree, and participate in religious rites and discourses carried out by Buddhist monks and learned scholars. Consequently, this pilgrimage will be a life-changing self-discovery and spiritual awakening through the peace walked step by step, just like how the Buddha walked in his journey to freedom from suffering.

While all these pilgrimage places of religious orientation had their distinguished features and special contexts, commonalities related to the spiritual quest, cultural heritage, and communal worship do seem to emerge. They provide pilgrims and tourists with transforming experiences that cut across faith, nationality, and linguistic barriers. In the background of a globalizing world, governance at such sites is related to challenges regarding overcrowding, environmental impacts, and the need for the preservation of site sanctity.

11.2 Queries That Might Justify This Following Case Study

1. What are some special rites or customs that are peculiar to pilgrimage sites like Camino de Santiago, Kumbh Mela, and Hajj by which their whole experience and spiritual journey get shaped?
2. What are the roles that religious tourism plays in protecting the culture of Bodh Gaya and Varanasi?
3. What are the general problems associated with managing a huge inflow of tourists to any pilgrimage destination, and how can such problems be resolved through sustainable practices?
4. What role has spiritual and personal transformation played in the motivation of pilgrims to such different religious shrines, and how are the experiences different across traditions?

12 Conclusion

Throughout this chapter, we have explored the complex interplay between religious motivations and touristic experiences, examining religious tourism's economic, social, cultural, environmental, and spiritual dimensions from a multidisciplinary perspective. We delved into the diverse motivations of religious travelers, ranging from spiritual quests and pilgrimage rituals to cultural exploration and personal transformation. Understanding these motivations is essential for destination managers to tailor marketing strategies, develop targeted experiences, and enhance visitor satisfaction while respecting the sacredness and authenticity of religious sites and traditions.

Moreover, religious tourism presents opportunities and challenges for destination management, requiring a balanced and holistic approach that integrates economic development, cultural preservation, environmental conservation, and social responsibility. By adopting multidisciplinary perspectives, fostering stakeholder collaboration, and embracing sustainable tourism practices, destination managers can harness the transformative power of religious tourism to create positive impacts for communities, travelers, and sacred sites alike. However, religious pilgrimage journeys continue to inspire, enlighten, and unite people from all walks of life for generations by upholding the principles of responsible and ethical tourism. Moreover, this chapter develops a model that reflects the digital transformation of religious tourism by which we can predict how our tourism industry looks like in the upcoming future.

References

Badone E, Roseman SR (2004) Approaches to the anthropology of pilgrimage and tourism. Intersecting journeys: The anthropology of pilgrimage and tourism, 1–23

Cànoves G, Romagosa F, Blanco A, Priestley GK (2012) Religious tourism and sacred places in Spain: old practices, new forms of tourism. Int J Tour Anthropol 2(4):282–298

Chowdhury A, Raj R, Griffin K, Clarke A (2017) Terrorism, tourism, and religious travelers. Int J Relig Tour Pilgr 5(1)

Collins-Kreiner N (2018) Pilgrimage-tourism: common themes in different religions. Int J Relig Tour Pilgr 6(1):3

Dubey DP (2003) Kumbha Mela: the festival of eternal bliss. Oriental Anthropol 3(2):164–173

Durán-Sánchez A, Álvarez-García J, del Río-Rama MDLC, Oliveira C (2018) Religious tourism and pilgrimage: bibliometric overview. Religions 9(9):249

Femenia-Serra F, Neuhofer B, Ivars-Baidal JA (2019) Towards a conceptualisation of smart tourists and their role within the smart destination scenario. Serv Ind J 39(2):109–133

Firdaus E, Rahmat M (2019, June) The impact of religious tourism in the formation of a peaceful and moderate society. In: In 3rd international seminar on tourism (ISOT 2018). Atlantis Press, pp 111–116

Hassan T, Carvache-Franco M, Carvache-Franco W, Carvache-Franco O (2022) Segmentation of religious tourism by motivations: a study of the pilgrimage to the city of Mecca. Sustain For 14(13):7861

Heidari A, Yazdani HR, Saghafi F, Jalilvand MR (2018) The perspective of religious and spiritual tourism research: a systematic mapping study. J Islamic Market 9(4):747–798

Iqbal A, Olariu S (2020) A survey of enabling technologies for smart communities. Smart Cities 4(1):54–77

Ivars-Baidal J, Casado-Díaz AB, Navarro-Ruiz S, Fuster-Uguet M (2024) Smart tourism city governance: exploring the impact on stakeholder networks. Int J Contemp Hosp Manag 36(2):582–601

Jongmeewasin, S. (2016). Religious tourism, pilgrimage, and cultural tourism. In 6th national and international graduate study conference: creative multi-disciplinary studies for sustainable development & the 1st national and international conference on creative multi-disciplinary studies for sustainable development (NCMSD & ICMSD 2016) (pp. 394–95)

Kartal B, Tepeci M, Atlı H (2015) Examining the religious tourism potential of Manisa, Turkey with a marketing perspective. Tour Rev 70(3):214–231

Kankanhalli A, Charalabidis Y, Mellouli S (2019) IoT and AI for smart government: a research agenda. Gov Inf Q 36(2):304–309

Kim B, Kim SS (2019) The effect of religious tourism experiences on personal values. Int J Relig Tour Pilgr 7(2):9

Kim B, Kim S, King B (2020) Religious tourism studies: evolution, progress, and prospects. Tour Recreat Res 45(2):185–203

Lochrie S, Baxter IW, Collinson E, Curran R, Gannon MJ, Taheri B et al (2019) Self-expression and play: can religious tourism be hedonistic? Tour Recreat Res 44(1):2–16

Luz N (2020) Pilgrimage and religious tourism in Islam. Ann Tour Res 82:102915

Olsen DH, Timothy DJ (2006) Tourism and religious journeys. In: Tourism, religion, and spiritual journeys. Routledge, pp 1–21

Olsen DH, Timothy DJ (eds) (2022) The Routledge handbook of religious and spiritual tourism. Routledge, Abingdon, UK

Rashid AG (2018) Religious tourism–a review of the literature. J Hosp Tour Insights 1(2):150–167

Reiter Y, Eordegian M, Khalaf MA (2000) Between divine and human: the complexity of holy places in Jerusalem. In: Jerusalem: points of friction-and beyond. Brill, pp 99–153

Rinschede G (1992) Forms of religious tourism. Ann Tour Res 19(1):51–67

Valeri M (2022a) Tourism risk. Crisis and recovery management. Emerald Publishing, UK

Valeri M (2022b) New governance and management in tourist destinations. IGI Global Publishing, Hershey PA, USA

Vantage Market Research (2023) Global relegious tourism market: growth trends and insight in 2032. https://www.vantagemarketresearch.com/industry-report/religious-tourism-market-2493

Vijayanand S (2012) Socio-economic impacts in pilgrimage tourism. Int J Multidiscip Res 2(1):329–343

Vukonić B (1998) Religious tourism: economic value or an empty box? Zagreb Int Rev Econ Bus 1(1):83–94

Exploring the Dynamics of Religious Tourism: A Bibliometric Analysis of Pilgrimage Sites and Modern Trends

Chandra Prakash and Manisha Paliwal

Abstract The study aims to explore the evolving landscape of religious travel within the global tourist economy, focusing on pilgrimage sites and the motivations behind modern society's inclination towards religious tourism. Utilising bibliometric analysis from 1989 to 2023, drawing from the Scopus database, the study examines 1368 papers, refining the selection to 838 articles. VOS viewer and Biblioshiny tools are employed to address four research objectives: publication patterns, citations, authorship and collaborations. The findings reveal diverse academic approaches and multidisciplinary perspectives on religious tourism, covering various methodologies, subjects, applications, geographical focuses and manifestations of the phenomenon. Notably, recent research extends beyond tourist motives to investigate identities, religious connections and religiosity. It acknowledges limitations such as the potential further research to comprehensively explore religious tourism in non-European contexts.

1 Introduction

Religious travel was not a recent occurrence. Religion has always been a fundamental motivator for conducting voyages and is generally regarded as the earliest type of non-financial travel (Smith and Eadington 1992). Religious tourism is a rising area within the cultural tourist industry. The worldwide religious tourism market is diverse (Abbate and Di Nuovo 2013). In recent years, the phenomenon of religious tourism has gained increased attention, prompting a growing body of literature on the subject (Eade 1992; Shuo et al. 2009). This chapter seeks to delve into religious tourism, which is a buzzing topic in today's era, with growing interest from researchers, policymakers and practitioners (Singh 2006; Ryan and Mizerski 2010; Shuo et al. 2009). The dynamics of this phenomenon, particularly the relationship between pilgrimages and tourism, have been extensively explored in the literature. Religious

C. Prakash · M. Paliwal (✉)
R&D cell at Sri Balaji University, Pune, Maharashtra, India

M. Valeri (ed.), *Contemporary Religious Tourism*, Tourism, Hospitality & Event Management, https://doi.org/10.1007/978-3-031-79074-4_7

tourism encompasses all travel to religious locations, as it combines religious experiences with travel (Bond et al. 2015). Definitions of religious tourism based just on the words 'religion' and 'tourist' are insufficient for comprehending the phenomena. In the same way, researchers have defined 'religion' as a commonly accepted set of beliefs in a higher entity, power or being which are held with strong levels of conviction, belief and dedication amidst a large number of people (Tilson 2005; Kumermanis and Rudzitis 2009). Growth in spiritually driven travel has corresponded along with the enlarging rise of tourism in the contemporary age today (Lloyd 2006) and this business of tourism along with its various ancillary behaviours are having an interaction which can be seen on a global scale, in reference with its interactions with the religious lives of the devout, other related people and organisations in the world today (Bremer 2005). Bremer (2005) emphasises that a large number of researchers have investigated the convergence of tourism and religion, categorising them into three different 'approaches' for study: (1) the location-based approach (pilgrims and other travellers occupying the same space with different geographic actions), (2) the conventional strategy (the connection between religious forms of travel and tourism) and (3) the perspective of culture (pilgrimage and tourism as contemporary actions in a post-modern world) (Fig. 1).

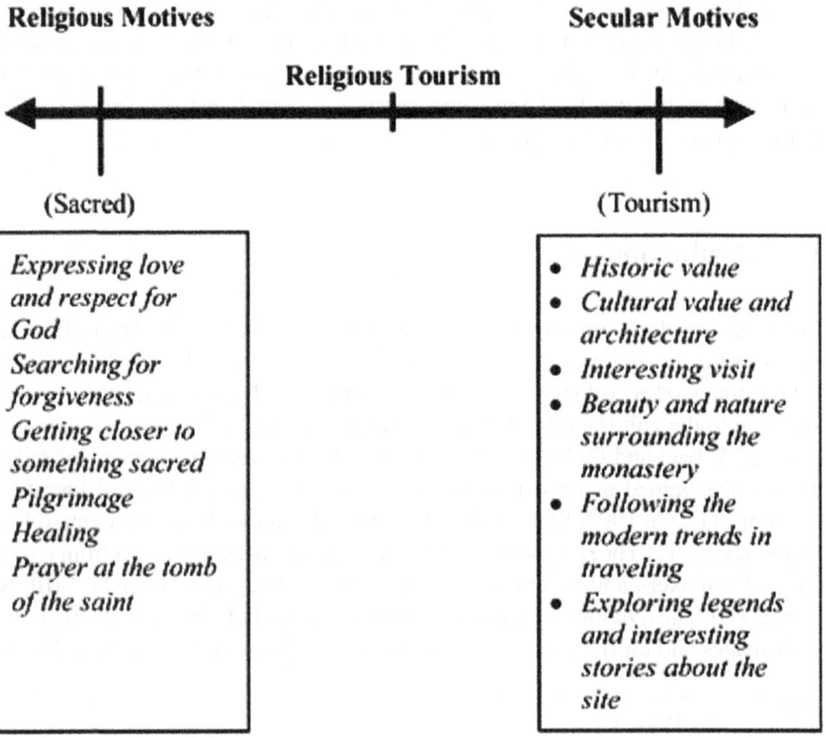

Source: Smith (1992) & Božic, et al. (2016)

Fig 1 Motives behind religious tourism

The majority of the current literature has perpetuated the belief that a sense of similarity can be found amidst tourism and pilgrimage journeys. The literature reveals that the primary focus of the visitor or tourists' engagement with the core objective or 'value' of the pilgrimage journeys undertaken by them along with the nature of their sentiments about the journey, which acts as a motivational factor in strengthening their faith or belief in the purpose for which they are undergoing the pilgrimage can act as an action factor which can get the visitors to invest in their own journey in life, so far to harvest something purposeful or meaningful benefit in their journey have been identified as considerations of pertinent value for the researchers in the study of religious tourism. It is to be noted that research is scarce in this area as very few empirical investigations have been done in this area of study.

Over the last decade, scientists have conducted several empirical investigations to comprehend these two notions better. However Scholars studying religious tourism phenomena have focused on understanding visitors, motivation and experiences (Collins-Kreiner 2010). Religious tourism's influence on modern civilisation poses various issues, including the monetisation of religion, which undermines the holiness of these locations and leads to the commercialisation of religious rites and symbols. Religious concepts can provide insight into life-death relationships and inspire people to visit sacred places for spiritual experiences, nevertheless they differ from cultural tourism. (Heidari et al. 2018)?

This suggests that pilgrimage and tourism are not entirely distinct phenomena but share a degree of overlap in the visitor experience and motivations. It is projected that more than 240 million individuals travel to religious sacred sites each year, with the majority being Hindus (Singh et al. 2021). Muslims, Christians and religious or spiritual tourism have been increasingly popular in recent decades (Kim et al. 2020). There are religious sites for tourists to visit almost anywhere in the world. Tourist activities can produce a location that is worthy of visitors wherever religious people have created centres of engagement with sacred forces (Verma and Sarangi 2019). In religious locations that receive large volumes of visitors, a simultaneous blending of spaces occurs, where the sacred and the touristic coexist within parallel geographies. Furthermore, the significance of these intersecting locations depends on identity narratives that strengthen the ties that both religious followers and tourists have to the location (Budovich 2023). Many of these individuals come to sacred places in the hope of being treated for various kinds of diseases or cleaning there their aura performing the ritual and paying their respects to God. In contrast, others simply want to view the sacred sites and attractions. People visit sacred locations for various reasons, including the desire for knowledge, enlightenment or a real encounter with the divine (Iliev 2020). It is also maintained that pilgrims are often not considered tourists or at least are perceived to be distinct from tourists. Furthermore, the literature highlights the growing complexity of pilgrimage motivations, proposing a typology of five layers that recognize tourists, pilgrims of need, pilgrims of duty, pilgrims of union, pilgrims of hope, and pilgrims of transformation, each driven by distinct spiritual, cultural, and personal objectives (Shinde 2022).

Hindu pilgrimage sites, in particular, have been the focus of scholarly attention, with researchers examining the interplay between religious tourism and tolerance. These pilgrimage sites have been found to exhibit a diverse range of combinations which showcase the traditions of sectarian nature amidst the religion followed by the Hindus, and a similar level of acceptance has been found amidst the interactions of the Hindu pilgrims with adherents of other faiths, including proponents of Abrahamic religions (Christianity et al., etc). and foreigners. The research studies have raised a suggestion that the ability of tolerance and acceptance within the Hindu sects, along with other believers and proponents of other religious faiths, can be identified as a core part of the values, practices and traditions witnessed within and near the various sites of pilgrimage (Das et al. 2024). However, the above investigations utilised a descriptive study approach that only examined a part of religious literature and did not reflect contemporary advancements in the subject. This study covers a research gap identified in Table 1's last column, distinguishing it from existing literature. Previous research (Collins-Kreiner 2020a, b; Suban et al. 2021) has not addressed the non-permanent aspects of the thematic evolution of the study area, revealing the scope for research which has been filled, in a form of this research study, which will contribute to the body of literature by performing an evaluation on the intellectual structure through a bibliographical analysis and will contribute to the rapidly increasing body of literature in religious tourism. This study also suggests promising areas for further investigation and will offer several insights of value through the methodology of bibliometric analysis, which will enhance the value of the body of literature in this field and will offer valuable insights into the scholarly landscape and trends in this field. By examining the patterns of publication, citation and collaboration, this study aims to provide a comprehensive understanding of the evolution and current state of research in religious tourism and pilgrimage (Duran-Sanchez et al. 2018). Additionally, it will identify the major trends and areas of focus within the field, highlighting gaps and opportunities for further research.

Table 1 Criteria for discovering documents with the keywords 'Religious tourism, Pilgrimage tourism'. Table 1: author compilation

	Criterion of inclusion	Criterion of exclusion
Source of database	'Scopus'	Further databases aside from Scopus
Timeframe	1989–2023	2024
Area of study	Religion, tourism, pilgrimage, business and management, social sciences	Any topics other than inclusive criteria
Document used in the study	Research article, book chapter	Conference paper, review paper
Publication stage	Published	Article in press
Language	English	Different languages apart from the studies

2 Methodology

In line with the aims, the study conducts a bibliometric analysis that identifies research clusters based on quantity and objectivity. Van Raan (2004) suggests that bibliometric reviews provide academics with defined quantitative markers to help their analysis. Qualitative outcomes and insights from academic literature provide a deeper understanding of underlying themes, patterns, and contextual nuances. This study included a bibliometric analysis. The goal is to identify the most productive research conducted on religious tourism in India. Between 1989 and 2023, there has been considerable scholarly production aimed at understanding distant perspectives, future possibilities, or long-term trends. A retrospective analysis was done using Bibliometrix. Aria and Cuccurullo developed the R package. (Aria and Cuccurullo 2017). The Bibliometrix R package makes it simple to do a comprehensive bibliometric study utilising approaches developed for both bibliometric and scientific metric purposes. Bibliometric analysis is a popular statistical approach used to understand the epistemology of a topic area (Singh et al. 2022), Prakash and Dwivedi (2022). Karamustafa et al. (2022) reviewed studies on bibliometric analysis in the tourist and hospitality industries. Bibliometric techniques are extensively used in sub-disciplines of tourism, including religion and sustainability (Niñerola et al. 2019) and ecotourism (Bi et al. 2020; Khanra et al. 2021). However, it is difficult to explain the most current advancements or gaps in terms of the research clusters. This method allows them to look at research publications that are relevant to the discovered clusters (Paliwal et al, 2023). This study uses this strategy to identify research clusters of considerable interest that are aligned with the objectives in a bias-free way. The search phrases used to get the closest comparable article contained 'Pilgrimage Tourism' and 'Religious tourism' keywords associated with the document. The criteria for the documents extracted by the researcher for the study are provided in Table 1.

Along with fulfilling the primary function of this study, the chapter also summarises the direction of contemporary research in the field. Previous literature evaluations have shown that this methodology has produced findings with better accuracy. The authors have defined the criteria of eligibility along with the criteria within which the research papers, articles and studies were included and the criteria of exclusion for the purpose of this study in congruence with the principles of systematic review (Fig. 2).

Pritchard 1969, has defined 'bibliometric methodology' as the methodology of research which seeks to integrate the quantitative approaches for searching (for example, analysis of the keywords used by the researchers) along with bibliometric data related to the literature (for example, the units published in the year, the year of publication of the literature and the citations used in the literature). The researchers scoured the Scopus database for all worldwide literature on 'religious tourism,' 'pilgrimage tourism,' 'Sustainable Religious Tourism,' and 'Tourism Development,' published between 1989 and 2023, covering 34 years of literature on the religious tourism discipline.

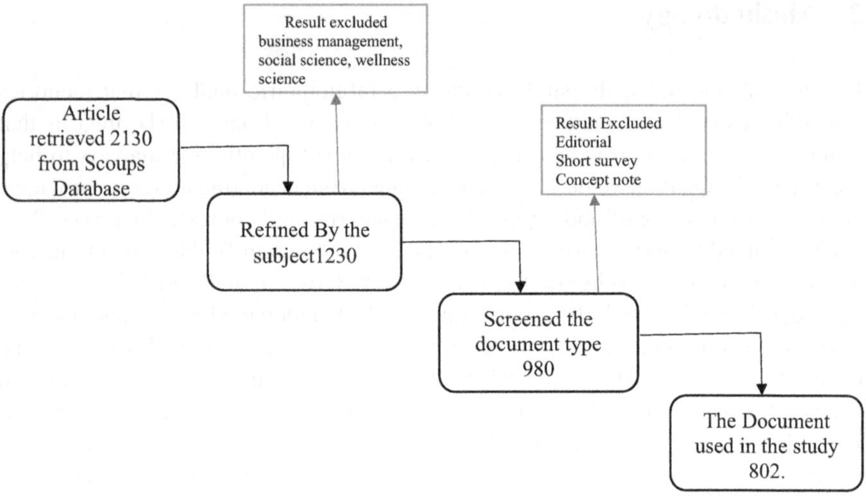

Fig. 2 A flowchart of the included publications on the dynamics of religious tourism. Source: authors' compilation

3 Result and Discussion

The research employs quantitative metrics to identify connections with various entities, including co-occurrence, bibliographic coupling, co-word and co-citation analysis. The article focuses on contemporary literature and the contributions of authors on the topic of religious tourism. As the quantity of research publications increases, obtaining information becomes more complex due to scattered distributions (Fig. 3).

From 1989 to 2023, the scholarly literature on religious tourism has increased at a 10.87 percent yearly pace. Every year, around 39 publications are published, which is quite decent when compared to other disciplines' production, leaving a large opportunity for scholars to write articles relating to this topic. Table 1 summarises the bibliometric data. As per the primary information summary, over 802 writers from 276 sources have contributed to this field of religious tourism research. There are also 66 single-authored articles and 575 multi-authored publications. It was also found in this study that the number of citations (on average) per document has been found to be 21.43.

3.1 The Prominent Authors

Table 2 and Fig. 4 show the top writers who contributed to religious tourism from 1989 to 2023. The parameters include highly cited authors, co-citation networks, productivity over time and author partnership networks. Figure 2 depicts the creators' long-term contributions, including the number of publications published and

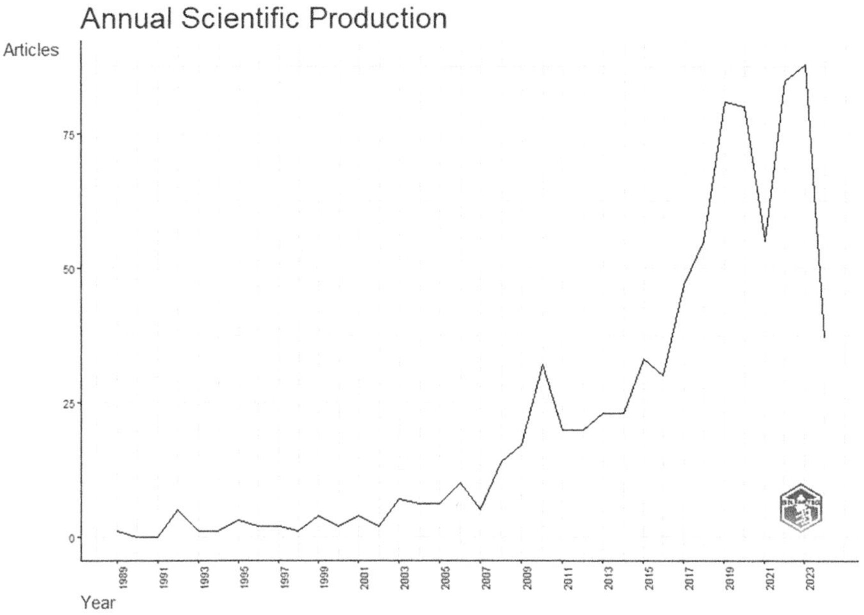

Fig. 3 Annual publication of the paper. Source: graph generated using Bibliometrix tool

Table 2 Summary of religious tourism bibliometric data from Scopus

Description	Results
Main information about data	
Timespan	1989:2023
Sources (journals, books, etc.)	276
Documents	802
Annual growth rate %	10.87
Document average age	7.06
Average citations per doc	21.43
Document contents	
Keywords Plus (ID)	1105
Author's keywords (DE)	2443
Authors	1547
Authors of single-authored docs	267
Single-authored docs	309
Co-authors per doc	2.24
International co-authorships %	17.58
Document types	
Article	764
Book chapter	38

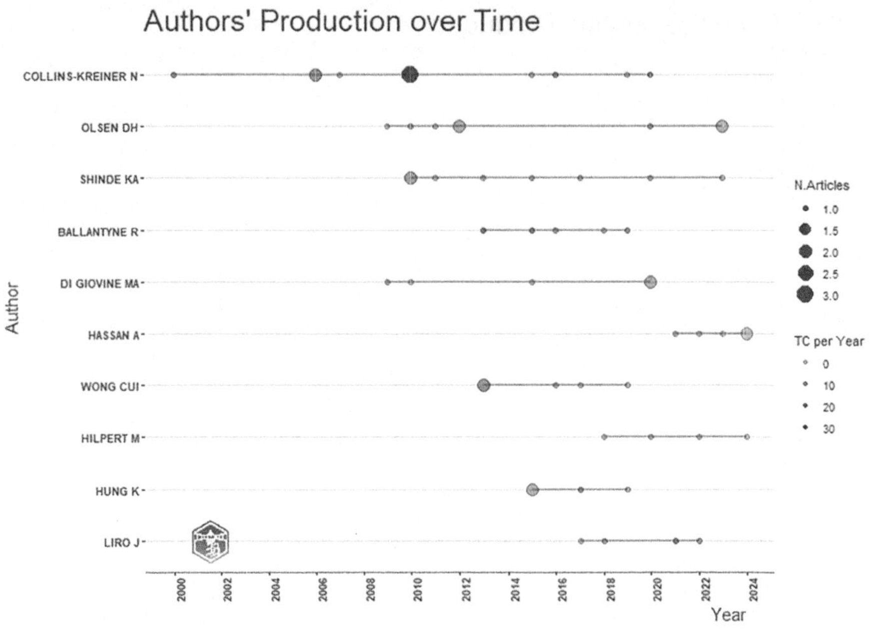

Fig. 4 The most prominent authors (Source: the author's compilation Bibliometrix tool)

total citations received in each year. The circle sizes determine the volume of the published document. The circle's tones represent the total number of citations, with deeper shades indicating more citations. The author 'Collins Kreiner' has been recognised as a great author over time, contributing to the field of religious tourism.

Table 3 shows author-level indicators, including g index, h index and m index (total citation), to assess productivity and academic output over time. The journal's success is evaluated using these indicators and measures. The h-index represents the number of citations a certain author's document has received. Citations are directly related to the quality of the journal. Higher h-index metrics indicate superiority over the journal. The g-list refers to top articles that have been referred to a certain number of times. The m-index is a version of the g-index, which works independently in this bibliometric study (Table 4) (Fig. 5).

Cluster Analysis

3.2 *Cluster Analysis*

Cluster 1 Pilgrimage tourism, hospitality (colour: red). Emphasise religion, spirituality, cultural tourism, temples and government. This cluster focuses on tourism motivated by religious and spiritual beliefs. The combination of terms like 'temples' and 'government' indicates that this cluster includes religiously significant sites, as

Table 3 The most prominent author

Element	h_index	g_index	m_index	TC	NP	PY_start
Collins-Kreiner N	11	11	0.44	1013	11	2000
Shinde KA	7	8	0.46666667	205	8	2010
Olsen DH	6	8	0.375	197	8	2009
Ballantyne R	5	5	0.41666667	337	5	2013
Di Giovine MA	5	5	0.3125	82	5	2009
Wong Cui	5	5	0.41666667	137	5	2013
Hung K	4	4	0.4	108	4	2015
McIntosh A	4	4	0.33333333	155	4	2013
Reichel A	4	4	0.18181818	342	4	2003
Ryan C	4	4	0.25	249	4	2009
Terzidou M	4	4	0.5	240	4	2017
Bond N	3	3	0.25	276	3	2013
Chun B	3	3	0.375	35	3	2017
Duda T	3	3	0.5	29	3	2019
Henderson JC	3	3	0.13636364	309	3	2003
Herstein R	3	3	0.42857143	80	3	2018
Huang K	3	3	0.33333333	112	3	2016
Li J	3	3	0.17647059	107	3	2008
Liro J	3	4	0.375	74	4	2017

Source: Bibliometrix—R tool; author's compilation

Table 4 Cluster-based co-occurrence of authors keywords

Cluster no	Colour	Name of cluster	No of keyword	Major keywords linked and normalised average citation
1	Red	Pilgrimage tourism, hospitality	19	Religious tourism, spirituality, cultural tourism, temples, government
2	Orange	Religious tourism, tourist behaviour	14	Visitor, motivation, experience, service quality
3	Green	Cultural identity, conservatism	16	Religious heritage, rural area, urban area, identity
4	Sky blue	Tourism development	14	China, Asia, Israel, south America
5	Navy blue	Sustainable tourism, tourism development,	13	Tourist satisfaction, destination image, tourism of economics

Source: Bibliometrix—R tool; the author's compilation

well as possible government participation in their management or promotion. The emphasis on 'hospitality' highlights the significance of services for pilgrims and spiritual tourists.

Cluster 2 Traveller behaviour and religious tourism (shade: orange) focusing on quality of service, experience, motivation and behavioural characteristics of travel-

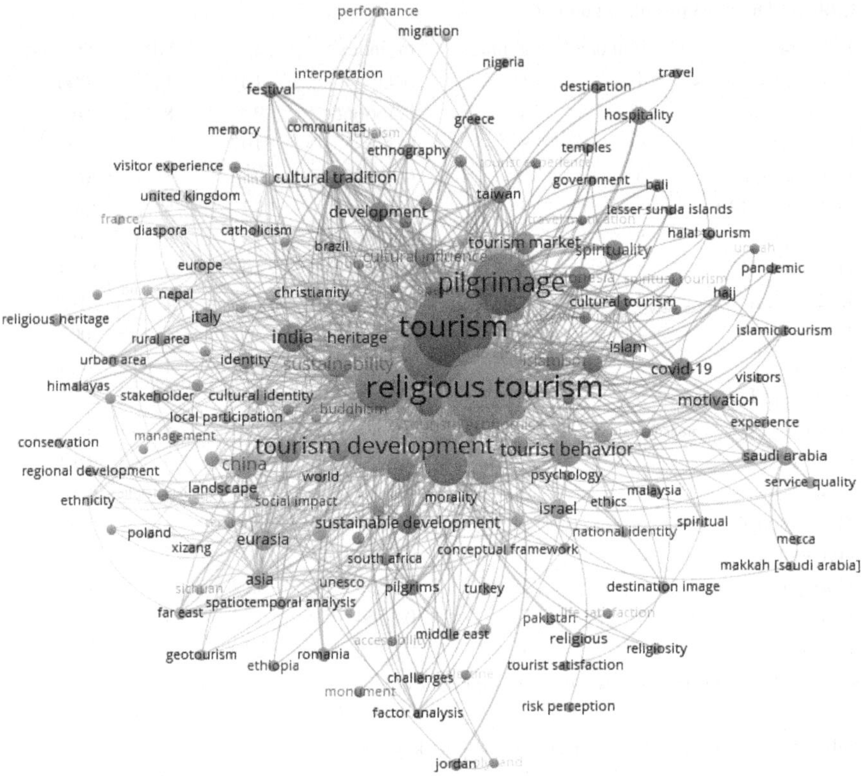

Fig. 5 Author keyword occurrences. Source: author compilation

lers who engage in religious tourism are the focus of this cluster. The emphasis is on comprehending visitor experiences and motives, which may be essential for raising the calibre of services. The goal is probably to improve the overall experience of visitors by attending to their individual requirements and motivations.

Cluster 3

Conservatism and cultural identity; green in colour emphasis.

Identity, rural/urban region, religious legacy. This cluster focuses on the relationship between religious heritage and cultural identity and tourism. The reference to both 'urban' and 'rural' locations alludes to a contrast or interaction between the two. The emphasis on 'identity' highlights how crucial it is to preserve religious and cultural identities in the face of tourism.

Cluster 4

Development of tourism; colour, sky blue; South America, China, Asia and Israel. This cluster focuses on the development of tourism in different locations, with a particular emphasis on geography. The mention of certain nations and continents suggests a comparison of tourist development plans and global trends. It can entail looking at how various areas create regulations and infrastructure related to tourism.

Cluster 5

Development of tourism, sustainable tourism; navy blue in colour.

Economic tourism, destination perception and visitor satisfaction This cluster emphasises the financial benefits of tourism, with a particular emphasis on sustainability. The focus on 'destination image' and 'tourist satisfaction' points to initiatives to ensure that tourism's growth does not jeopardise the destinations' ecological or cultural integrity. It is likely that the cluster studies methods for striking a balance between social and environmental sustainability and economic growth.

3.3 *Country-Wise Analysis*

The map illustrates how, in many regions, only a small number of countries—such as the United States in North America, Israel and India in Asia and the United Kingdom in Europe—have displayed an active interest in collaboration on research projects related to religious tourism. South America has not developed a regional leadership in cooperation, while South Africa has emerged as a key leader in regional integration and pan-African initiatives. Religious tourism research has been found to originate from developed nations such as the United States and the United Kingdom, as well as emerging countries and economies of significance such as China, India and Taiwan (Tkaczynski and Arli (2018). In addition, several nations like Israel, China, India and Germany, have significantly impacted pilgrimage research more as measured by their average normalised citation numbers. The United States and Israel have established a great deal of cross-national collaboration to conduct research. However, new countries like Hong Kong, Japan, Italy, Canada, Magnolia, Cyprus and New Zealand have also begun religious tourism research, indicating a geographical expansion beyond Asia and North America (Table 5) (Fig. 6).

Table 5 Most cited countries

Country	TC	Average article citations
Israel	1921	76.8
USA	1551	35.2
United Kingdom	1496	37.4
Australia	1302	34.3
China	717	18.9
Singapore	424	70.7
New Zealand	408	68
India	321	8.7
Canada	294	24.5
Italy	291	15.3
Hong Kong	266	29.6
Spain	229	12.7
Japan	189	21
South Africa	184	12.3

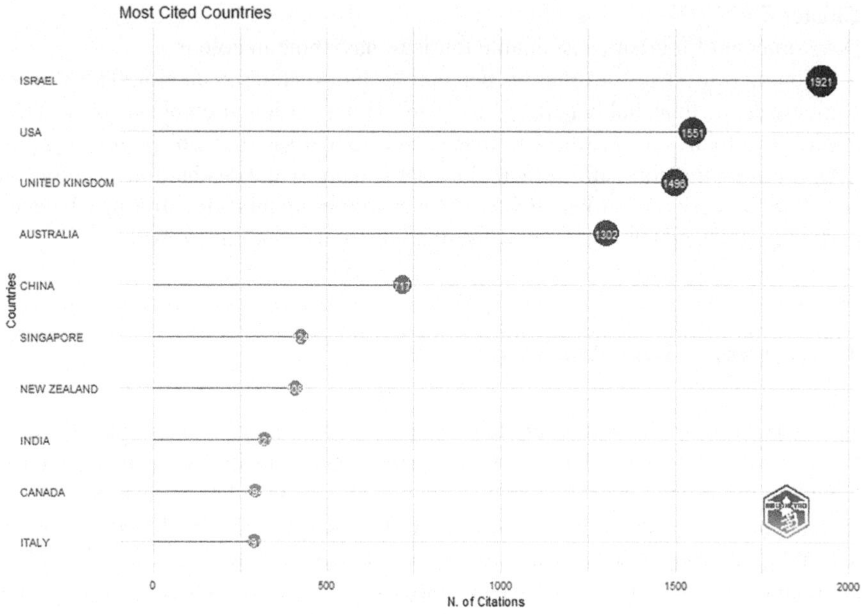

Fig. 6 Most cited countries. Source: author compilation

4 Conclusion and Implications

Religious tourism is starting to emerge as a serious business of its own as the number of people engaging in tourism activities is on a rapid scale rise today, with travelling, touring and journeying becoming more popular than it was before. This study will benefit the researchers and academics working in religious tourism and its associated disciplines. In the initial step, the researchers whose focus for search is knowledge or development might feel a need to cite the most active and productive authors, writers and key research papers of interest (Table 2) in the available body of information in this research area or they can make use of the literature as a source of reference for the identification of the appropriate journals for publication of their research papers and articles. In addition, the researchers and academics who are interested in collaboration for the development of possible quality research articles can refer to Fig. 5, which depicts the countries with the most collaboration so as to enhance the quality of their research output. In addition, this present investigation will also be very useful for the researchers and academics looking to enhance the body of literature in studies related to religious tourism to extend their understanding of this topic of research. The author, N. Collins-Kreiner from Israel, associated with Haifa University, has been cited as the most prominent researcher in this area of study, and his research papers, articles and studies have been cited by several researchers across the world. The evidence also indicates that the authors' research based in the United States has been recognised as the researchers with the largest volume of publications in this discipline, followed by authors based in China and the United Kingdom. With reference to this study, the majority of our sample's study has focused on Christianity.

This suggests that tourism is tied to other religions' more scholarly research, such as Hinduism, Judaism and Buddhism (Liro 2021). Religious tourism may be a true gift for everybody engaged in numerous ways. It may have a significant influence on local communities, corporations and governments alike. The flood of pilgrims creates economic possibilities, bolstering local businesses ranging from modest souvenir stores to pleasant bed and breakfasts. Governments profit from higher tax income, which allows for infrastructure upgrades and public services tailored to tourists' requirements. Furthermore, religious tourism promotes cultural interaction, understanding and tolerance—the ultimate road to world concord.

The following paragraph provides various themes for the researchers to carry out research in this area:

4.1 **Heritage and Cultural Preservation:** Protecting cultural assets is a top priority for religious tourism. These sites are frequently ancient and have substantial historical importance, necessitating efforts to maintain and preserve them for future generations. This includes physical preservation and the preservation of intangible cultural activities related to the locations. Balancing tourist expansion and historical protection involves collaboration among government agencies, local communities and tourism providers.

4.2 **Opportunity for Sustainable Tourism Development:** Sustainable development in religious tourism provides a unique chance to provide economic advantages while also protecting cultural and environmental resources. As religious tourism grows, the problem is to balance tourist demand with the need to protect sacred places and local populations' well-being. Implementing sustainable tourism practices is critical to attaining this balance and maintaining long-term viability. Religious tourism has the potential to considerably benefit the local economy by creating jobs, developing infrastructure and increasing income for tourism-related enterprises. However, these advantages might result in environmental damage and cultural loss without sustainable methods. Countries concentrating on sustainable tourism development stress responsible tourist management, which includes restricting visitor numbers to alleviate resource demand and constructing eco-friendly infrastructure to minimise the environmental impact of.

4.3 **Innovative Technologies in Pilgrimage Tourism:** Countries can use new technology to encourage sustainable religious tourism. This involves leveraging digital platforms for tourist management, such as online booking systems, to govern visitor flow and virtual reality to provide immersive experiences without requiring physical presence. By putting sustainability first, governments can guarantee that religious tourism remains a beneficial driver for economic growth, cultural preservation and environmental conservation, allowing future generations to continue benefiting from these priceless resources.

4.4 **Understanding Visitor Motivations:** Visitor motivations are often deeply personal and vary widely. For some, religious tourism is driven by a desire for spiritual growth or religious obligations, while others may be motivated by cultural curiosity or the aesthetic appeal of religious sites. Understanding these motivations allows destinations to create customised experiences that resonate with visitors. For example, offering guided spiritual retreats or educational tours about the history and significance of religious sites can cater to different interests and enhance visitor engagement.

5 Limitation

While researching religious tourism and pilgrimage is beneficial in understanding the cultural and spiritual components of travel, it is critical to recognise the possible negative effects of growing tourism on sacred places and local populations. Commercialising pilgrimage destinations can lead to exploitation, cultural commodification and environmental degradation. Additionally, the influx of tourists may disrupt the peaceful and contemplative atmosphere that pilgrims seek, ultimately diminishing the authenticity of the spiritual experience. Balancing the preservation of sacred sites with the promotion of tourism is a complex challenge that requires careful consideration of ethical and sustainable practices. Moreover, the emphasis on scholarly analysis and academic discourse may overlook the lived experiences of local communities and indigenous peoples, whose perspectives and rights should be integral to any discussion of religious tourism and pilgrimage. It is essential to approach the study of religious tourism with a critical lens, taking into account the potential conflicts between commercial interests, cultural heritage and spiritual significance. An in-depth examination of the social, environmental and economic implications of religious tourism is imperative to ensure responsible and sustainable management of pilgrimage sites and the well-being of the communities involved.

References

Abbate CS, Di Nuovo S (2013) Motivation and personality traits for choosing religious tourism. A research on the case of Medjugorje. Curr Issue Tour 16(5):501–506

Aria M, Cuccurullo C (2017) bibliometrix: An R-tool for comprehensive science mapping analysis. J Informet 11(4):959–975

Bi JW, Liu Y, Li H (2020) Daily tourism volume forecasting for tourist attractions. Ann Tourism Res 83:102923

Bond N, Packer J, Ballantyne R (2015) Exploring visitor experiences, activities and benefits at three religious tourism sites. Int J Tour Res 17(5):471–481

Bremer TS (2005) Tourism and religion. Encyclopedia of Religion 13:9260–9264

Budovich LS (2023) The impact of religious tourism on the economy and tourism industry. HTS Teol Stud/Theol Stud 79(1). https://doi.org/10.4102/HTS.V79I1.8607

Collins-Kreiner N (2010) Researching pilgrimage: Continuity and transformations. Ann Tour Res 37(2):440–456

Collins-Kreiner N (2020a) Religion and tourism: A diverse and fragmented field in need of a holistic agenda. Ann Tour Res 82:102892. https://doi.org/10.1016/j.annals.2020.102892

Collins-Kreiner N (2020b) A review of research into religion and tourism launching the annals of tourism research curated collection on religion and tourism. Ann Tour Res 82:102892

Duran-Sanchez A, Álvarez-García J, del Río-Rama MDLC, Oliveira C (2018) Religious tourism and pilgrimage: bibliometric overview. Religions 9(9):249

Das, A., Kondasani, R. K. R., & Deb, R. (2024). Religious tourism: a bibliometric and network analysis. In Tourism review 79, 3, pp. 622–634. Emerald Publishing. doi:https://doi.org/10.1108/TR-08-2022-0387

Eade J (1992) Pilgrimage and tourism at Lourdes, France. Ann Tour Res 19(1):18–32

Heidari A, Yazdani HR, Saghafi F, Jalilvand MR (2018) The perspective of religious and spiritual tourism research: a systematic mapping study. J Islamic Market 9(4):747–798. https://doi.org/10.1108/JIMA-02-2017-0015

Iliev D (2020) The evolution of religious tourism: concept, segmentation and development of new identities. J Hosp Tour Manag 45:131–140. https://doi.org/10.1016/j.jhtm.2020.07.012

Liro J (2021) Visitors' motivations and behaviours at pilgrimage centres: push and pull perspectives. J Herit Tour 16(1):79–99. https://doi.org/10.1080/1743873X.2020.1761820

Karamustafa K, Ülker P, Çalhan H (2022) Do level of tourism development and its type make a difference in residents' perceptions? Learning from Turkish cases. J Hosp Tour Ins 5(1):138–165

Khanra S, Dhir A, Kaur P, Mäntymäki M (2021) Bibliometric analysis and literature review of ecotourism: toward sustainable development. Tour Manag Perspect 37:100777

Kim B, Kim S, King B (2020) Religious tourism studies: evolution, progress, and future prospects. Tour Recreat Res 45(2):185–203. https://doi.org/10.1080/02508281.2019.1664084

Kumermanis M, Rudzitis J (2009) Microtopography model of the rough surface. Solid State Phenom 147:686–689

Lloyd K (2006) Catering to the backpacker: The transition of backpacker enclaves in Vietnam. Tour Recreat Res 31(3):65–73

Niñerola A, Sánchez-Rebull MV, Hernández-Lara AB (2019) Tourism research on sustainability: a bibliometric analysis. Sustainability 11(5):1377

Paliwal M, Chatradhi N, Tripathy S, Jha S (2023) Growth of digital entrepreneurship in academic literature: a bibliometric analysis. Int J Susble Dev Plan 18(6)

Prakash C, Dwivedi S (2022) Trends in Political Campaigning Research: A Bibliometric Literature Analysis. J Sci Res 11(2):262–271. https://doi.org/10.5530/jscires.11.2.28

Pritchard A (1969) Statistical bibliography; An Interim Bibliography

Ryan MM, Mizerski K (2010) Place branding for sustainable futures: a case study. Place Brand Public Dipl 6:49–57

Shinde K (2022) The spatial practice of religious tourism in India: a destinations perspective. Tour Geogr 24(4–5):902–922. https://doi.org/10.1080/14616688.2020.1819400

Shuo YSS, Ryan C, Liu GM (2009) Taoism, temples and tourists: the case of Mazu pilgrimage tourism. Tour Manag 30(4):581–588

Singh RP, Rana PS, Olsen DH (2021) Environment as a sacred space: religious and spiritual tourism and environmental concerns in Hinduism. In: The routledge handbook of religious and spiritual tourism. Routledge, pp 135–151

Singh R, Singh AK, PS, S. (2022) Journal of human resources in hospitality and tourism: a bibliometric overview. J Hum Resour Hosp Tour:1–22

Singh S (2006) Tourism in the sacred Indian Himalayas: an incipient theology of tourism? Asia Pacific J Tour Res 11(4):375–389

Smith VL, Eadington WR (eds) (1992) Tourism alternatives: potentials and problems in the development of tourism. University of Pennsylvania press

Suban SA, Madhan K, Shagirbasha S (2021) A bibliometric analysis of halal and Islamic tourism. Int Hosp Rev 37:219. https://doi.org/10.1108/IHR-05-2021-0038

Tilson DJ (2005) Religious-spiritual tourism and promotional campaigning: a church-state partnership for St. James and Spain. J Hosp Leisure Market 12(1–2):9–40

Tkaczynski A, Arli D (2018) Religious tourism and spiritual leadership development: Christian leadership conferences. J Hosp Tour Manag 35:75–84

van Raan AFJ (2004) Measuring science. In: Hand-book of quantitative science and technology research. Springer, Netherlands, pp 19–50. https://doi.org/10.1007/1-4020-2755-9_2

Verma M, Sarangi P (2019) Modeling attributes of religious tourism: a study of Kumbh Mela, India. J Conv Event Tour 20(4):296–324. https://doi.org/10.1080/15470148.2019.1652124

A Profile Analysis of Travellers Participating in a Mass Gathering Religious Event

Bárbara Sofía Pasaco-González, Francisco Sánchez-Cubo, Ana María Campón-Cerro, and Ricardo Gouveia Rodrigues

Abstract Given the importance of religious events as a cultural expression, literature has emphasised the need to identify their characteristics. This study aims to define the profile of tourists participating in religious events. Using data from a questionnaire distributed at the 2023 World Youth Day, Latent Class Analysis (LCA) was performed. The findings reveal that religious events attract visitors who are primarily interested in having a unique religious experience and growing spiritually. Attendees also consider their participation in the event as an opportunity to learn about the destination and the local culture, improving their knowledge about themselves and the world. These results position religious events as a holistic experience framed by the search for benefits derived from participation in the event and knowledge of the destination.

B. S. Pasaco-González (✉)
Institutos Universitarios de Investigación, Universidad de Extremadura, Cáceres, Spain
e-mail: sofiap@unex.es

F. Sánchez-Cubo
Department of Applied Economics (Economic Structure), University of Málaga, Málaga, Spain
e-mail: fsanchezcubo@uma.es

A. M. Campón-Cerro
Facultad de Empresa, Finanzas y Turismo, Universidad de Extremadura, Cáceres, Spain

Universidade da Beira Interior & NECE–Research Centre for Business Sciences, Covilhã, Portugal
e-mail: amcampon@unex.es

R. G. Rodrigues
Universidade da Beira Interior & NECE–Research Centre for Business Sciences, Covilhã, Portugal
e-mail: rgrodrigues@ubi.pt

M. Valeri (ed.), *Contemporary Religious Tourism*, Tourism, Hospitality & Event Management, https://doi.org/10.1007/978-3-031-79074-4_8

1 Introduction

Today, religious tourism is one of the most significant market segments of the tourism industry and generates millions of trips worldwide each year (Collins-Kreiner 2020; Das et al. 2024). In fact, there are between 300 and 330 million travellers visit the world's major religious sites annually. In addition to the growing number of trips generated by religious tourism, new niche markets have also emerged due to the wide range of providers offering religious tourism and pilgrimage trips (Iliev 2020). Religious tourism has thus become a highly diversified market segment, which, as a consequence, has also generated a diversification in the customer segments attracted by this form of travel.

Within the categorisation of religious tourism resources, events are positioned as one of the most expanding religious tourism activities (Tsai 2021) and compose one of the main categories within the event-tourism context (Getz and Page 2016). Because of this, the religious event experience has become an area of interest in the tourism literature (Gallarza et al. 2009). However, it is still a growing area of research, so there are still few studies that have addressed this topic (Wang et al. 2020).

Given the importance that religious events have acquired (Sánchez et al. 2017), it is relevant to know the motivations of the individuals who participate in them. That is because, as is the case in other areas of tourism, in religious tourism, there are also specific groups interested in particular experiences (Hassan et al. 2022; Liro 2021; Tsai 2021). Kamath et al. (2023) argue that the motivations for attending religious events may be associated with their religious beliefs, or, in other cases, they may be motivated by an interest in visiting the destination. Liro (2021) indicates that research on religious motivations has been growing; however, complementary motivations have only recently been considered. Therefore, there is still a lack of further discussion on the actual motivations that drive a pilgrim or tourist to attend a religious activity (Wang et al. 2020).

In addition to motivations, it is relevant to identify participants' characteristics or the existence of specific market niches. However, given the remarkable diversification of tourists' needs (Sánchez et al. 2017), it is still a challenge to determine the types of participants in religious events (Wang et al. 2020). Typologies of religious tourists have been identified in the tourism literature. However, these studies have primarily been associated with the experience of visiting sacred sites. In contrast, in the context of religious events, few studies have addressed the motivations and profiles of attendees (Wang et al. 2020).

Based on the above considerations, this study proposes to identify potential profiles of religious event participants based on their religious and non-religious motivations. On a theoretical level, this study contributes to the body of knowledge of religious tourism literature by identifying a potential classification of religious event participants. On a practical level, the results provide information that will enable event organisers and destination managers to formulate marketing strategies to position this type of event among diverse consumers.

2 Theoretical Background

2.1 Religious Events and Profile of Attendants

In the tourism literature, religious events have been classified as a religious tourism resource category (Abad et al. 2016; Nolan and Nolan 1992). Religious events are essential to people's culture (Gupta and Basak 2018) and have motivated travel to sacred sites since ancient times (Ohlan and Ohlan 2024). Even today, thousands of people continue to participate, interact and socialise during these gatherings (Verma and Sarangi 2019). That has led to religious events or festivals becoming relevant tourism resources for destinations (Nolan and Nolan 1992). In fact, religious events constitute one of the most expanding forms of religious tourism due to their potential to attract visitors (Tsai 2021). That means religious events can contribute to developing destinations and local communities by bringing economic, social and cultural benefits (Lee et al. 2015).

The growing popularity of religious events can be attributed to the holistic experience they provide. Kasim (2011) noted that religious events can be perceived as both a tourist experience and a religious experience that provides spiritual healing and holiness. Kim and Kim (2019) indicate that finding oneself represents a critical element people seek when engaging in religious tourism experiences or visiting religious tourism destinations. Kamath et al. (2023) pointed out that religious events allow visitors to achieve incredible spiritual growth and learn about the local culture of the place where they are held. Thirugnanasambantham et al. (2023) highlighted that participation in religious events generates feelings of holiness, devotion and consecration, allowing individuals to feel fulfilled and delighted. According to these studies, religious events involve both the experience of the event itself and the experience of the destination. Therefore, the potential for attracting diverse visitor profiles is broader, especially when it comes to events with an enormous international scope.

The experience that attendees seek from an event determines their motivations greatly. Therefore, one of the recurring themes in event studies is attendee motivation (Thirugnanasambantham et al. 2023). That is because motivation often makes people decide to make a trip or visit a specific destination (Abad et al. 2016). Studies on religious tourism have found that religion prevails as the main motivation for religious tourism trips. However, as the needs of individuals are constantly evolving, nowadays, motivations in the field of religious tourism are also of a cultural and leisure nature (Iliev 2020; Kim et al. 2019). Therefore, nowadays, participants in religious events, in addition to expressing their religious interests, also play the role of traditional visitors interested in different aspects unrelated to the sacred (Abad et al. 2016). In this way, motivations for religious tourism can become quite diverse (Iliev 2020).

Given that the choice to participate in one religious tourism experience or another is individual (Kamath et al. 2023) and that people visit religious sites for different motivations (Tsai 2021), it can be assumed that different visitor profiles can be identified, each with their own preferences and interests. Certain visitor profiles in the religious tourism literature have been identified based on motivational dimensions. For example, Wang et al. (2020), in the context of a religious event, identified four types of participants according to their motivations: leisure travellers, devout believers, cultural enthusiasts and religious pragmatists. Tsai (2021), in a religious festival, explored the types of participants in a polytheistic society. The results yielded four distinct groups: sacred pilgrims, believers, experience companions and secular polytheists. Hassan et al. (2022) defined the types of pilgrims visiting the city of Mecca. The results showed three segments of demand: multiple motives, passive tourists and religious tourists. As can be seen, the results of these previous studies confirm that religious tourists cannot be treated as a homogeneous group, not only because of the individual interests of each person but also because each type of religious attraction (i.e. event, sacred place, temples, etc.) provides a unique experience. Therefore, for each type of religious tourism experience, the types of market segment it attracts should be identified.

Based on the above considerations, this study aims to better understand the types of visitor profiles attending religious events. For this purpose, motivational indicators categorised into religious and non-religious will be used. According to Liro et al. (2017), while religious motivations are still present as the essence of religious tourism, it has become increasingly common for this type of travel to also cater to non-religious needs, which may affect the diversity of the visitor profile. Liro (2021) indicates that non-religious motivations are more related to cultural heritage, leisure, social life and other non-spiritual factors. These non-religious motivations are also important because they function as magnets attracting visitors and often prolonging their stays. Thus, for the purposes of this study, this classification can provide further insight into the profiles of religious event attendees who travel for one reason or another. Even considering the two categories are not mutually exclusive, verifying what kind of religious and non-religious motivated profiles prevail would be possible.

3 Methodology

3.1 Data Collection

Data was collected through a face-to-face survey for which a questionnaire structured in three sections was developed. The first section included questions on the type and frequency of participation in WYD. The second section was aimed at gathering information about the participants' motivations. Fifteen items adapted from previous research on religious tourism (e.g. Amaro et al. 2018; Hassan et al. 2022; Tsai 2021; Wang et al. 2020) were used to measure motivations. All indicators were measured using a 5-point Likert scale, where 1 is 'not very important' and 5 is 'very important'. Finally, the third section covered questions on the socio-demographic profile of the participants. The questionnaire was developed in the five official WYD languages (i.e. English, French, Italian, Portuguese and Spanish). Prior to the launch of the survey, a pre-test and a pilot test were carried out to verify the correct functioning of the survey.

The non-probabilistic convenience sampling method was used to select the sample. The sample consisted of WYD participants aged 18 years or older. Data collection took place in Lisbon from 1 to 6 August, the official days of the event. Participants were contacted at the main points of concentration of attendees. All participants gave their consent to participate in the study. As a result of the fieldwork, 234 responses were obtained. However, for the purposes of this study, only fully completed questionnaires were considered. Therefore, those records with missing values were discarded from the sample. Thus, the final sample consists of 225 valid questionnaires.

3.2 Data Analysis

In order to achieve the objective of this research, the application of a methodology appropriate to both the purpose and the nature of the data is required. In this sense, knowing the religious nature of the event, the classification of WYD 2023 attendees into homogeneous groups must be carried out according to a criterion that qualifies this generic grouping of religious tourists. To this end, information on the motivations of those attending the event will be used, divided into two groups following previous contributions from the literature (e.g. Liro 2021; Liro et al. 2017). On the one hand, we consider those motivations strictly religious and encompass a dimension exclusively associated with the event. On the other hand, we group together those motivations that may take place at the event but are not unequivocally linked to the celebration of a religious event. Since the subdivision of these motivations does not make them mutually exclusive, the same sample is applied for the entire study.

Considering the above, the methodology that most closely meets the needs of this study is Latent Class Analysis (LCA). This method allows individuals to be classified according to their responses on a Likert scale of 1 (not very important) to 5 (very important) for each motivation indicator. It also offers several goodness-of-fit tests—AIC, BIC, G^2, X^2 and MLE. The R (R Core Team 2021) *poLCA* package (Linzer and Lewis 2011) was used to run this methodology and obtain the graphical outputs. Also, although the age variable towards younger values skews the sample, this is due to the nature of the event. Hence, bias is ruled out by the mere definition of the object of the study. Below are some descriptions of the sample.

4 Results

4.1 Profile and Motivations of Attendee

According to the results of the socio-demographic profile, 55.11% of the sample comprises female participants and 44.89% of male participants. About the age of the individuals, it can be seen that the most representative groups are made up of individuals aged 18 to 25 years (43.56%) and 26 to 35 years (39.11%), which demonstrates the prevalence mentioned above of young participants in the sample. Likewise, the declared level of education stands out, with 86.22% of the sample being university graduates. As for their place of residence, participants were found to be from the Americas, Europe and Africa. The American and European continents show relatively equal percentages (49.33% and 49.78%, respectively), while the African continent shows a less significant percentage (0.89%). Between the three continents, there are a total of 33 countries, of which from the American continent, the presence of participants from Mexico stands out (19.82%); from the European continent, they come mainly from Spain (77.68%), and from the African continent only participants from Zimbabwe are present in the sample.

Regarding participants' motivations, Table 1 presents a descriptive analysis of the motivational indicators analysed and their correspondence with each category. The results show that the 'religious motivations' category is the most important, followed by non-religious motivations. Within religious motivations, participants are more interested in gaining a unique religious experience and achieving spiritual growth. On the other hand, among the non-religious motivations, the attendees valued the opportunity to live the event as a unique travel experience.

Table 1 Motivation indicators analysed

Motivation indicators by category	Mean
Religious motivations	**4,50**
MOT1_Personal development	4,44
MOT2_Religious devotion	4,54
MOT4_Spiritual growth	4,65
MOT7_Visiting religious pilgrimage sites	4,15
MOT11_Reflecting on the meaning of life	4,47
MOT14_To be at an event with the pope	4,60
MOT15_Living a unique religious experience	4,66
Non-religious motivations	**4,07**
MOT3_Learning about local culture	3,68
MOT5_Participate in programmed activities	4,20
MOT6_Experience a cultural exchange	4,15
MOT8_Meet new people	4,31
MOT9_Escape from the daily routine	3,80
MOT10_Having a unique experience	4,64
MOT12_Visit new tourist destinations	3,55
MOT13_Enjoying a multicultural environment	4,32

Source: Authors

4.2 Latent Class Analysis (LCA)

For the LCA, socio-demographic data were not considered as no consistent and statistically significant relationships were found through their inclusion as covariates. Therefore, only the inclusion of the motivation indicators subdivided into the abovementioned categories is considered. Thus, Table 2 shows the goodness-of-fit indices of the LCA for both groups of motivations—religious and non-religious—running models for 2, 3 and 4 classes. This methodology requires manually inserting the number of classes and comparing the model fit through the classical indicators included in the table. The lower values, comparatively speaking, will indicate the optimal number of classes. This study finds minor and variable differences between methods when comparing the results. However, for both cases, it appears that the optimal solution is around three classes, as this is the one with the best interpretability, as opposed to the inclusion or deletion of an additional class for a minimum improvement of the values of the information and fit criteria contained in Table 2.

On the other hand, it should be remembered that the individuals in the classes are the same for both groups of motivations, so the three classes in these are different. For ease of understanding and interpretation of the results, the classes about religious motivations are listed as Class 1, Class 2 and Class 3, and the non-religious motivations as Class A, Class B and Class C. Table 3 lists the Likert response probability 1 to 5 for each of the religious motivations, while Fig. 1 graphically represents these results. All in all, it is observed that the three classes are easily assimilated to three types of religious tourists, similar to Wang et al. (2020). On the one hand, Class 1 contains individuals less interested in the religious part of WYD. Given the nature of WYD 2023, this class accounts for only 7.4%. Class 2, on the other hand, contains the most devout believers, representing 50.7% of the individuals. Finally, Class 3 contains those who, although very interested in these aspects, moderate their response (religious pragmatists), accounting for 41.9%.

Then, the results for non-religious motivations, shown in Table 4 and Fig. 2, are slightly more heterogeneous. Thus, while in all three classes, the indicator

Table 2 Goodness of fit for religious and non-religious motivations' LCA models

		Religious motivations			Non-religious motivations	
Goodness of fit	4 classes	3 classes	2 classes	4 classes	3 classes	2 classes
AIC	2687.106	2667.011	2698.421	3990.601	4001.839	4095.018
BIC	3079.957	2960.796	2893.138	4438.11	4336.617	4317.065
G^2	730.7335	768.6388	858.0482	1516.357	1593.595	1752.774
X^2	281,159	22328.12	3,708,529	183468.3	194996.7	647501.7
MLE	−1228.553	−1247.506	−1292.21	−1864.301	−1902.92	−1982.509
Param.	115	86	57	131	98	65
d.f.	110	139	168	94	127	160
n	225	225	225	225	225	225

Source: Authors

Table 3 Types of religious tourists by religious motivations

	Pr(1)	Pr(2)	Pr(3)	Pr(4)	Pr(5)
[MOT_1] personal development					
Class 1	0.1205	–	0.3743	0.2463	0.2589
Class 2	–	–	–	0.0444	0.9556
Class 3	–	0.0636	0.1886	0.4543	0.2935
[MOT_2] religious devotion					
Class 1	–	0.4219	0.2433	0.1969	0.1379
Class 2	–	–	0.0081	0.0518	0.9402
Class 3	–	–	0.1065	0.4541	0.4394
[MOT_4] spiritual growth					
Class 1	–	0.1747	0.3736	0.2513	0.2005
Class 2	–	–	–	–	1
Class 3	–	0.0117	0.0509	0.4329	0.5045
[MOT_7] visiting religious pilgrimage sites					
Class 1	0.1333	0.2241	0.5593	0.0834	–
Class 2	0.0173	0.0088	0.0081	0.2177	0.7481
Class 3	0.0086	0.1196	0.2311	0.3476	0.2931
[MOT_11] reflecting on the meaning of life					
Class 1	0.2184	–	0.2642	0.3266	0.1908
Class 2	–	0.0173	0.0147	0.0661	0.9018
Class 3	0.0146	0.0215	0.1583	0.3397	0.4659
[MOT_14] being at an event with the pope					
Class 1	0.0603	0.4219	0.0924	0.1216	0.3038
Class 2	–	–	–	0.0683	0.9317
Class 3	–	–	0.1110	0.3307	0.5583
[MOT_15] living a unique religious experience					
Class 1	–	0.3022	0.2964	0.3300	0.0714
Class 2	–	–	–	0.0492	0.9508
Class 3	0.0106	0.0104	0.0327	0.2959	0.6504

Source: Authors

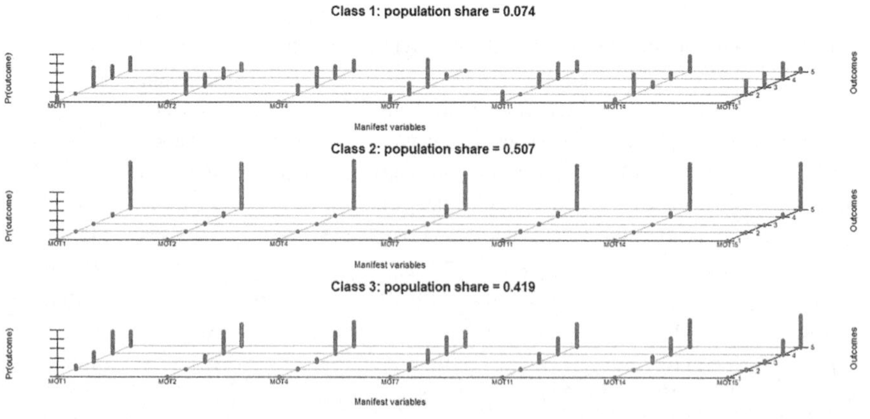

Fig. 1 Visual representation of the probabilities of each outcome for each given religious motivation. Source: Authors

Table 4 Types of religious tourists by non-religious motivations

	Pr(1)	Pr(2)	Pr(3)	Pr(4)	Pr(5)
[MOT_3] learning about local culture					
Class A	0.2725	0.3405	0.2986	0.0679	0.0205
Class B	0.0207	0.0870	0.1395	0.1469	0.6059
Class C	–	0.0139	0.2920	0.5655	0.1286
[MOT_5] participation in planned activities					
Class A	–	0.1014	0.2663	0.3858	0.2465
Class B	0.0085	0.0254	0.0693	0.3130	0.5839
Class C	0.0175	0.0167	0.1503	0.5227	0.2928
[MOT_6] experiencing a cultural exchange					
Class A	0.0254	0.2011	0.6311	0.0801	0.0623
Class B	0.0168	–	0.0645	0.1324	0.7863
Class C	0.0132	–	–	0.8303	0.1565
[MOT_8] meeting new people					
Class A	0.0201	0.1821	0.3068	0.2977	0.1933
Class B	–	0.0165	0.0509	0.0787	0.8539
Class C	–	–	0.1880	0.5415	0.2705
[MOT_9] leaving the daily routine behind					
Class A	0.2235	0.1851	0.2439	0.1817	0.1658
Class B	0.1219	0.0090	0.0223	0.1039	0.7428
Class C	0.0608	0.1004	0.2846	0.3976	0.1566
[MOT_10] having a unique experience					
Class A	0.0402	0.0605	0.1053	0.1675	0.6265
Class B	–	–	0.0084	0.0442	0.9474
Class C	–	0.0174	0.1011	0.4111	0.4704
[MOT_12] visiting new tourist destinations					
Class A	0.3135	0.2464	0.4126	0.0275	–
Class B	0.0627	0.0532	0.1465	0.1800	0.5576
Class C	–	0.1307	0.3010	0.4269	0.1414
[MOT_13] enjoying a multicultural social environment					
Class A	–	0.2011	0.3689	0.3171	0.1129
Class B	–	–	–	0.0815	0.9185
Class C	0.0175	–	0.1693	0.6591	0.1540

Source: Authors

'MOT10_Having a unique experience' is key for all, the rest show more significant variability. Again, the first class, Class A, shows relatively low values for the motivations studied. Specifically, the subjects included in this category seem less interested in the destination's cultural and tourism issues while positively valuing both the experience and the multicultural social environment and meeting new people. This group accounts for 22.1% of respondents. Class B, at 52.6%, again includes those individuals who give very high ratings to all non-religious motivations. Finally, Class C includes those participants who, although their interest is high, are more spread out among the scores given, grouping 25.4% of the individuals.

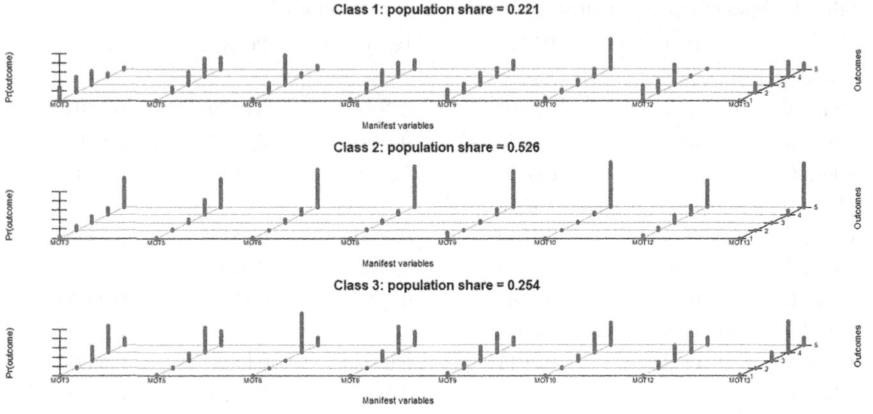

Fig. 2 Visual representation of the probabilities of each outcome for each given non-religious or spiritual motivation. Source: Authors. Note: Class numbers associated with letters as 1, A; 2, B; 3, C

5 Discussion

This study applies to the context of religious events and has had the main objective of identifying potential profiles of participants of religious events. In this study, the results indicate that attendees of religious events have religious and non-religious motivations. These findings are consistent with previous studies that have found that religious tourism is undertaken to satisfy both religious and non-religious interests (Abad et al. 2016; Hassan et al. 2022; Wang et al. 2020). Therefore, both types of motivations coexist among the participants of such events. When analysing the importance of the two types of motivations, the results indicate that religious motivation is the most important, followed by non-religious motivations. These findings align with those obtained by Abad et al. (2016) and Hassan et al. (2022). Thus, it is confirmed that religious aspects remain the most important motivation for religious tourism. However, given the evolution of demand, these are not the only ones; non-religious motivations also accompany them.

About the category of religious motivations, the results show three classes of participants, each with different interests:

- The first refers to a small percentage of attendees attributing little importance to the majority of motivations related to the religious connotation of the event. Similar to Abad et al. (2016), in their study on visitors' motivations for the Ignatian Way, they identified that a small percentage of visitors showed secular and cultural, rather than religious, motivations.
- The second refers to a highly devout and religiously involved participant. They intend to deepen their religious beliefs and achieve maximum subjective benefits, such as increased spirituality and personal development. According to Abad et al. (2016), these characteristics could correspond to those of a traditional pilgrim inspired primarily by religion and spiritual quests.

- Finally, the third corresponds to a visitor who is religiously involved but who attends the event primarily for the religious experience as an opportunity for individual reflection and a search for meaning in life. This finding is in line with Nilsson (2018), who, in the context of the Camino de Santiago, found that many pilgrims undertake this journey to reflect on aspects of their lives and reach a deeper understanding of themselves.

Regarding the category of non-religious motivations, the participants' interests are divided into three classes:

- The first corresponds to a type of participant who, while positively valuing the event as a unique experience in their lives, shows little interest in the possibility of visiting the destination and experiencing the local culture. That could be because, in line with the results of the category of religious motivations, this type of participant is the most religiously involved and, therefore, understands their participation in the event more as a religious practice than a tourist one.
- The second refers to a type of participant who values all indicators of non-religious motivations as very important. It could indicate a participant interested in cultural exploration with a strong tourist inclination. That may be because, according to Olsen and Timothy (2006), more and more people are taking religious trips for cultural interests in order to learn about the history of the place and understand its religious tradition. In this way, religious events function as a catalyst for tourism activity in the destination.
- Lastly, the third type of attendee gave medium importance to all non-religious motivations. However, the motivations 'experiencing a cultural exchange' and 'enjoying a multicultural social environment' are the most highly rated among these. This implies that for this type of attendee, the event represents an opportunity for social interaction. In this regard, Maeng et al. (2016) found that socialisation represents one of the most important motivational factors in the context of events because events foster social cohesion. Religious events, therefore, provide an environment that enables the building of a community among people who share a common interest.

Overall, these results demonstrate that while religious events provide a context for people to connect with their religious beliefs and have a meaningful spiritual experience (Abad et al. 2016), they also attract visitors with multifaceted interests that are not precisely linked to religion (Nilsson 2018). Therefore, the symbolic meaning attributed to religious events depends on the individual's own interests.

6 Conclusions

Given the large number of trips generated by events worldwide, it is necessary to investigate the characteristics and motivations of the attendees. This is because the potential to attend an event depends on the individual's own interests

(Özdemir-Bayrak 2011). In the case of religious events, the results of this study have shown that attendees' motivations evolve, and, therefore, new visitor segments constantly emerge (Iliev 2020). In fact, multiple attendee profiles can be found at the same event, each with different expectations about the experience and the benefits they expect to achieve from the event. Thus, this study has sought to elucidate the characteristics of these different profiles based on their motivations. As indicated in the literature, these motivations have been categorised as religious and non-religious, further enhancing these profiles' heterogeneity. The results have confirmed the coexistence of both types of motivations among the preferences of those attending religious events and the increasingly evident link between religion and tourism.

On a theoretical level, given that little research has addressed the issue of participants' motivation for religious events (Thirugnanasambantham et al. 2023), this study broadens the scope of previous research by focusing specifically on religious events. Furthermore, according to Iliev (2020), identifying visitor motives is relevant to understanding religious tourism in contemporary society. Therefore, this study contributes to the body of knowledge of religious tourism literature by examining the motivations that drive people to attend religious events. This study also contributes to the literature by identifying a potential classification of religious event participants based on religious and non-religious motivations. While previous research has mentioned these two types of motivations, this study goes beyond identifying motivations within each category and explores how these may contribute to a segmentation of religious tourists. Finally, it contributes to the literature on events in tourism by providing information on an event typology that has been less studied.

On a practical level, the results of this study provide a better understanding of attendees' behaviour at religious events. It contributes to the decision-making process of event organisers and destination managers in terms of striving for balanced event management, that is, trying to preserve the essence and authenticity of the event (Kasim 2011) and avoiding marketing the event as a tourism product (Ohlan and Ohlan 2024). At the same time, proper planning of these events should be managed, taking into account the needs and possible impacts on residents. On the other hand, the identification of typologies of participants according to their motivational behaviour indicates that religious events do not encompass homogenous groups of people. Therefore, the challenge in designing the event and destination experience lies in formulating marketing strategies to position the event among diverse consumers (Pegg and Patterson 2010). Furthermore, the activities undertaken by participants must align with the attendees' motivations (Buzinde et al. 2014).

The limitations of this study are related, first, to the use of the non-probabilistic convenience sampling method, which limits the generalisability of the results. Second, due to the event's subject matter, the most significant percentage of the sample comprises a young market segment. Further research could focus on profiles of individuals with higher age ranges or carry out comparative studies between different age groups. Third, motivational indicators were used to identify visitor typologies. Future studies could use additional variables such as religious affiliation, geographical region, form of travel organisation and length of stay, among others.

Comparisons could also be made between classes of tourists according to perceived experience, satisfaction and behavioural intentions. Finally, this study has focused on religious events as a specific type of religious tourism resource. Future studies could apply this analysis to other types of religious tourism experiences or compare experiences of different categories.

Acknowledgments Project developed within the framework of the Margarita Salas grant of the Universidad de Extremadura (Ref. MS-23), awarded within the actions for the concession of subsidies to public universities for the requalification of the Spanish university system (Ministerio de Universidades, R.D. 289/2021).

The NECE-UBI, Research Centre for Business Sciences, Research Centre and this work are funded by FCT—Fundação para a Ciência e a Tecnologia, IP—project UIDB/04630/2020 and DOI identifier 10.54499/UIDP/04630/2020.

References

Abad M, Guereño B, Makua A, Iriberri J (2016) Pilgrimage as tourism experience: a preliminary approach to profiling pilgrims on the Ignatian way. Int J Relig Tour Pilgr 4(4):48–66. https://doi.org/10.21427/D7KT5N

Amaro S, Antunes A, Henriques C (2018) A closer look at Santiago de Compostela's pilgrims through the lens of motivations. Tour Manag 64:271–280. https://doi.org/10.1016/j.tourman.2017.09.007

Buzinde C, Kalavar J, Kohli N, Manuel-Navarrete D (2014) Emic understandings of Kumbh Mela pilgrimage experiences. Ann Tour Res 49:1–18. https://doi.org/10.1016/j.annals.2014.08.001

Collins-Kreiner N (2020) Religion and tourism: a diverse and fragmented field in need of a holistic agenda. Ann Tour Res 82:102892. https://doi.org/10.1016/j.annals.2020.102892

Das A, Kondasani R, Deb R (2024) Religious tourism: a bibliometric and network analysis. Tour Rev 79(3):622–634. https://doi.org/10.1108/TR-08-2022-0387

Gallarza M, Arteaga F, Floristán E, Gil I (2009) Consumer behavior in a religious event experience: an empirical assessment of value dimensionality among volunteers. Int J Cult Tour Hosp Res 3(2):165–180. https://doi.org/10.1108/17506180910962168

Getz D, Page S (2016) Progress and prospects for event tourism research. Tour Manag 52:593–631. https://doi.org/10.1016/j.tourman.2015.03.007

Gupta S, Basak B (2018) Exploring pilgrim satisfaction on facilities for religious events: a case of Ratha Yatra at Puri. Asia Pacific J Tour Res 23(8):765–779. https://doi.org/10.1080/10941665.2018.1486864

Hassan T, Carvache-Franco M, Carvache-Franco W, Carvache-Franco O (2022) Segmentation of religious tourism by motivations: a study of the pilgrimage to the city of Mecca. Sustainability 14(13):7861. https://doi.org/10.3390/su14137861

Iliev D (2020) The evolution of religious tourism: concept, segmentation and development of new identities. J Hosp Tour Manag 45:131–140. https://doi.org/10.1016/j.jhtm.2020.07.012

Kamath V, Ribeiro M, Woosnam K, Mallya J, Kamath G (2023) Determinants of Visitors' loyalty to religious sacred event places: a multigroup measurement invariance model. J Travel Res 62(1):176–196. https://doi.org/10.1177/00472875211057600

Kasim A (2011) Balancing tourism and religious experience: understanding Devotees' perspectives on Thaipusam in Batu caves, Selangor, Malaysia. J Hosp Market Manag 20(3–4):441–456. https://doi.org/10.1080/19368623.2011.562437

Kim B, Kim S (2019) The effect of religious tourism experiences on personal values. Tour Pilgr 7(2):85–93. https://doi.org/10.21427/5gvy-hh90

Kim B, Kim S, King B (2019) Religious tourism studies: evolution, progress, and future prospects. Tour Recreat Res 45(2):185–203. https://doi.org/10.1080/02508281.2019.1664084

Lee T, Fu C, Chang P (2015) The support of attendees for tourism development: evidence from religious festivals, Taiwan. Tour Geograp 17(2):223–243. https://doi.org/10.1080/14616688.2014.997280

Linzer D, Lewis J (2011) poLCA: an R package for Polytomous variable latent class analysis. J Stat Softw 42(10):1–29. https://www.jstatsoft.org/v42/i10/

Liro J (2021) Visitors' motivations and behaviours at pilgrimage centres: push and pull perspectives. J Herit Tour 16(1):79–99. https://doi.org/10.1108/ijse-08-2021-0481

Liro J, Sołjan I, Bilska-Wodecka E (2017) Visitors' diversified motivations and behavior–the case of the pilgrimage center in Krakow (Poland). J Tour Cult Chang 16(4):416–435. https://doi.org/10.1080/14766825.2017.1332068

Maeng H, Jang H, Li J (2016) A critical review of the motivational factors for festival attendance based on meta-analysis. Tour Manag Perspect 17:16–25. https://doi.org/10.1016/j.tmp.2015.10.003

Nilsson M (2018) Wanderers in the shadow of the sacred myth: pilgrims in the 21st century. Soc Cult Geogr 19(1):21–38. https://doi.org/10.1080/14649365.2016.1249398

Nolan M, Nolan S (1992) Religious sites as tourism attractions in Europe. Ann Tour Res 19(1):68–78. https://doi.org/10.1016/0160-7383(92)90107-Z

Ohlan R, Ohlan A (2024) Religious tourism scholarship: current state and future research directions. J Islamic Market 15(3):800–818. https://doi.org/10.1108/JIMA-05-2023-0152

Olsen D, Timothy D (2006) Tourism and religious journeys. In: Olsen T, Olsen D (eds) Tourism, religion and spiritual journeys. Routledge, pp 1–21

Özdemir-Bayrak G (2011) Festival motivators and consequences: a case of Efes Pilsen blues festival, Turkey. Anatolia 22(3):378–389. https://doi.org/10.1080/13032917.2011.634916

Pegg S, Patterson I (2010) Rethinking music festivals as a staged event: gaining insights from understanding visitor motivations and the experiences they seek. J Conv Event Tour 11(2):85–99. https://doi.org/10.1080/15470141003758035

R Core Team (2021) R: a language and environment for statistical computing. R Foundation for Statistical Computing, Vienna, Austria. https://www.R-project.org/

Sánchez V, Fernández M, Lara J (2017) Economic impact of a religious and tourist event: a holy week celebration. Tour Econ 23(6):1255–1274. https://doi.org/10.1177/1354816616675996

Thirugnanasambantham K, Rajasekharan P, Patwardhan V, Raghavendra G, Rao S (2023) Determinants of visitors' festival satisfaction and subjective Well-being: tracing the roles of place attachment and satisfaction. Int J Event Festiv Manag 14(4):441–459. https://doi.org/10.1108/IJEFM-07-2022-0061

Tsai H (2021) Exploring the motivation-based typology of religious tourists: a study of welcome Royal Lord Festival. J Destin Mark Manag 21:100623. https://doi.org/10.1016/j.jdmm.2021.100623

Verma M, Sarangi P (2019) Modeling attributes of religious tourism: a study of Kumbh Mela, India. J Conv Event Tour 20(4):296–324. https://doi.org/10.1080/15470148.2019.1652124

Wang K, Kasim A, Yu J (2020) Religious festival marketing: distinguishing between devout believers and tourists. Religions 11(8):1–12. https://doi.org/10.3390/rel11080413

Religious Tourism Phenomena Evolution and Future Research Perspectives

Kamarun Muhsina

Abstract The objective of this study is to examine the historical development, present situation, and future potential of religious tourism, with a focus on its economic conditions, various phenomena, and strategies for progress. The research develops a thorough literature review methodology to track the historical development of religious tourism, starting from ancient pilgrimages to contemporary organized tours. The investigation of current trends and challenges in religious tourism is conducted through the analysis of scholarly articles, reports, and case studies. The study emphasizes notable advancements in religious tourism, such as the influence of technology on visitor experiences, the trend of integrating spirituality with recreational pursuits, and the economic consequences for destination regions. It identifies significant obstacles, such as the impact of global politics and concerns regarding authenticity and commercialization.

1 Introduction

Religious tourism has been a topic of interest over the years given the growing interest in the same. This can be categorically described as religious tourism; it is traveling with the aim of carrying out certain religious activities, for instance, going for pilgrimage, visiting certain religious shrines, or attending specific religious ceremonies. The following are some of the factors that have boosted religious tourism in the world today: globalization, technology, and change in social cultural values and religious tourism as the current trend of people traveling for spiritual purposes. Therefore, much work has been done regarding this subject in order to understand various approaches and concerns. A relatively new field of research interest that has been receiving much attention of scholars is the investigation of the impact of religious tourism on host communities and their development. This involves

K. Muhsina (✉)
Department of Tourism and Hospitality Management, Daffodil Institute of IT, Dhaka, Bangladesh

M. Valeri (ed.), *Contemporary Religious Tourism*, Tourism, Hospitality & Event Management, https://doi.org/10.1007/978-3-031-79074-4_9

assessing the impacts of the contact between the interdependent entities of the tourism system, that is, culture, economy, society, and environment. Another important element is the evaluation of the following aspect of how technology supports the improvement of the religious tourism experience and how it is capable of addressing the challenge of overcrowding especially in the most frequent visited places.

Furthermore, there is also growing attention of the world towards religious tourism for wellness where physical activities like yoga, trekking, spirituality, etc. are conducted. Existing studies have also begun exploring the possibilities of creating economic development based on health and spirituality. The aims of the chapter are as follows:

1. To analyze the historical evolution of religious tourism, tracing its origins from ancient times to the present day
2. To assess the current state of religious tourism by investigating its economic conditions and tourist motivational factors
3. To categorize the diverse phenomena associated with religious tourism
4. To analyze technological advancements, explore the connection between religion and recreation, and address current concerns and challenges including the influence of global politics
5. To propose the strategies of the prospect of religious tourism

2 Literature Review

2.1 Evolution of Religious Tourism

The theme of religious tourism has been a long and diverse one, including such forms of travel as pilgrimages and other religious jaunts with the sacral significance for the believers. Over the years religious tourism has evolved, and today's pilgrims are no longer only seeking the religious experience (Sharpley 2009). They also search for interest in the role of the religion and history of the religious places. This means that religious tourism has evolved gradually towards a more structured type of tour offerings that are religious tours and pilgrimage packages provided by travel agents and religious bodies (Griffin and Raj 2017). These tours are often followed by visits to various religious places where a guide in the form of religious scholar describes the place to the tourists. Furthermore, the participants can have an opportunity to join in the group prayers and perform self-analysis. There is a clear change in religious tourism where technology has been integrated into the tourism sector (Campbell 2012). A number of religious places of interests have thus come up with the virtual tours and other forms of participation to meet the needs of those who cannot physically access the places. This has given rise to new occasions for engagement and accessibility to all those who earlier could not indulge in religious tourism (Fig. 1).

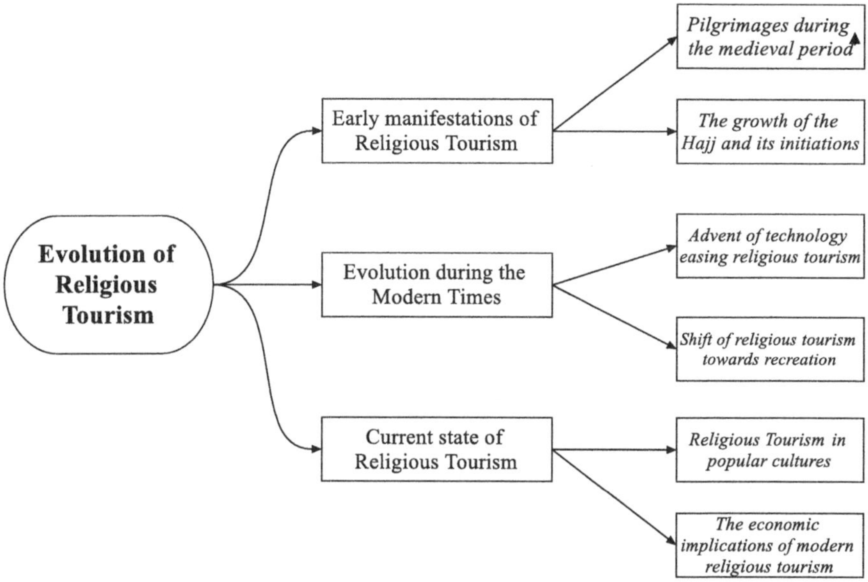

Fig. 1 Evaluation of religious tourism (developed by authors)

2.1.1 Early Manifestations of Religious Tourism

Religious tourism can be defined as traveling with the aim of taking blessings from the divine power at various shrines. The religious tourism has expanded over the years as a means of not only satisfying the religious demands of the pilgrims but also to improve the cultural and historical experience of the visited region (Griffin and Raj 2017). Religious tourism enhances the experience of pilgrimage journeys, cultural interchange, and helps to foster economic trade routes. They would come back from the journey with stories, wares, and gains, and the pilgrims would come back with new insights and ideas from the world to their own. Also, the early religious tourism was related to the public ceremonies and meetings, which meant that the pilgrims could find like-minded people, therefore join them, and celebrate together devotion. These practices are still common today and religious tourism has not been an exception, because the pilgrims still seek for not only the. It is therefore important for the readers to appreciate early forms of religious tourism so as to identify the role of pilgrimage and religious journeys in shaping individual faith and world societies today (Griffin and Raj 2017).

I. ***Pilgrimages During the Medieval Period***

The actual movement to holy places is a key part of faith, as believers continue to attempt to reinforce their beliefs (Olsen and Timothy 2006). These pilgrimages were strenuous and time consuming because the pilgrims were to encounter many difficulties in the course of the journey. The pilgrims wished to do penance in order to be closer to God and be spiritually reborn in some way. Most of the people undertake a strenuous and a time-consuming process in search of spiritual transformation or to atone for their sins (Kyalo 1998; Shekhar 2023). These kind of pilgrimages were great for the followers for they got a chance to have divine experience and a way of being absolved for sins they made.

II. ***The Growth of the Hajj and Its Initiations***

This is why the hajj pilgrimage and its rituals are increasing in size. However, the following cannot be overlooked; transport has also been realized to be becoming more readily available to make the pilgrimage easily accessible to Muslims from all over the world (Fullilove 2008). Furthermore, the expansion of the area of the Islamic religion to the areas that have not been reached before has also played a very important role in enhancing the hajj pilgrimage. This is because more and more people are turning to Islam and have the strong feeling that they should fulfill this very important commandment of Islam. These tendencies are seen from the growth in the number of people who perform hajj which evidence the involvement of Muslims in the world and the importance of the process. In addition, because of the improved and enhanced technology and communication, people's awareness has been increased and the organization of hajj pilgrimage has been made easy. It has therefore become a useful tool in linking the pilgrims, sharing of experience and sharing information on the journey (Gretzel 2019). The communication has not only made the journey easy to undertake but also has fostered unity of the Muslims in the whole world. This is because the hajj is an ever-evolving and continuously strong institution because it serves as the link that unites all Muslims around the globe.

2.1.2 Evolution During the Modern Times

I. *Advent of Technology Easing Religious Tourism*

The phenomenal growth of technology has significantly enhanced religious tourism by changing how the pilgrims are approaching the sacred places around the world. Whereas in the past pilgrims would have limited easy access to guides, maps, and historical information about other religious sites, it is now possible for them to download these and many more materials directly on their mobile devices. Some religious engagement technologies include virtual reality and augmented reality, which present virtual world views that allow worshippers to spiritually travel to religious places and markers before it is possible to physically make a pilgrimage (Amaro et al. 2021). Besides, in terms of organization, such sites facilitate the process of organizing accommodation and travel and transport linked with the assistance of guided tours associated with religious tours. Sharing of experiences, communication with fellow travelers, and seeking advice from a community of followers in whatever part of the world are made easier through the social networks (Tilson 2005). In general, technology has not only impacted the aspect of reaching sacred spaces but also positively impacted the perception of undertaking religious pilgrimages by offering deepened connected substantive religious beliefs.

II. *Shift of Religious Tourism Towards Recreational Purposes*

Over the last few years, the trend in religious tourism has changed markedly, featuring an expanding movement that unites spiritual exploration with leisure pursuits, as travelers look for more diverse and rich experiences (Iliev 2020). An increasing number of tourists want to integrate relaxation and adventure into their cultural experiences, instead of just concentrating on standard religious rituals or classic pilgrimages. It is not difficult to associate this change with multiple influencing elements, such as demographic shifts, changed travel preferences, and the initiative to deliver comprehensive experiences. At present, a variety of religious pilgrimage sites are now offering visitors a wide selection of accommodations, gourmet food, and entertainment targeted towards today's travelers' expectations and preferences. In addition, the expansion of wellness tourism has encouraged the foundation of spiritual retreats and wellness facilities in places of religious importance, welcoming guests wanting personal growth, energy renewal, and quiet. Thanks to the heightened visibility of religious tourism in current society, it is progressively seen as a relaxation, revitalization, and spiritual exploration technique, reflecting the transformation of travel (Boniface 2001).

2.1.3 Current State of Religious Tourism

Locations throughout the world are implementing changes to their tourism strategies in response to the changing tastes of travelers and current trends observed in religious tourism. Despite the fact that traditional pilgrimage destinations still pull in large crowds of worshippers and spiritual searchers each year, there is also a strengthening interest in experiential and specialized religious tourism alternatives. Current travelers are interested in abundant experiences that overlap historical, cultural, and spiritual themes (Stausberg 2012). Consequently, we have seen the emergence of functional travel packages, the launch of current digital tools, and techniques in sustainable tourism. The COVID-19 pandemic has also been monumental for religious tourism, motivating locales to adopt health and safety regulations, adjust to changing travel rules, and think of online alternatives to traditional pilgrimages. In light of the difficulties, the ongoing presence of religious tourism is a consequence of an important human requirement for connection, deep mental engagement, and spiritual development (Ptasznik 2015). There is a remarkable effect happening due to it on a worldwide scale for travel and cultural exchange.

I. *Religious Tourism in Popular Cultures*

The relationship between religious tourism and society, along with entertainment media, is expanding in popularity. To define characters, storyline settings, and character development, a range of media, particularly film, television, literature, and video games, increasingly makes use of religious tourism (Reijnders 2016). "The Da Vinci Code" and "Indiana Jones" franchisees create characters that deal with matters connecting religious history, artefacts, and symbolism. People around the world become interested and curious about sacred sites due to this. Similarly, travel documentaries and reality television shows routinely incorporate travels to well-known religious sites, assuring viewers a view into a number of cultural practices and spiritual traditions. Also, the launch of virtual reality experiences and narrative platforms centered on immersion has enabled users to rapidly connect with important religious sites and sacred places from any location they happen to be (Mansour 2019). This outcome has produced a confluence of the lines that separate fiction from reality. Within the purview of popular culture, religious tourism has grown in importance and respect, becoming an important tool for improving dialogue, understanding, and appreciation of the global religious heritage.

II. *The Economic Implications of Modern Religious Tourism*

Today's religious tourism has important economic impacts that attract diverse sectors and stakeholders to the locales in which these experiences transpire. The increase in religious tourism dramatically improves local economic situations by having tourists expend their dollars on lodging, dining, transportation, mementos, and other goods and services offered. A rise in what tourists invest creates jobs and encourages businesses, resulting in economic improvement and better off for the communities facilitating these services (Mbaiwa 2005). As a side note, investment in infrastructure and facilities linked to tourism, featuring hotels, transportation networks, and visitor centers, continues in order to serve the requirements of religious tourists. This action helps to increase both economic productivity and increase the standard of living of people in the local area. Tourism focused on religious faith may energize heritage conservation, cultural protection, and the reinvigoration of cities. Places are leveraging their religious heritage to attract tourists and to obtain a competitive benefit in the international tourism industry (Buhalis 2000). The priority of confronting actual obstacles, including strong commercialization, environmental worsening, and cultural socio impacts, continues to develop religious tourism in a sustainable and effective manner while also maximizing the lasting economic profits for both tourists and local communities.

2.2 Phenomena Associated with Religious Tourism

Religious tourism is inextricably linked to a range of phenomena that have an impact on both traveler experiences and the relationships between communities at destinations. Definable as a phenomenon, pilgrimage is the act of people who undertake journeys to locales with religious importance, to showcase their loyalty and strive for spiritual enlightenment, healing, or satisfaction (Barber 1991). Normally, pilgrimages involve rituals, praying, and honoring esteem, all which create a connection between participants. An additional phenomenon is the act of turning experiences derived from religion into products for the market. This situation emerges when localities utilize their religious heritage to provide guided tours, sell sacred artifacts, and run souvenir stores (Olsen and Timothy 2006). Even with

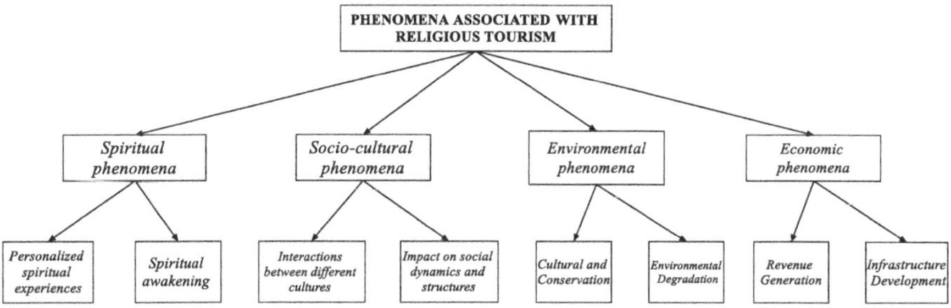

Fig. 2 Phenomena associated with religious tourism (self-developed)

potential economic rewards, it similarly reveals weaknesses related to authenticity and the commercial exploitation of those experiences. To that end, religious tourism overlaps with cultural exchange, compelling travelers from a multitude of backgrounds to research special faith traditions, customs, and rituals, which increases mutual knowledge and exchange. In religious tourism, some of the problems may be excessive volumes of visitors, the nature of cultural interactions, and conflicting insights into sacred spaces. This informs us of the important role that having responsible and sustainable management strategies play in safeguarding religious sites and supporting the prosperity of local communities (Quang et al. 2023). Fundamentally, these episodes investigate the detailed connections that exist between spirituality, culture, and commerce related to religious tourism (Fig. 2).

2.2.1 Spiritual Phenomena

Signals of spiritual occurrences indicate a wide range of interactions that transcend the physical plane and encourage awe for a greater power or being. Events can show diversity through experiences like spiritual openings, important transcendental insights, or serious deep comprehensions. Persons seeking answers regarding importance, intent, and the divine commonly do so because of the significant relaxation, awe, and wonder they associate with their spiritual experiences (Pond 2013). Historically, churches and temples plus practices such as contemplation, prayer, or pilgrimage facilitate the generation of this level of expression. On top of that, the phenomena of a spiritual nature are subject to personal beliefs, community context, and personal interpretations. An interpretation of spiritual experiences in a religious sense occurs with some, but a significant majority generally understands them as indications of our shared human urge for connection and transcendence (Rothman 2009). The significance of spiritual phenomena in the process of individual belief, value, and identity formation is consistent, independent of insight. They increase understanding about the mysteries of life and our status in the universe.

(a) *Personalized Spiritual Experiences*

Individualized spiritual experiences indicate very specific interactions with the sacred or transcendent, designed to correspond with the unique overall needs, beliefs, and preferences of individuals (Russo-Netzer 2017). Opposite to standard religious rituals or practices, individualized spiritual experiences do not face boundaries set by recognized dogma or institutional structures. This autonomy lets people examine their spirituality so it powerfully relates to them on a personal level. A variety of experiences can emerge, including individual reflective exercises in scenic places and participation in involved therapeutic rituals backed by spiritual mentors (Lindahl et al. 2017). As well, individualized spiritual encounters regularly give precedence to self-exploration, individual growth, and the elaboration of a connection to either inner wisdom or increased awareness. According to Winkleman (2010), these kinds of methods can connect mindfulness with either original energy management styles, spiritual practices, or unique artistic endeavors. They deliver individuals various techniques and methods to improve their ability to be fully present, truthful, and in synchronization with their spiritual walk. Being part of a secular and varied society enables individuals to enjoy personalized spiritual experiences that help them easily advance their spirituality in ways that represent their personal inclinations. This enhances an awareness of empowerment, personal importance, and spiritual linkedness to the divine inside and outside of one's being.

(b) *Spiritual Awakening Resulting from Religious Tourism*

The religious tourism of visiting sacred sites can produce deep, remarkable spiritual awakening along with transformative elements for individuals. Grandly speaking, these experiences provide opportunities for people to participate in vital analysis, form personal beliefs, and create spiritual links, while also communally confronting the complexities related to religious history, tradition, and symbolism (Davids and Waghid 2018). Participation in events that involve sacred places often produces powerful emotions of loyalty, respect, and a call to take initiative, which subsequently reinforce vital moments of awareness, recognition, and individual spiritual development. Also, the complete enthusiasm of fellow pilgrims together with the sacredness of the environment supports an ideal atmosphere for personal development and educational spirituality. Getting involved in the world can truly improve by following rituals, praying, and performing reflective practices (Pryce 2018).

2.2.2 Sociocultural Phenomena

Societies' visible patterns, behaviors, and events come directly from their cultural norms, values, and practices. These developments affect the way people engage, deliver messages, and build organization in community frameworks. Cultural phenomena represent a range of labels defined by rituals, traditions, customs, and social movements that stand for group beliefs and identities (Salman and Assies 2017). The examples of sociocultural phenomena include the celebrations of cultural festivals, events that recognize rites of passage, and prayer rituals. Using this methodology, the group's cohesiveness is sustained, its cultural beliefs are protected, and anniversaries important to individual biographies receive proper recognition. Furthermore, movements facilitated by globalization, migration, and technological evolution have important effects on society, affecting cultural environments while deepening intercultural relationships and inquiring into longstanding cultural ways of life. Societies are perpetually dynamic and evolving, and sociocultural phenomena provide important understandings of the details surrounding human behavior, social transformation, and the relation of identity, social structures, and culture (Levin and Mamlok 2021). A complete understanding of these phenomena is important for enhancing social coherence, forging a range of cultures, and taking on disparities, biases, and the defense of cultural heritage in an ever more connected society.

I. *Interactions Between Different Cultures During Religious Tourism*

The dynamics and layers of interactions among different cultures in religious tourism are clear. Usually, these exchanges produce essential moments for cultural exchange, advanced communication, and a better insight into each other (Reisinger and Turner 2012). The experience benefits from the unique insights, ritual practices, and cultures of participants from different heritages meeting at places of spiritual or religious pilgrimage. Engagement activities increase engagement in cultural conversations and fully appreciate the elaborate diversity of international religious beliefs and practices. Also, the social relationships that take place between pilgrims and nearby communities serve to better cultural integration, because visitors become acquainted with local customs, dining conventions, and hospitality, positioning them to observe local life with greater depth. In return, religious tourism acts as a compelling motivation for worldwide partnership and collaboration, bringing together diverse stakeholders to defend cultural sites, support sustainable tourism methods, and deal with mutual challenges. Despite the existing circumstances, disputes regarding religious tourism resulting from cultural differences could produce tensions and confusion due to obstacles in communication, multiple interpretations of religious symbols that could heighten tensions or misunderstandings (Stausberg 2012). Developing confidence, respect, and the talent of openness within these relationships implies that they work more efficiently than cultural differences, which results in harmony and a greater understanding of multiple people.

II. *Impact on Social Dynamics and Structures*

Tourism related to religion plays a critical role in the dynamics and frameworks observed in society, impacting both community views of themselves and of each other. In local conditions, an increase in tourist arrivals may generate economic opportunities, support infrastructure growth, and offer job opportunities, causing changes in community classifications and the power relationships among those tourist hotspots. The techniques by which tourists connect with local residents could hold the ability to change social expectations and cultural beliefs as well as cultural behaviors. This occurs, of course, when people from different backgrounds convene and trade their concepts, values, and practices. On top of that, religious tourism could strengthen the efforts of cultural heritage restoration and secure intangible cultural assets, while enhancing local pride, identity, and shared memories among the community (Qiu et al. 2022). In any event, the possibility exists for it to initiate conflicts and tensions, especially when the aspirations of tourists differ significantly from those of local communities, or when cultural authenticity and sacred places are imperiled by excessive commercialization or unsustainable tourism. Religious tourism can have an impact on the operation of major social systems and processes at both national and global levels. There is potential for its advancement in transnational networks, politics which arise from identity, and religious movements extending across different geographical regions (Vertovec 2001). A serious examination of the social dynamics and social structures influenced by religious tourism is important, since such exploration of its social, cultural, and economic outcomes affects both local populations and tourists.

2.2.3 Economic and Environmental Phenomena

The economic and environmental factors around religious tourism consist of numerous interrelated variables that presage the enduring viability and impact of pilgrimages and spiritual voyages on the places they address. Viewing it through an economic perspective, religious tourism often serves an important purpose in income production, resulting in a growth of the hospitality, transportation, and retail sectors together. Tourist spending growth can stimulate job growth, improved infrastructure, and a rising community investment (Zaei and Zaei 2013). One must strive to keep intact a unique relationship between the rewards of economic development and the liabilities of too much dependency on tourism, irregular revenues, and declining natural resources. Also, environmental concerns greatly endanger the sustainability of religious travel considering that the locations for pilgrimages and consecrated landscapes typically deal with problems that include huge visitor flows, contamination, and the decline of natural ecosystems. For these environments, keeping ecological stability together with an exciting guest experience calls for the application of sustainable management approaches, including waste management innovations, conservation plans, and environmental transportation techniques (Mihalič 2000). In addition, climate change poses important hurdles for religious tourism, resulting from increasing temperatures, greater frequency of severe weather, and the growth of sea levels that imperil both the endurance and accessibility of sacred destinations and pilgrimage paths. Tackling these problems in economics and sustainability brings together different organizations, comprising governments, local groups, faith-based organizations, and tourists. It is important for them to construct strategies that maintain sustainable development, safeguard cultural heritage, and conserve the environment in places related to religious tourism (Xiao et al. 2018).

2.3 *Economic Boons and Challenges of Religious Tourism*

Religious tourism delivers economic advantages and challenges to the locations that act as tourist destinations. On one side, it stands out by being a primary revenue maker, lending important support to economic growth, job growth, and infrastructure formation. Pilgrims and spiritual travelers, in their allocation of resources for lodging, travel, dining, souvenirs, and expert guided tours, drive local commerce and assist in keeping a thriving economy (Aruljothi and Ramaswamy 2019). Also, religious tourism offers an opportunity to improve economic diversity, particularly in places with modest economic opportunities, by attracting tourists to cultural heritage and sacred sites. There are different kinds of challenges stemming from religious tourism. Overemphasizing dependence on tourism brings economic risks, as we respect that fluctuations or disruptions in tourism might negatively affect local firms and their communities. On top of all this, the sale of religious experiences packaged together with the commercialization of holy sites can lessen cultural authenticity, lower spiritual value, and exploit conflicts between tourists and residents. Also, religious tourism regularly experiences the challenges posed by environmental destruction, heavy crowds, and strained infrastructure (Shinde 2012). It is necessary for sustainable management strategies to exist in order to keep a positive balance between economic profits, the care of environmental resources, and the community's welfare. In essence, religious tourism is able to advance economic growth in the locations it explores. Efficient planning, well-regulated operations, and increased collaboration are key to capitalize on its advantages and to avert unwanted consequences for all major participants.

I. *Environmental Impact of High-Density Religious Events*

Ritualistic gatherings of major religious denominations, which encourage a vast crowd for spiritual reasons, can instigate a great deal of damage to the environment. Since participants typically hail from numerous locales, these situations often require significant transportation, which adds to air pollution and carbon emissions. When a vast group shows up, area resources begin to struggle, increasing waste production and consumption (Satterthwaite 2021). In addition, the establishment of interim buildings and facilities to support the significant number of people can bring about habitat damage and a deterioration of the land, especially within ecologically fragile areas. On top of that, the management of plastic waste, food packaging, and alike that do not compost contributes to a higher level of environmental pollution. Programs that focus on supporting sustainable transport alternatives, carrying out waste reduction and recycling, and the responsible control of natural resources must be launched by us (Iqbal 2020). Informational initiatives directed at religious communities regarding the important value of environmental stewardship, in addition to merging sustainable practices into religious traditions, eases environmental concerns tied to densely populated religious celebrations.

2.4 Impact of Religious Tourism on Local Economies

The financial success of tourism activities over the long run, with respect to their effects on the destination, community, and stakeholders, is an important aspect of a study focused on economic sustainability in religious tourism. Assessing different economic metrics is part of this, which includes income generation, employment generation, investment prospects, and the distribution of benefits (Stiglitz et al. 2009). Achieving sustainable economic advancement in religious tourism requires finding a precise equilibrium between the profits derived from tourist spending and the preservation of cultural and environmental assets. A case in point is that sustainable tourism endeavors to cut down on money outflow by elevating the participation of local people in tourism activities and promoting locally produced goods and services (Ashley et al. 2007). As a further point, economic sustainability requires broadening tourism choices to go beyond religious destinations in order to reduce the risks associated with variable seasons and dependency on certain markets. In addition, it demands insight into the social and ecological prices associated with tourism development, covering crowding, pollution, and the commercialization of cultural ways of life, and adopting strategies to mitigate these results (Taiminen 2018). Participants in religious tourism have to strive for sustainable economic success by taking into account a broad range that includes economic, social, and environmental elements. This strategy assures that economic advantages are shared fairly and that future generations will inherit a whole cultural and natural legacy.

2.5 Evolving Nature of Religious Cognitive and Affective Experiences

The changing characteristics of religious cognitive and affective experiences evince the lively interaction of personal beliefs, social norms, and cultural elements over time. According to cognitive experiences, people interpret, make sense of, and perceive religious concepts and doctrines, in contrast, affective experiences consist of the emotional feelings and reactions generated by religious practices, rituals, and symbols (Barsalou et al. 2005). Religious experiences modify and adapt in response to technological growth and cultural changes. General support for digital technologies has made it easier to form new religious practices and strengthen religious communities using web platforms, digital ceremonies, and lively social media engagement. In addition, the phenomenon of globalization has produced greater understanding of a variety of religious customs, which in turn promotes intercultural engagements and the arising of fused spiritual rituals (Cruz et al. 2018). Stated trends in secularization over set regions have triggered a reassessment of traditional religious beliefs and practices, causing a growth in a variety of alternative spiritual movements and practices. Developments in the ways people experience their religious emotions and thoughts reflect the myriad forms of human spirituality as well as the persistent dream for meaning and relatedness in a continuously shifting context.

2.6 Personal Religious Experiences

A large, interdisciplinary field of study explores personal religious experiences, trying to understand the different methods individuals use to engage with religion and spirituality. These experiences range from major spiritual interactions to routine evidence and images of faith. Psychologists, sociologists, anthropologists, and religious studies scholars look into the emotional, cognitive, and social dimensions of a variety of cultural, traditional, and belief systems associated with religious experiences (Beit-Hallahmi 2014). This work reviews the range of aspects that impact the evolution and comprehension of religious experience, including personal thoughts, cultural impacts, interactions within society, and neurological interactions. And at the same time, researchers are evaluating how religious experiences affect individual health, build personal identity, create moral principles, and direct social conduct. In order to more deeply understand the detailed and varied nature of personal religious experiences, researchers combine qualitative and quantitative research methods, which include surveys, interviews, ethnographic studies, and experimental psychology (Yoshikawa et al. 2008). This illustrates our improved knowledge of human spirituality and the part religion plays in the society.

2.7 Emotional Connection with Religious Sites

The ties created with religious sites examine the intense and complicated ways individuals relate to and experience sacred spaces. Both believers and visitors find that religious sites have a considerable weight which evokes feelings of awe, reverence, serenity, and transcendence (Abdrakhmanova et al. 2023). Community members tend to develop commitments to a site influenced by their personal beliefs, heritage, and their life experiences. These components are critically important in defining their understanding and actions towards the site. A psychological relationship is present within faith sites that evokes both a sense of community and calm, as well as reinforcing their individual and community identities as members of their faith. To add, the design details, artwork, and natural features of spiritual sites can maximize powerful emotional reactions, portraying each the religious and the humanity in creativity, plus the charm of nature (Kellert 2003). The strong emotional connections we have to religious areas go further than our private experiences to encompass group gatherings, pilgrimages, and celebrations that foster both greater social cohesiveness and support existing cultural beliefs and values. Experts from religion, anthropology, and psychology are using qualitative interviewing, surveys, and ethnographic research to understand emotional components of visits to sacred sites. To truly value the key contribution that religious sites make in formulating both public and private religious identities, experiences, and practices, researchers have to be knowledgeable about the emotional connections that exist with these sites (Knott 2010).

2.8 Influence of Global Politics on Religious Tourism

The ramifications of worldwide politics for religious tourism are considerable and complicated, because political influences at the global, national, and local dimensions inform the experience of religious travel. Seyfi and Hall (2020) find that religious tourism is likely affected by complications, tensions, and global partnerships. Evidently, this evidence consists of the application of travel limitations, visa strategies, security frameworks, and the regular pressures in diplomatic interactions. Should travel restrictions be in effect, if there is political unrest, or if individuals doubt safety in defined areas, both holidaymakers and those journeying for spiritual reasons might pass over holy sites, which could affect tourism and have unfortunate economic ramifications for the local region. According to Assembly et al., activity within diplomacy and intergovernmental partnerships plays a positive role in religious tourism by easing international travel, fostering cultural conversations, and maintaining heritage sites common to multiple nations. Both the development and management of religious tourism infrastructure and promotion activities, as well as interfaith relations, are molded by politics within the country, government policies, and specific religious regulations. How both the availability of and representation for religious places develop is shaped by national political ideologies and power dynamics, which ultimately affect the views and experiences of those who visit. The growth of religious tourism is dependent on political dynamics at the local level, the way the community functions, and differences in socio-economics. Municipal stakeholders, local political leaders, and a variety of interested organizations frequently affect decisions about land management, infrastructure development, and tourist management (Skog 2018).

3 Conceptual Model

This model is developed through the information of literature Review. Here shows the way of the development of religious tourism by different factors of tourist motivation (Fig. 3).

4 Challenges of Religious Tourism

Religious tourism encounters various obstacles that can affect its long-term viability, ease of access, and overall quality. Several primary obstacles faced in religious tourism are as follows (Fig. 4):

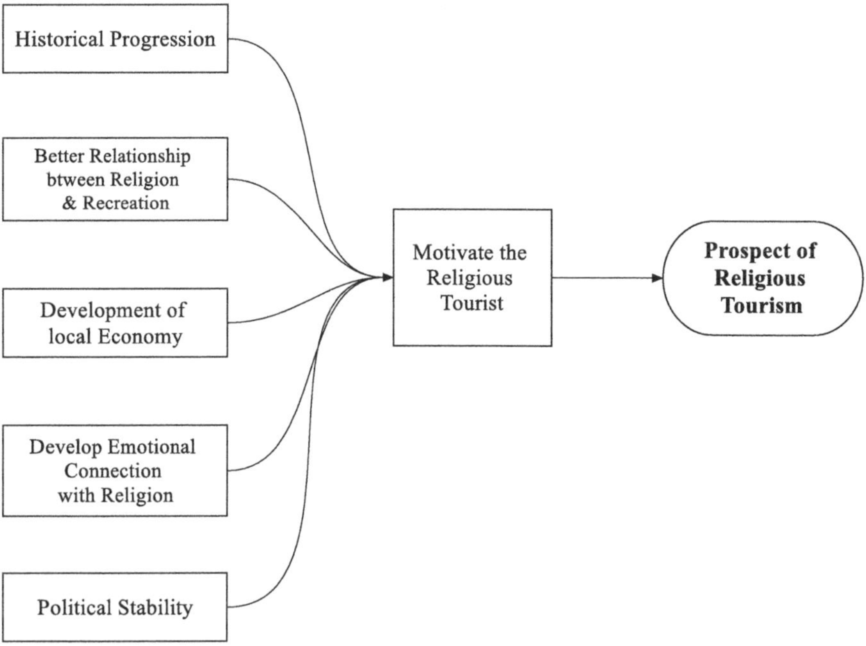

Fig. 3 Prospect of religious tourism (self-developed)

4.1 Security Concerns

Religious tourism involves the numerous individuals at important religious locations, making them vulnerable to security risks, acts of terrorism, or social disorder. Ensuring the safety and security of tourists and pilgrims is a crucial and difficult task for destinations and consultants.

4.2 Infrastructure Limitations

In remote or historically important locales, a vast number of religious sites find themselves with deficient infrastructure, such as transportation, lodging, and sanitation services. At times of peak pilgrimage, there may be greater difficulties for visitors to gain both access and convenience because of an insufficient infrastructure.

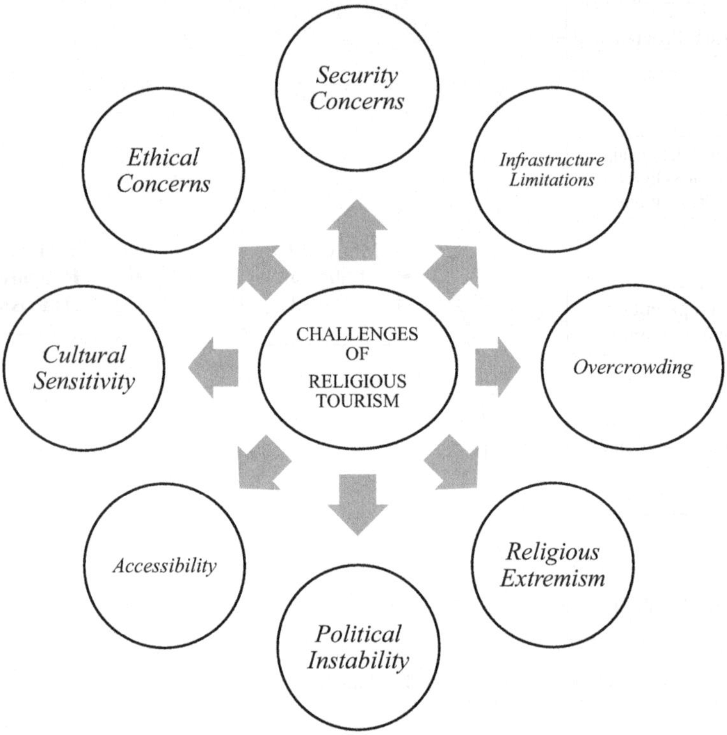

Fig. 4 Challenges of religious tourism (self-developed)

4.3 Overcrowding and Environmental Impact

With large numbers of pilgrims, the most revered religious sites usually become overly congested, resulting in environmental injury, an increase of waste, and a strain on community resources. The conservation of these sites requires the direction of visitor flows and the execution of sustainable tourism approaches.

4.4 Cultural Sensitivity and Conflict

Religious tourism entails the engagement of visitors from various cultural backgrounds with local communities that hold significant religious and cultural importance. Continuously addressing the challenges of promoting mutual respect, fostering understanding, and effectively managing potential conflicts that may arise due to cultural differences is an ongoing task.

4.5 Commercialization and Authenticity

The commercial ambition of making money from religious tourism may sometimes endanger the integrity and sanctity inherent to these spaces in the introduction of souvenir shops alongside hotels and tour operators. Interweaving religious integrity and commercial ambitions proves to be a difficult task to do.

4.6 Ethical Concerns

There is an opportunity for difficult conversations in religious tourism, influenced by the monetization of theological positions and changes in sacred traditions along with the risks of cultural appropriation. Maintaining respect and real authenticity is critically dependent on applying ethical principles in religious tourism.

4.7 Accessibility and Inclusivity

Barriers that are both discriminatory and physical in access planning might make it difficult for people from oversight communities and those with disabilities to reach religious tourism sites. Reaching fair access to religious experiences is fundamental to supporting both accessibility and inclusiveness in the field of religious tourism.

4.8 Political Instability and Travel Restrictions

Incursions into the tourism sector from political upheaval, strained partnerships, or travel constraints in several areas can interrupt the movement of religious tourists and convince people not to travel to important spiritual lands. Maintaining religious tourism requires both the commitment to political stability and the resolution of any travel problems.

4.9 Religious Extremism and Intolerance

An increase in intolerance or religious extremism might impact the fellowship of varied faith visitors and cause disruption while they enjoy their quiet experiences at religious tourism sites. The key to religious tourism is to develop interfaith discussions and increase mutual comprehension so as to create harmony.

To tackle these challenges, it is necessary for governments, religious institutions, local communities, tour operators, and visitors to work together.

5 Impact of Religious Tourism Phenomena on Well-Known Destinations

The influence of religious tourism on popular destinations has garnered growing attention in recent years.

Notable effects of this phenomenon encompass the following:

I. ***Boosting Local Economy***

On their journeys to both international nations and urban areas, pilgrims regularly set aside a lot of money for accommodations, food, souvenirs, and different services. As a result, such tourism can foster a productive economic situation by facilitating income for local industries and forming job opportunities.

II. ***Preservation of Cultural Heritage***

The enhanced tourism at important religious destinations can spur the focus of authorities on the essential requirement for maintenance approaches, which in turn supports the conservation of historical sites.

III. ***Encouraging Interfaith Dialogue***

Those who seek religious experiences while on vacation generally encounter varying cultural and religious journeys that improve their awareness of people from several cultural backgrounds.

IV. ***Negative Environmental Effects***

Poor tourist management at these areas might produce severe environmental impacts, including pollution and damage to natural environments.

V. ***Conflict over Resources Among Locals***

From time to time, friction can develop between residents and tourism activities as a result of rising demand for common facilities or amenities, for example, water usage. The contest may produce stress in relationships within the community as people strive to steer these resources (Rasoolimanesh et al. 2023).

6 Future Perspectives on Religious Tourism

The future outlook for religious tourism is expected to be influenced by a blend of technological progress, shifting travel preferences, and evolving sociocultural dynamics. Due to the growing availability of information on digital platforms, religious tourists are inclined to search for customized and immersive experiences that cater to their spiritual interests. Virtual reality and augmented reality technologies have the potential to greatly facilitate remote exploration of religious sites and rituals, potentially revolutionizing the way pilgrimage is perceived (Steinicke 2016). Furthermore, there is an increasing recognition of the importance of sustainable travel practices, resulting in a change towards environmentally friendly pilgrimage routes and accommodations. Additionally, there is a stronger focus on responsible tourism initiatives in religious destinations. Furthermore, with the ongoing process of globalization erasing cultural distinctions, religious tourism is anticipated to become increasingly varied and all-encompassing, accommodating a wider range of religious traditions and convictions.

6.1 Technological Advances Impacting Religious Tourism

Technological changes have greatly influenced religious tourism in the sense that it has changed the way that pilgrims have an encounter with the divine and perform their rites. Another important improvement has been the use of social networks and mobile applications that have elements like tours, maps, and video and audio files containing a lot of information about the religious objects and rituals (Ramos et al. 2016). AR and VR technologies enable the user to visit and make offerings to the religious places, as well as get information about the history of such places from any location. Also, improved transport systems such as building of high-speed rail and purchase of electric cars has also helped the pilgrims gain easy access to the shrines. This has facilitated advanced transport to areas with a difficult terrain as well as reducing the damaging impacts on the environment (Bailey and Kurland 1999).

6.1.1 Role of Virtual Reality and the Future Technologies

That is why virtual reality and other modern technologies can change the link between people and religiosity. Virtual reality (VR) recreates real-life religious places, events, activities, and experiences. This makes it possible for people to have a feel of the virtual pilgrimages without having to move an inch. This is fascinating. This technology takes into account geographical barriers where people are able to access religious centers from distant communities and experience diverse forms of spiritual activities (Barzilai-Nahon and Barzilai 2005). Moreover, the application of virtual reality (VR) may also include interactive options, for example, haptic feedback and real-time communication. This would enhance the feeling of being there which in turn would enhance the possibilities to actually interact with religious content. New possibilities in the near future for augmented reality (AR) and mixed reality (MR) may offer even more convincing ways of enabling people to engage with the religious and thus bring about a state of affairs where the divide between the real and the virtual disappears. These innovations can help to give equal opportunities for receiving information about religion and for the communication between different confessions to improve the attitudes of the adherents of different faiths (Edwards 2016).

6.1.2 Role of Transport Development for the Possibility of Visiting Distant Holy Places

As a result, the development of transport has been vital in the realization of the pilgrims' ability to access the distant shrines as well as in the discovery and utilization of sacred artefacts all over the world. Historically, travels to the sanctuaries were not easy, time consuming, and expensive affair that could take years if one is to do it on foot, on a horse, or even by ship. However, advances in transportation have altered this context to ensure that movement times have been greatly reduced and accessibility has been enhanced (Banister and Stead 2004). Development of railway systems, roads, and highways has enhanced mobility through facilitating movement to and within various difficult to access areas and hence enable the faithful to access religious shrines Gelbman (2019).

6.2 The Relationship Between Religion and Recreation

Religion and recreation are two aspects of human life that are linked in a manner which the two have related impacts on human experience. Many of the religions of the world have recognized the role of play and recreation in their rites and ceremonies and festivals to ensure that happiness and celebrations are part of the spiritual being (Olsen and Timothy 2006). Festivals, pilgrimages, and meals are times when believers come together to observe their faith and strengthen their togetherness. Recreation likes music, dance, and art is employed in a number of ways to reflect adherence and worship in religions (Segler and Bradley 2006).

6.2.1 The Concept of Religious Tourism Experience Phenomenology

It consists of a comprehensive journey that is both spiritual, cultural, and historical but all under one umbrella of religious tourism. The activity comprises visiting the shrines, doing the worship, and taking part in religious activities and ceremonies (Iftikhar et al. 2023). The religious tourism experience refers to the experience of the sacred which is a hedonic experience of the sublime and the reverent and which basically entails a feeling of being other than the self and a feeling of unity with the divinity. Using rituals, prayers, and meditations in the sacred space, participants can experience his/her transformational process. The experiences may create different feelings of peacefulness and happiness while at the same time bring about thinking and letting out feelings. But religious tourism enhances also cultural and religious interrelation to make people get to know each other and be more tolerant of the other faiths (Phan and Ray 2014). The distinctive feature of religious tourism is that it contributes to the formation of the spiritual quest, enhances the sense of unity with the large family, and adds to the value of human relationships.

7 Conclusion

This research has shed light on the reasons of religious tourism and strategies of its sustainable development, further research should. To this end, it will be possible for the people involved to have a better understanding of these new features so that they can make right decisions on the strategic development, marketing, and successful implementation of successful religious tourism business. More empirical studies should be done on the effects of religious tourism and pilgrimage on the economic, social, and economic development of the receiving communities. This will entail assessment of the effects of religious tourism to the economic, social cultural, and physical environment. Therefore, one has to look at other social impacts and indicators associated with religious tourism so that all the stakeholders could gain from the tourism and its impacts as well as tourism being sustainable. It is in these areas of knowledge gap that researchers can provide useful information. The following are the implications of the findings that can be useful in developing right and sound strategies and policies.

References

Abdrakhmanova S et al (2023) Exploring the emotional dimensions of religious and spiritual tourism: a study of visitors to the Ascension Cathedral in Almaty, Kazakhstan. Int J Tour Cult Spirit 6(2):123–143

Amaro S et al (2021) The use of information and communication technologies in religious tourism. In: The routledge handbook of religious and spiritual tourism. Routledge, pp 372–384

Aruljothi C, Ramaswamy S (2019) Pilgrimage tourism: socio-economic analysis. MJP Publisher

Ashley C et al (2007) In: John F (ed) The role of the tourism sector in expanding economic opportunity. Kennedy School of Government, Harvard University

Bailey N, Kurland NB (1999) The advantages and challenges of working here, there, anywhere, and anytime. Organ Dyn 28(2):53–68

Banister D, Stead D (2004) Impact of information and communications technology on transport. Transp Rev 24(5):611–632

Barber R (1991) Pilgrimages. Boydell & Brewer Ltd.

Barsalou L et al (2005) Embodiment in religious knowledge. J Cogn Cult 5(1–2):14–57

Barzilai-Nahon K, Barzilai G (2005) Cultured technology: the internet and religious fundamentalism. Inf Soc 21(1):25–40

Beit-Hallahmi B (2014) Psychological perspectives on religion and religiosity. Routledge

Boniface P (2001) Dynamic tourism: journeying with change. Channel View Publications

Buhalis D (2000) Marketing the competitive destination of the future. Tour Manag 21(1):97–116

Campbell HA (2012) Introduction: the rise of the study of digital religion. Routledge, pp 1–31

Cruz AGB et al (2018) Religion as a field of transcultural practices in multicultural marketplaces. J Bus Res 91:317–325
Davids N, Waghid Y (2018) Teaching and learning as a pedagogic pilgrimage: cultivating faith, hope and imagination. Routledge
Edwards, S. (2016). Critical conversations about religion: promises and pitfalls of a social justice approach to interfaith dialogue., IAP
Fullilove M (2008) World wide webs: diasporas and the international system.
Gelbman A (2019) Tourism, peace, and global stability. In: Handbook of globalisation and tourism, pp 149–160
Gretzel U (2019) The role of social media in creating and addressing overtourism. Overtourism: Issues, realities and solutions 1:62–75
Griffin KA, Raj R (2017) The importance of religious tourism and pilgrimage: reflecting on definitions, motives and data, Int J Relig Tour Pilgr
Iftikhar A et al (2023) Sacred seduction: unveiling the allure of religious activities as a tourist magnet. The Islamic culture as-Saqafat-ul Islamia الثقافة الإسلامية-Research Journal-Sheikh Zayed Islamic Centre, University of Karachi 48(2)
Iliev D (2020) The evolution of religious tourism: concept, segmentation and development of new identities. J Hosp Tour Manag 45:131–140
Iqbal Q (2020) The era of environmental sustainability: ensuring that sustainability stands on human resource management. Glob Bus Rev 21(2):377–391
Kellert SR (2003) Kinship to mastery: biophilia in human evolution and development. Island Press
Knott K (2010) Religion, space, and place: the spatial turn in research on religion. Religion and Society 1(1):29–43
Kyalo B (1998) Franciscan Penitential Spirituality.
Levin I, Mamlok D (2021) Culture and society in the digital age. Information 12(2):68
Lindahl JR et al (2017) The varieties of contemplative experience: a mixed-methods study of meditation-related challenges in Western Buddhists. PLoS One 12(5):e0176239
Mansour N (2019) Virtually sacred: effect of light on the spiritual experience in virtual sacred architecture
Mbaiwa JE (2005) Enclave tourism and its socio-economic impacts in the Okavango Delta, Botswana. Tour Manag 26(2):157–172
Mihalič T (2000) Environmental management of a tourist destination: a factor of tourism competitiveness. Tour Manag 21(1):65–78
Olsen DH, Timothy DJ (2006) Tourism and religious journeys. In: Tourism, religion and spiritual journeys. Routledge, pp 1–21
Phan PC, Ray J (2014) Understanding religious pluralism: perspectives from religious studies and theology. Wipf and Stock Publishers
Pond MF (2013) Investigating climbing as a spiritual experience. Ohio University
Pryce P (2018) The monk's cell: ritual and knowledge in American contemplative Christianity. Oxford University Press
Ptasznik A (2015) Thru-hiking as pilgrimage: transformation, nature, and religion in contemporary American hiking novels." (Master's thesis). University of Colorado, Boulder
Qiu Q et al (2022) Intangible cultural heritage in tourism: research review and investigation of future agenda. Land 11(1):139
Quang TD et al (2023) Can tourism enhance inclusitivity for indigenous peoples? Cham perspectives on tourism benefit sharing at living heritage sites in Vietnam. Equal Divers Incl: Int J 43:959
Ramos CM et al (2016) Augmented reality for smart tourism in religious heritage itineraries: tourism experiences in the technological age. In: Handbook of research on human-computer interfaces, developments, and applications. IGI Global, pp 245–272
Rasoolimanesh SM et al (2023) Tourists' perceptions of the sustainability of destination, satisfaction, and revisit intention. Tour Recreat Res:1–20
Reijnders S (2016) Stories that move: fiction, imagination, tourism. Eur J Cult Stud 19(6):672–689

Reisinger Y, Turner L (2012) Cross-cultural behaviour in tourism. Routledge
Rothman J (2009) Spirituality: what we can teach and how we can teach it. J Relig Spiritual Soc Work Soc Thought 28(1–2):161–184
Russo-Netzer P (2017) “Sometimes I don’t even know where I am going”: what supports individualized personal spiritual change? Psychol Relig Spiritual 9(1):82
Salman T, Assies W (2017) Anthropology and the study of social movements. In: Handbook of social movements across disciplines, pp 57–101
Satterthwaite D (2021) Sustainable cities or cities that contribute to sustainable development? The Earthscan reader in sustainable cities. Routledge, pp 80–106
Segler FM, Bradley R (2006) Christian worship: its theology and practice. B&H Publishing Group
Seyfi S, Hall CM (2020) Political transitions and transition events in a tourism destination. Int J Tour Res 22(4):493–506
Sharpley R (2009) Tourism, religion and spirituality. In: The Sage handbook of tourism studies, pp 237–253
Shekhar VM (2023) Modelling the effect of spiritual tourism motivators on spiritual tourism. Tour: Int Interdiscip J. https://doi.org/10.37741/t.71.3.13
Shinde K (2012) Policy, planning, and management for religious tourism in Indian pilgrimage sites. J Policy Res Tour, Leis Events 4(3):277–301
Skog KL (2018) How do policies and actors’ attitudes, interests and interactions influence farmland conversion outcomes in land-use planning? Sustain For 10(6):1944
Stausberg M (2012) Religion and tourism: crossroads, destinations and encounters. Routledge
Steinicke F (2016) Being really virtual. Springer
Stiglitz JE et al (2009) Report by the commission on the measurement of economic performance and social progress., The Commission Paris
Taiminen S (2018) The negative impacts of overtourism on tourism destination from environmental and socio-cultural perspectives.
Tilson DJ (2005) Religious-spiritual tourism and promotional campaigning: a church-state partnership for St. James and Spain. J Hosp Leis Mark 12(1–2):9–40
Vertovec S (2001) Religion and diaspora, University of Oxford. Transnational Communities Programme
Winkleman M (2010) Shamanism and the origins of spirituality and ritual healing. J Study Relig Nat Cult 3(4):458
Xiao W et al (2018) Geoinformatics for the conservation and promotion of cultural heritage in support of the UN sustainable development goals. ISPRS J Photogramm Remote Sens 142:389–406
Yoshikawa H et al (2008) Mixing qualitative and quantitative research in developmental science: uses and methodological choices. Dev Psychol 44(2):344
Zaei ME, Zaei ME (2013) The impacts of tourism industry on host community. Euro J Tour Hosp Res 1(2):12–21

Part II
Case Studies: Intersections Between Tourism and Religion

Tourist Destination Placement and Supported Architectural Digital Tools for Cultural Pilgrimage Tourism: Lefkada Island

Tilemachos K. Koliopoulos and Marco Valeri

Abstract In the post-COVID world, cultural religious tourism on Lefkada Island is poised to undergo significant transformation, leveraging digital innovation to enhance visitor engagement and ensure safety. The integration of advanced digital tools, particularly in the digital representation and analysis of 3D architectural design and virtual reality, presents a valuable opportunity to showcase the island's rich historical and cultural heritage while adhering to health and safety protocols. Lefkada island demonstrates heritage religious buildings, and cultural monuments from Enetic times inspired by Italian Architecture as well as historical structures, buildings, monuments from ancient times until later times during the Byzantine Epirous Empire and Ottoman local Empire in past times. Useful conclusions are presented in terms of historical heritage monuments, environmental health and public health within safe green tourism, based on the examining study for stakeholders and promotion of safe sustainable ecological religious tourism destinations.

1 Introduction

Nowadays, alternative safe travel destinations are needed for cultural heritage tourism destinations. The aim of this investigation study is to promote safe travel destinations, landscape upgrade, support of renewable resources and regional goods for public health protection at post COVID-19 era, and in climate change management. Leukada island at Ionian sea in Greece due to its geological strata and

T. K. Koliopoulos (✉)
Telegeco Research and Development. Collaborator University of West Attica, Faculty of Health, Athens, Greece

M. Valeri
Faculty of Economics, Niccolò Cusano University, Rome, Italy

Faculty of Social Sciences and Leisure Management, School of Hospitality, Tourism and Events, Taylor's University, Subang Jaya, Selangor, Malaysia
e-mail: marco.valeri@unicusano.it

M. Valeri (ed.), *Contemporary Religious Tourism*, Tourism, Hospitality & Event Management, https://doi.org/10.1007/978-3-031-79074-4_10

geo-morphology belongs to a seismological zone with a lot of earthquakes with big magnitudes in the past, where proper building standards for particular constructions facilities should be applied for public health protection (Hwang and Yoon 1981; Paliwal 2006; Szigeti et al. 2011; Katsoni and Stratigea 2015; Latinopoulos and Vagiona 2013; Hsiao 2016; Chang and Dong 2016; Zakarya et al. 2021; Vardopoulos 2023; Carballada 2021).

When it comes to planning a cultural pilgrimage to Lefkada, there are several architectural digital tools that should be taken into account so as to support tourists in their exploration of the island's cultural heritage. Leukada island due to its geographical location can be visited either by sea or by car, traveling bus, or van due to the existence of a bridge linking it with the mainland Aitoloakarnania Prefecture at Eipiros where more churches, heritage monuments, and Byzantine churches can be visited by tourists. Tourists from several countries from the Adriatic sea as well as from Mediterranean sea can visit the Lefkada island. Also either Thessaloniki's international airport or Athens' one could be used for tourists traveling by air (Zakarya et al. 2021; Vardopoulos 2023; Carballada 2021; Contoes et al. 2000; Contoes et al. 2000; Manola 2022; Peeters and Dubois 2010; Pongthanaisawan et al. 2018; Zhang and Zhang 2020).

Leukada island demonstrates heritage religious buildings and cultural monuments from Enetic times inspired by Italian Architecture as well as historical structures, buildings, and monuments in later times during the Byzantine Epirous Empire and Ottoman local Empire in past times. Several old heritage monuments have been maintained at several stages on the island. Useful solutions to support and promote sustainable safe religious tourism are analyzed in next sections. All the relative analysis that is analyzed in next sections in terms of placement and supporting architectural digital tools for cultural pilgrimage tourism promotes useful sustainable tourism services for stakeholders and tourists that should be taken into account for Lefkada island as a unique travel destination.

2 Promotion of Sustainable Tourism at Lefkada Island in Greece

Climate change is indeed one of the most pressing challenges facing humanity today, and its multifaceted impacts are increasingly evident across various sectors. The rise in greenhouse gas emissions, primarily from human activities such as fossil fuel combustion, deforestation, and industrial processes, has led to a range of environmental consequences including rising global temperatures, extreme weather events, and sea-level rise.

Given the stark reality of climate change, it is crucial for governments, organizations, and individuals to prioritize both mitigation and adaptation strategies. Mitigation efforts focus on reducing or preventing the emission of greenhouse gases, while adaptation involves making adjustments in ecological, social, or economic practices to minimize the damage from climate impacts (Pongthanaisawan et al. 2018; Zhang and Zhang 2020; Metropolitan church Evagelistrias and architectural elements at Lefkada 2007; Ahmad et al. 2021; Baggio and Valeri 2020; Valeri and Baggio 2020a; Koliopoulos and Valeri 2022; Koliopoulos and Papakonstantinou 2024; INSETE 2021; Reihanian et al. 2012; Chemli et al. 2020).

Key strategies for addressing climate change within religious tourism facilities include:

Transitioning to renewable energy: Shifting to solar, wind, hydro, and other renewable energy sources can significantly reduce emissions from the energy sector.

Enhancing energy efficiency: Improving energy efficiency in buildings, transportation, and industries can lower energy consumption and reduce emissions.

Sustainable land use: Implementing sustainable agricultural practices, reforestation, and avoiding deforestation can help sequester carbon and protect biodiversity.

Innovation and technology: Investing in research and development of new technologies, such as carbon capture and storage, can play a vital role in reducing emissions.

International cooperation: Climate change is a global issue that requires coordinated efforts. International agreements, such as the Paris Agreement, are essential for establishing targets and fostering collaboration among countries.

Raising public awareness: Educating communities about the impacts of climate change and promoting sustainable practices can empower individuals to make a difference.

Strengthening climate resilience: Developing infrastructure and systems that are resilient to climate impacts—such as severe storms, flooding, and droughts—will help communities adapt and thrive in the face of change.

By prioritizing these strategies and fostering a culture of sustainability, we can address the challenges posed by climate change and work towards a more stable and secure future for both the planet and its inhabitants.

Several digital tools are necessary to promote sustainable religious tourism for the promotion of past heritage historical monuments, like part of old church architectures, maintenance projects at old ruins, and maintenance of old churches.

Furthermore, some of these digital tools should include: Virtual tours: With the help of digital drawing reality technology, tourists can take virtual tours of the island's historical sites and ruins before actually visiting them in person or due to probable disabilities for elderly people or other people with disabilities can have remote access to several touristic places.

This can help visitors plan their trip and get a better understanding of the island's rich cultural heritage. 3D reality proper apps: Three-dimensional representation reality apps for ruins of churches and digital representation of elements at building architectures can provide tourists with interactive and informative experiences while exploring the island's cultural attractions. These apps can overlay information, images, and videos onto the real-world environment, enhancing visitors' understanding of the historical significance of the sites they are visiting.

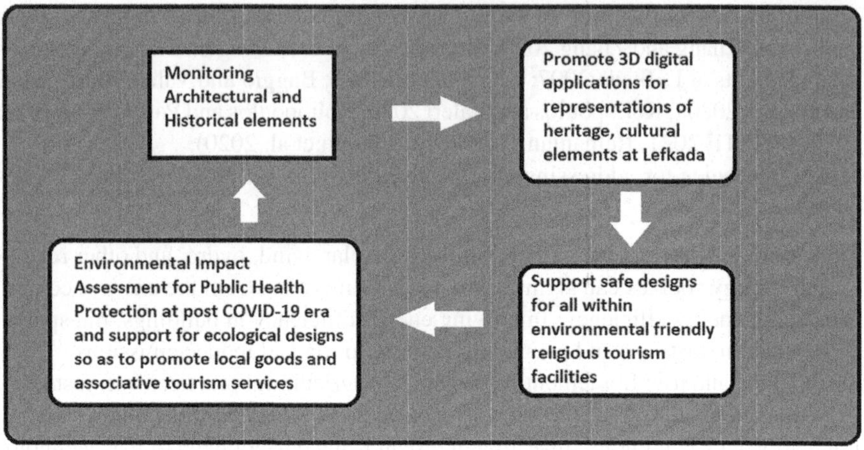

Fig. 1 A useful framework on the use of architectural digital tools that can enhance the cultural pilgrimage tourism experience on Lefkada island and results for safe designs for all

Moreover, it can be applied properly for people with disabilities. Audio guides: Audio guides can be a helpful tool for tourists exploring Lefkada's cultural sites virtually or by foot. These guides provide visitors with information about the history, culture, and significance of each site, allowing them to delve deeper into the island's cultural heritage at their own pace. 3D printing and reconstructions: 3D reconstructions of historical sites and buildings on Lefkada can provide tourists with a more immersive and realistic experience of the island's cultural heritage. These digital models can help visitors visualize how the sites would have looked in their original state, bringing history to life in a new and engaging way. Overall, Fig. 1 presents a useful framework in using architectural digital tools that can enhance the cultural pilgrimage tourism experience on Lefkada island and providing results for safe designs for all tourists with a deeper understanding and appreciation of the island's rich history and heritage (Chemli et al. 2020; Björk and Kauppinen-Räisänen 2019; Elmo et al. 2020; Valeri and Baggio 2020b; Sklavenitis 2019; Amer et al. 2013; Paliwal 2006; Hwang and Yoon 1981; Vagiona and Doxopoulos 2017; Vagiona and Palloglou 2021; Cho et al. 2016).

The proper use and application of digital architectural tools, virtual learning techniques, sanitary digital drawings should exist for safe travel destinations supporting proper designs for all so as to promote local products based on religious tourism applications, activities, support bazaars, and festivals at environmentally friendly comfortable facilities at indoors and outdoors. Educational tourism, cultural within religious tourism, promote local goods, foods, and applied associative geoinformation technologies and associative computer-aided design applications promoting particular services between tourists and stakeholders (Paliwal 2006; Hwang and Yoon 1981; Vagiona and Doxopoulos 2017; Vagiona and Palloglou 2021; Cho et al. 2016; Yang et al. 2021).

3 Planning Safe Religious Tourism Historical Destinations and Ecological Designs for All

Developing useful digital resources that support sustainable tourism and promote cultural and religious tourism can greatly enhance the experiences of tourists while also serving to educate staff and stakeholders (Carballada 2021; Contoes et al. 2000; Contoes et al. 2000; Manola 2022; Peeters and Dubois 2010; Pongthanaisawan et al. 2018; Zhang and Zhang 2020; Hwang and Yoon 1981; Vagiona and Doxopoulos 2017; Vagiona and Palloglou 2021; Cho et al. 2016; Yang et al. 2021).

Sustainable religious tourism growth refers to the development of tourism in a way that maximizes benefits for local communities and the environment while minimizing negative impacts. This approach seeks to ensure that tourism can be sustained over the long term, preserving the natural and cultural resources that make destinations attractive to visitors. Here are some key aspects of sustainable tourism growth and its impacts:

Positive impacts of sustainable tourism growth.

Economic benefits.

Job creation: Sustainable tourism can create jobs in areas such as hospitality, transportation, and guided tours, often prioritizing local employment.

Diversification of economy: It can help diversify local economies, reducing dependence on traditional industries like agriculture or fishing.

Cultural preservation.

Community engagement: Encouraging participation from local communities helps preserve cultural heritage, traditions, and practices.

Cultural exchange: Sustainable tourism fosters interactions between tourists and locals, promoting understanding and appreciation of different cultures.

Environmental conservation.

Protection of natural resources: Funds generated from tourism can be reinvested into conservation efforts, protecting natural habitats and biodiversity.

Promoting eco-friendly practices: Sustainable tourism encourages the adoption of environmentally friendly practices in the tourism industry.

Infrastructure development: Improved infrastructure benefiting both tourists and locals, such as transportation, sanitation, and healthcare services.

Education and awareness: Tourists may become more aware of environmental issues and cultural sensitivity, potentially fostering greater appreciation and advocacy for sustainable practices.

Negative impacts if not managed properly.

Overcrowding and overuse: Popular destinations may experience overcrowding, leading to strain on local resources, infrastructure, and ecosystems.

Environmental degradation: If not managed sustainably, tourism can lead to pollution, habitat destruction, and resource depletion (e.g., excessive water use in arid regions).

Cultural erosion: Increased tourism can lead to commodification of culture and loss of authenticity, as local traditions and practices are altered to cater to visitors.

Economic leakage: Profits from tourism may not always benefit local economies if large companies dominate the sector and send revenues back to their home countries.

Social disruption: An influx of tourists can disrupt local communities, leading to conflicts over resources, changes in social norms, and increased cost of living.

Key strategies for sustainable tourism growth.

Regulation and planning: Developing policies that manage tourist numbers and protect vital resources.

Community involvement: Encouraging local communities to take part in decision-making processes and benefit directly from tourism.

Education and training: Providing education for both tourists and the local population about sustainable practices and conservation.

Sustainable certifications: Promoting eco-labels and certifications for businesses that adhere to sustainable practices.

Technology and innovation: Utilizing technology to manage tourism flows, reduce environmental impact, and enhance visitor experiences.

However, in Fig. 2 a useful framework for the promotion of local religious tourism architectural, cultural, and heritage elements investigating indicative 3D digital representation is presented for several architectural elements based on investigated old structures, ruins, and building architectures that can be found in Lefkada island. Such heritage religious old monuments can be found at Lefkada main city, south part of Lefkada island, other villages, and ruins on Lefkada island (Metropolitan

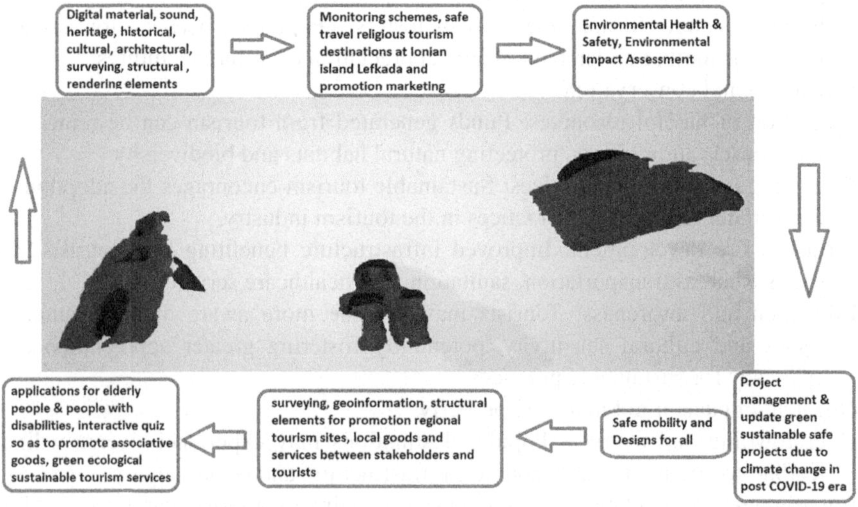

Fig. 2 A useful framework for the promotion of local religious tourism cultural architectural elements based on 3D digital representation for associative heritage elements within religious tourism at Lefkada island

church Evagelistrias and architectural elements of churches at Lefkada 2007; INSETE 2021; Sklavenitis 2019). A proper digital trip of cultural knowledge for elderly people and people with disabilities could be established for people that had an impact from COVID-19 and are afraid to visit these sites at close environments where a lot of people may exist or climate change situations are making difficult for them to visit such places. Proper sustainable designs, within ecological mechanics, environmental biotechnologies and safe indoor and outdoor facilities should exist supporting relative digital twin tools in project management, e-commerce, training services and structures due to climate change in terms of health and safety in protecting public health, green circular economy, improving both impacts based on UN's sustainable goals.

Based on the above new opportunities are coming up between stakeholders and tourists for the growth of religious tourism on Lefkada island. Proper digital repositories could be set up in the web supporting associative websites for people with disabilities. The relative material could be useful for training and educational purposes and should be enhanced by particular additional future projects processing relative historical, cultural, heritage, and architectural elements for particular interactive applications for stakeholders, research scholars, elderly people, and people with disabilities supporting safe proper applications on the post-COVID-19 era and in climate change management. All the above presented framework will be useful for the growth of sustainable religious tourism in similar case studies. However, Lefkada presents its competitive advantage combining the above with safe green particular tourism infrastructures and its ecological beauties, topographical views, and fresh local food products which are good for public health and ecological tourism sites.

Sustainable religious tourism growth is a dynamic and multifaceted approach to developing the tourism industry that seeks to balance economic, social, and environmental interests. The challenge lies in effectively managing tourism growth to maximize its benefits while minimizing adverse environmental impacts. Successful sustainable tourism requires collaboration among governments, businesses, communities, and tourists themselves.

Moreover, based on the findings of current working study, below are some suggestions for future research supporting proper digital tools and resources that can be created or improved upon:

Digital resources for tourists
Interactive mobile apps
Cultural guides: Apps that provide in-depth information about cultural and religious sites, traditions, and practices. Include AR features to enhance the learning experience
Itinerary planners: Personalized itinerary building tools that suggest sustainable transport, accommodations, and activities
Local community interaction: Platforms for tourists to connect with local guides or communities, allowing for immersive experiences that respect cultural practices
Virtual tours and experiences: Create VR or 360-degree tours of significant cultural and religious sites, allowing tourists to explore before visiting

Provide access to live-streamed events and festivals that may be culturally or religiously significant

Sustainability checklists: Interactive checklists or guidelines available through a website or app that help tourists make sustainable choices during their travels

Cultural etiquette guides: Digital resources detailing cultural norms and etiquette in various regions, important for religious tourism to promote respectful engagement

Crowdsourced eco-friendly reviews: Platforms where tourists can share their experiences about sustainable accommodations, restaurants, and activities, creating a peer-driven resource for others

Digital training resources for staff and stakeholders

E-Learning Modules: Develop comprehensive e-learning courses focusing on sustainable tourism practices, cultural sensitivity, and management of religious tourism

Webinars and workshops: Regularly scheduled online sessions featuring experts in sustainable tourism, cultural heritage preservation, and religious tourism.

Digital resource libraries: Create an online repository of case studies, best practices, and research on sustainable tourism models and successful cultural tourism initiatives

Collaborative platforms: Online forums and networks where stakeholders can share insights, challenges, and successes related to sustainable and cultural tourism.

Crisis management kits: Develop digital toolkits that provide guidance on managing crises in tourism (e.g., natural disasters, pandemics) with a focus on maintaining sustainability and community relations.

Community and stakeholder engagement tools

Stakeholder collaboration platforms: Create digital platforms where local businesses, communities, and stakeholders can collaborate on tourism initiatives and share resources

Feedback mechanisms: Digital surveys and feedback tools for tourists and local communities to gauge the impact of tourism on cultural sites and their concerns

Cultural mapping projects: Online mapping tools that identify and promote cultural and religious points of interest, integrating local narratives and histories

Social media campaigns: Utilize social media to raise awareness about sustainable tourism practices and cultural preservation efforts, engaging both tourists and locals

Digital reporting tools: Tools that help stakeholders report on the impact of tourism (economic, social, and environmental), which can be used to inform policy and practice

By developing these digital resources, the tourism industry can foster a more sustainable approach that respects cultural and religious practices while enhancing the overall experience for tourists.

4 Conclusions

The study effectively highlights the potential benefits of integrating digital architectural tools for virtual learning techniques so as to enhance the relationship between tourists and stakeholders in the realm of religious tourism. By leveraging sanitary digital drawings and focused safe associated environmental health applications, stakeholders can better promote local products and services that align with religious tourism endeavors.

Key outcomes from the study suggest that by adopting sustainable practices, stakeholders can create a multifaceted approach to religious tourism. This includes the promotion of heritage monuments, cultural architecture, and educational opportunities, all of which contribute to a richer tourist experience. Additionally, cultural activities that intersect with religious tourism can facilitate a deeper understanding and appreciation of local traditions, further enriching the visitor experience.

The application of associative geoinformation technologies and three-dimensional computer-aided design (CAD) tools provides innovative avenues for showcasing local goods, foods, and services while enhancing the visibility of cultural and historical sites. By doing so, stakeholders can not only attract more visitors but also foster a greater sense of community engagement and support for safe local economies.

In conclusion, the findings of this study underline the importance of collaboration and the strategic use of digital tools in promoting sustainable religious tourism. Stakeholders can utilize these insights to bolster the growth of religious tourism, ensuring that it remains both an enriching experience for tourists and a viable economic opportunity for local communities.

The research examined the opportunities for interactions between tourists and stakeholders in the context of religious tourism, providing valuable insights for both parties. By utilizing advanced digital architectural tools alongside virtual learning techniques and sanitary digital representations, the study aims to enhance the promotion of local products related to religious tourism.

The findings suggest that by integrating these digital tools, stakeholders can create various activities within sustainable religious and proper applied associative tourism frameworks. This approach not only highlights heritage sites and cultural monuments but also fosters educational tourism and cultural activities. In turn, this promotes local goods and foods while leveraging geoinformation technologies and three-dimensional computer-aided design applications.

Overall, the materials and insights presented in this study have the potential to support stakeholders in the development and growth of religious tourism, ensuring that it is sustainable, culturally enriching, and economically beneficial for local communities protecting public health and in climate change management.

References

Ahmad A, Jamaludin A, Shaliza N, Zuraimi M, Valeri M (2021) Visit intention and destination image in post-Covid-19 crisis recovery. In: Proceedings of the international crisis and risk communication conference, volume 4 (pp. 16-18). Nicholson School of Communication and Media, Orlando Fl. https://doi.org/10.30658/icrcc.2021.04

Amer M, Daim T, Jetter JA (2013) A review of scenario planning. Futures 46:23–40

Baggio R, Valeri M (2020) Network science and sustainable performance of family businesses in tourism. J Family Bus Manag 12:200. https://doi.org/10.1108/JFBM-06-2020-0048

Björk P, Kauppinen-Räisänen H (2019) Destination foodscape: a stage for travelers' food experience. Tour Manag 71:466–475. https://doi.org/10.1016/j.tourman.2018.11.005

Carballada AM, Balsa-Barreiro (2021) J. Geospatial analysis and mapping strategies for fine-grained and detailed COVID-19 data with GIS. ISPRS Int J Geo Inf 10:602

Chang YZ, Dong SC (2016) Evaluation of sustainable development of resources-based cities in Shanxi Province based on unascertained measure. Sustain For 8:585

Chemli S, Toanoglou M, Valeri M (2020) The impact of Covid-19 media coverage on tourist's awareness for future traveling. Curr Issue Tour. https://doi.org/10.1080/13683500.2020.1846502

Cho YJ, Wang Y, Hsu LLI (2016) Constructing Taiwan's low-carbon tourism development suitability evaluation indicators. Asia Pac J Tour Res 21:658–677

Contoes CC, Raptis V, Lautner M, Oberstadler R (2000) The potential of kernel classification techniques for land use mapping in urban areas using 5m-spatial resolution IRS-1C imagery. Int J Remote Sens 21:3145–3151

Elmo GM, Arcese G, Valeri M, Poponi S, Pacchera F (2020) Sustainability in tourism as an innovation driver: an analysis of family business reality. Sustainability 12(15):6149. https://doi.org/10.3390/su12156149

Hsiao T-Y (2016) Developing a dual-perspective low-carbon tourism evaluation index system for travel agencies. J Sustain Tour 24:1604–1623

Hwang CL, Yoon K (1981) Methods for multiple attribute decision making. In: Multiple attribute decision making, Lecture notes in economics and mathematical systems, vol 186, 1st edn. Springer, Berlin/Heidelberg, Germany, pp 58–191

INSETE (2021) Marketing plan tourism product 2015–2020. Available online: https://insete.gr/wp-content/uploads/2020/02/marketing_plan_ionian_islands.pdf

Katsoni V, Stratigea A (2015) A strategic policy scenario analysis framework for the sustainable tourist development of peripheral small Island areas–the case of Lefkada-Greece Island. Eur J Futures Res 3:5

Koliopoulos T, Papakonstantinou D (2024) Allied architectural digital solution Design for all at Landscape & Safe Community Health Facilities at post COVID era and in climate change. Afr J BioSc 6(8):374–380. https://doi.org/10.33472/AFJBS.6.8.2024.374-380. https://www.afjbs.com/uploads/paper/b5d9878154a9ef57c2ab46777265b7f2.pdf

Koliopoulos T, Valeri M (2022) Efficient healthcare policy and engineering management facilities for planning sustainable tourism development in post-COVID-19 crisis recovery. In: Valeri M (ed) Tourism risk. Emerald Publishing Limited, Bingley, pp 155–162. https://doi.org/10.1108/978-1-80117-708-520221011

Latinopoulos D, Vagiona D (2013) Measuring the sustainability of tourism development in protected areas: an indicator based approach. Int J Innov Sustain Dev 7:233–251

Manola M (2022) Contribution of the venetian monuments of Rhodes to cultural tourism and the local development of the Island. Open Journal for Research in Economics 5(2):35–42. https://doi.org/10.32591/coas.ojre.0502.02035m

Metropolitan church Evagelistrias and architectural elements of churches at Lefkada (2007) Publications metropolitan church Evagelistrias. Lefkada, Greece

Paliwal R (2006) EIA practice in India and its evaluation using SWOT analysis. Environ Impact Assess Rev 26:492–510

Peeters P, Dubois G (2010) Tourism travel under climate change mitigation constraints. J Transp Geogr 18:447–457
Pongthanaisawan J, Wangjiraniran W, Chuenwong K, Pimonsree L (2018) Scenario planning for low carbon tourism city: a case study of Nan. Energy Procedia 152:715–724
Reihanian A, Mahmood NZB, Kahrom E, Hin TW (2012) Sustainable tourism development strategy by SWOT analysis: Boujagh National Park. Iran Tour Manag Perspect 4:223–228
Sklavenitis K (2019) A walk through Lefkada's old town. Fagotto Publications, Athens, Greece
Szigeti H, Messaadia M, Majumdar A, Eynard B (2011) STEEP analysis as a tool for building technology roadmaps. In proceedings of the conference: eChallenges, Florence, Italy
Vagiona D, Doxopoulos G (2017) The development of sustainable tourism indicators for the islands of the northern Sporades region in Greece. Fresenius Environ Bull 26:1301–1309
Vagiona DG, Palloglou A (2021) An indicator-based system to assess tourism carrying capacity in a Greek Island. Int J Tour Policy 11:265–288
Valeri M, Baggio R (2020a) Italian tourism intermediaries: a social network analysis exploration. Curr Issue Tour 1–14:1270. https://doi.org/10.1080/13683500.2020.1777950
Valeri M, Baggio R (2020b) Social network analysis: organisational implications in tourism management. Int J Organ Anal 29(2):342–353
Vardopoulos I (2023) Adaptive reuse for sustainable development and land use: a multivariate linear regression analysis estimating key determinants of public perceptions. Heritage 6:809–828
Yang X, Zhao C, Xu H, Liu K, Zha J (2021) Changing the industrial structure of tourism to achieve a low-carbon economy in China: an industrial linkage perspective. J Hosp Tour Manag 48:374–389
Zakarya YM, Metwaly MM, Abdel Rahman MAE, Metwalli MR, Koubouris G (2021) Optimized land use through integrated land suitability and GIS approach in west El-Minia governorate. Upper Egypt Sustain 13:12236
Zhang J, Zhang Y (2020) Assessing the low-carbon tourism in the tourism-based urban destinations. J Clean Prod 276:124303

Religious Tourism and Local Development: Case Study of Samos Island

Panoraia Poulaki, Theodoros Rachiotis, and Nikolaos Vasilakis

Abstract This chapter examines the influence of religious tourism on the development of local communities, with a particular focus on the island of Samos in Greece. The objective of the investigation is to assess the influence of religious tourism on regional affluence and economic development. This study employs qualitative methodologies, such as interviews with stakeholders from the religious and tourism sectors, to investigate the current state and potential of religious tourism in Samos. The results suggest a lack of strategic planning, insufficient infrastructure, and insufficient coordination among stakeholders. Religious tourism presents a significant opportunity for sustainable growth, provided that comprehensive policies are implemented. These findings underscore the necessity of a methodical approach to the utilisation of Samos' cultural and religious assets for the purpose of tourism development.

1 Introduction

The primary catalyst for religious tourism is religious feeling, which compels individuals to travel in order to explore religious monuments and customs of a religious nature. Religious tourism capitalises on cultural legacy and serves as a focal point for cultural visitors, while for religious tourists, it is a manifestation of their

P. Poulaki
Department of Economics and Management of Tourism, University of the Aegean, Chios, Greece
e-mail: panoraia@aegean.gr

T. Rachiotis (✉)
Department of Physical Education and Sport Science, National and Kapodistrian University of Athens, Athens, Greece
e-mail: theorax@phed.uoa.gr

N. Vasilakis
Ionian University, Kerkira, Greece

M. Valeri (ed.), *Contemporary Religious Tourism*, Tourism, Hospitality & Event Management, https://doi.org/10.1007/978-3-031-79074-4_11

everyday life. Religious tourism undoubtedly has an economic aspect, since it significantly contributes to the economic prosperity of the regions that receive religious travellers.

The global religious travel and tours industry is thriving, reinvigorating the practice of pilgrimage from past times (Rachiotis and Poulaki 2022, 2024a, b, c, d). Engaging in diverse religious practices has always been a significant incentive for human migration throughout history. Religious tourism, as defined by Poulaki (2018), refers to the engagement in various tourist activities that are centred on the fulfilment of religious obligations. Religious travel, which has been practiced since ancient times, is considered the oldest and most prevalent kind of travel in human history (Kaelber 2006; Poulaki 2018). It continues to be a significant aspect of tourism worldwide, both in terms of the number of people involved and the frequency of such trips. Anthropologists and archaeologists have identified the presence of pilgrimage expeditions in Europe, Asia, America, and Australia dating back to ancient eras. Religious tourists are motivated by a desire to connect with the divine, perform a vow, or take part in religious ceremonies and activities. Religious tourism refers to a kind of tourism when individuals travel, either partly or solely, for religious purposes (Poulaki 2018).

Religious tourism, characterised by a stronger Christian influence and a preference for solitude, refinement, and environmental consciousness, can be seen as a significant component of high-quality tourism and cultural heritage. It is being promoted as an essential alternative to the prevailing sun and beach tourism (Dubish 2000; Gassiot et al. 2015). Religious tourism is a widespread occurrence worldwide. Religious excursions and expeditions around the globe are thriving, reinvigorating the practice of pilgrimage from past times. Religious tourism, which incorporates secular components, has become a prominent kind of special and alternative tourism. It is rapidly replacing organised mass tourism, thanks to the many religious groups it encompasses (Rachiotis and Poulaki 2022, 2024a, b, c, d).

The primary goal of this study is to facilitate the application of empirical research results on the growth of religious tourism on the border island of Samos. If the recommendations and policy measures derived from empirical research are included into a comprehensive and specialised tourist development program, they may be progressively implemented and make a significant contribution to the tourism development of the North Aegean Region, particularly on the island of Samos. In order to achieve this objective, it is necessary to have the collaboration of all competent entities in the Region to ensure the optimal utilisation of the natural and cultural treasures found on its islands. By implementing meticulous planning and strategic elements, religious tourism has the potential to serve as the primary catalyst for the economic growth and municipal development of Samos.

2 Religious Tourism

Religious tourism is a form of tourism in which the participants are either entirely or partially motivated by religious considerations. The term religious tourism should include all types of travel which are motivated by religion in combination with other motives and are intended for a religious site but for which the route itself is not a religious practice. Religious tourism is not intended to cause the loss of religious meaning or sanctity (Vukonic 2002). It reflects more of another form of tourism, which is placed between the sacred and the secular (Kasim 2011; Poulaki 2018).

While the religious tourism business is seeing rapid growth, it is crucial that its fundamental objective of promoting devotion and spiritual restoration remains unchanged (Raj and Bozonelos 2015; Poulaki et al. 2015). This form of travel is directly related to the beliefs of participants in religious experiences (Yeoman 2008; Kasim 2011; Kumar 2016). Common types of religious destinations are (a) emblematic places of pilgrimage or trips to religious centres (holy sites open to all believers or to specific categories of persons, such as clergy, teachers, monks), (b) religious destinations linked to historical events or cultural achievements or trips to festivals, and (c) religious festivals or trips to festivals (Lucarno 2016; Polyzos 2010).

Religious travel is intrinsically connected to the practice of religion. According to Santos, religion may be succinctly defined as a belief system centred on a supreme entity, characterised by strong faith and devotion. All faiths have a common belief in the existence of a higher entity. (Timothy and Olsen 2006). It is a determinant that functions as a catalyst in the conduct and mindset of individuals. Religion encompasses the belief and expression of information, thoughts, emotions, intentions, and behaviours about the influence of superior beings, whether they are individual or impersonal. The word "religion" is often employed in ordinary public discourse to describe phenomena that extend beyond the scope of conventional religious institutions (Poulaki 2013; Poulaki 2016). A religious monument is directly associated with the following: (a) Adherence to religious regulations regarding holiness, (b) Consideration for the religious aspects and requirements of visitors, (c) Conservation of the natural landmark and its promotion with a focus on sustainability, (d) Dual status as both a functioning religious monument and a cultural attraction open to visitors, (e) Addressing the commercial requirements of businesses, (f) Ensuring social equilibrium among the local population. Contemporary pilgrimages often integrate a spiritual objective with customary landmarks. The primary focus of this trip is on the spiritual advantages it offers, including the rejuvenation of one's spirit, the development of faith, a greater understanding of the Bible and theology, and transformative life experiences (Munro 2017; Poulaki 2018).

Throughout history, sacred locations and religious rites have attracted travellers seeking spiritual experiences. In these locations, the devout seek to establish a connection with the divine. The interest in religious tourism is steadily increasing, and this may be attributed to the following factors: (a) The evolution of global religions, (b) The quest for spirituality and genuine encounters via spiritual and cultural customs, (c) The advancement of niche markets, (d) Targeting well-travelled customers, (e) Faith in religion, (g) Personal fulfilment, and (h) Volunteering (Yeoman 2008).

Currently, the act of fulfilling religious obligations is connected to a diverse array of tourist activities, together known as religious tourism. The primary goals of religious tourism are to emphasise and advertise locations or features of religious significance, in order to attract and accommodate travellers who have religious or pilgrimage intentions. Historical events have a significant role in shaping a religious movement. Religious tourism has emerged as a significant catalyst for the conservation and use of several religious and historical landmarks (Cohen 2003). Furthermore, when visitors engage with other religious beliefs, it provides them with a distinctive and exclusive experience that can only occur when there is complete reverence for the local community hosting them (Poulaki 2018; Rachiotis and Poulaki 2024a, b, c, d). Religious tourism encompasses the act of visiting religious locations by those who are motivated by either partially or solely religious grounds. Religious tourism is not subject to seasonal variations, and it does not discriminate or show a preference for well-known tourist attractions. Religious tourists exhibit a wide range of age, economic, social, and professional diversity, including all social strata. This is due to the universal and all-encompassing nature of religious feeling. Religious tourism encompasses two groups of travellers: pilgrims and tourists, as well as devout and knowledge-seeking leisure travellers (Poulaki et al. 2021a; Raj and Morpeth 2007a).

The reasons for religious tourism include the following: (a) The ubiquitous spiritual sentiment that compels travellers to visit locations of religious activities in order to fulfil their religious requirements. (b) The classification of significant Christian holidays as public holidays, allowing workers the opportunity to be absent from work, as well as enabling parents with school-age children to take time off. (c) The widespread distribution of religious monuments and places of worship in various locations, apart from the residential areas of the faithful. (d) The increasing inclination to go back to one's origins is a prevailing tendency that has gained momentum in recent years. (e) The inclination to evacuate and elude metropolitan areas, which is intensified by the demanding, stressful, and exhausting lifestyle of the city and the challenges of the urban environment (Poulaki et al. 2021b; Raj and Morpeth 2007b).

The religious tourist exhibits characteristics of both a conventional tourist and a devout pilgrim. He is a religious traveller who lacks the knowledge to effectively pursue his spiritual objective. Despite being surrounded by professional travel advisors, adhering to predetermined itineraries, participating in organised group meals, and following conventional routes, he loses sight of his genuine religious-spiritual goal. The traveller's expedition may be referred to as a pilgrimage, yet, in actuality, they often experience isolation and transform from a pilgrim into a mere tourist (Rachiotis and Poulaki 2024a, b, c, d). The opulence, convenience, cosmopolitan atmosphere of travel and lodging in upscale hotels, the presence of cameras and recording devices carried by many travellers, the commercialisation of religious artefacts, the need for lodging, food, and organised activities, etc. They eliminate the spiritual aspect from the journey and convert it to just a tourist attraction (Cerutti and Piva 2016). Once the religious tourist has satisfied his religious obligations, he then assumes the role of a typical tourist. Consequently, it relies on tourist services to fulfil its requirements for transportation, lodging, sustenance, amusement, acquisition of souvenirs, and other goods (Poulaki et al. 2021c; Poulaki et al. 2022).

The concept of religious tourism, without diminishing its intrinsic significance and holiness, is a distinct kind of tourism that exists between the realms of the holy and the secular. Commuters, regardless of whether they are solely pilgrims, tourists, or a combination of both, have fundamental requirements that must be fulfilled. Religious tourism contributes significantly to the economic prosperity of the areas that host it. The advantages include higher earnings, the generation of employment opportunities, the promotion of regional growth, and the establishment of connections with other sectors of the local economy (Rinschede 1992; Poulaki et al. 2021d). The endeavour to establish religious tourism is a worldwide and all-encompassing undertaking. Religious tourism, which incorporates secular components, has become a prominent aspect of special and alternative tourism. It is progressively replacing organised mass tourism (Poulaki et al. 2015; Poulaki 2018).

Several factors significantly influence the demand and growth of religious tourism. These include: (a) The inclusion of short-term religious travel as a means for individuals in consumer societies to pursue a higher quality of life, (b) The combination of leisure time for older individuals with religious or pilgrimage journeys, (c) The advancement of technology, communication, and information systems, which particularly aid in the development of tourism as a whole, including religious tourism, (d) The strong emphasis placed on short-term trips, brief vacations, and weekend getaways, and (e) The pursuit of aesthetic appeal and excellence in travel, which is directly linked to the unique spiritual experience of embarking on a pilgrimage (Rachiotis and Poulaki 2024a, b, c, d).

Religious tourism is linked to the protection of the environment and the preservation of local culture. Religious tourism takes on particular importance as an institutional form in the following religions: Hinduism, Buddhism, Christianity, and Islam. Modern society has broadened its definition of sacredness by encouraging the establishment of new sanctuaries. These destinations are considered pilgrimages, and individuals travel to them in pursuit of a location or state that they believe will exemplify a valuable ideal (Poulaki et al. 2021e; Santos 2002).

Religious tourism facilitates the sharing of cultural accomplishments and fosters opportunities for people to unite. Religious tourism has a crucial role in fostering and enhancing social and cultural connections via the mobility of people (Griffin and Raj 2012). Religious tourism is a component of sustainable tourist development that focusses on alternative forms of tourism, specific interests, and the preservation of holy places. It is closely connected to the overall goal of sustainable tourism development.

Long-term religious tourism refers to the practice of visiting religious centres for extended periods of time, often lasting many days or weeks. Religious tourism includes many travel activities that are not considered pilgrimages, such as trips to Christian places renowned for their art and architecture (Poulaki 2018; Ron 2009).

The economic advantages linked to religious tourism surpass those linked to other market sectors due to the enthusiastic purchasing behaviour of pilgrims and other religious visitors towards religious items. Furthermore, the expansion of religious tourism may contribute to the conservation and safeguarding of cultural assets (Fleisher 2000; Poulaki 2018).

This kind of tourism also provides the chance to diminish prejudice, fosters a shift in mindset, and may overcome cultural disparities via interaction with various cultures. It is important to regulate the excitement for financial gain in this field in order to preserve the authenticity of religious tourism (Kasim 2011; Cassar and Munro 2016; Poulaki 2018). Religious visitors undoubtedly enjoy the pleasures that come with every journey, as they explore unfamiliar destinations and acquire new experiences.

Religious tourism has an impact on population growth and the economy in religious hubs. Indeed, in Islam and other faiths, the concept of "religious tourism" is seen inappropriate. In several faiths, including Islam and Buddhism, those who do not adhere to the faith are prohibited from accessing certain locations, such as the revered Kaaba stone in Islam. This feature diminishes the appeal of the holy places for tourists, but simultaneously fosters more reverence and a heightened feeling of religious devotion among the devout (Lagos and Poulaki 2014; Poulaki 2018).

3 Religious Tourism and Pilgrimage

The scientific community has been concerned with the differentiation between pilgrims and religious tourists for a few years. The individual's wants, aspirations, and behaviours are influenced by the environment, whether it is religious or secular, in which they reside. This shaping process occurs throughout their journey (Timothy and Olsen 2006; Metreveli and Timothy 2010; Poulaki 2018). Religious tourism may be categorised into two categories, which are determined by the objectives and the mode of transportation. One version is motivated by religious feeling, namely by pilgrims, whereas the other form is driven by the pursuit of knowledge, known as religious tourism. We categorise individuals into two groups: (a) those who are just pilgrims, and (b) religious tourists.

A pilgrimage entails the obligation for a believer to go to a location that is deemed sacred. It is a religiously inspired trip to a holy site, attributed to the extraordinary intervention of God or a deity in that particular area. Believers engage in acts of devotion and piety inside a holy location. The trip has two destinations: one is a physical trek towards a holy spot, and the other is an inward and metaphorical journey (Kasim 2011; Poulaki 2018; Yeoman 2008). The pilgrimage has always been and continues to be an expedition into uncharted territory. The pilgrim entrusts mankind to the divine, seeking divinity via this act of faith, detached from the constraints of ordinary existence (Cohen and Hattab 2010; Stausberg 2011). The pilgrim attempts to attain spiritual redemption by embarking on a journey/ritual characterised by a profound spiritual connection, a profound devotion to the divine, and a deliberate detachment from worldly desires. The primary motivations for doing a pilgrimage often revolve on the pursuit of spiritual enlightenment, deepening one's faith, seeking blessings, and finding answers to profound problems. Undoubtedly, pilgrimage, including both its religious and non-religious manifestations, has seen significant growth worldwide in recent years (Poulaki 2018).

Pilgrims dedicate their time to deep reflection and supplication, engaging in religious rituals, while on their trip they visit designated "sanctuaries". The majority of these pilgrims lack knowledge about the historical or cultural (i.e. non-religious) importance of the cities and villages they pass through on their journey. Their primary objective is to undertake a "special pilgrimage" to a place of worship where it is feasible to get healing for an ill individual or redemption for the soul, leading to admittance into heaven. Conversely, religious tourists are people who primarily travel with the intention of gaining information. They acquire knowledge and encounters by means of the path and interaction with the people, areas, and urban centres they traverse. Religious visitors often frequent the city centre and engage in some of the ceremonies (Yeoman 2008; Kasim 2011; Poulaki 2018).

Throughout history, holy locations and religious rites have attracted travellers from throughout the globe. These locations are where individuals strive to establish a connection with the divine. According to Dillon (1997) and Sfakianakis (2000), pilgrimage travels have been recorded since ancient times. These excursions were undertaken to take part in spiritual events and diverse rites. In addition, they journeyed to seek prophecies or seek remedies for certain ailments (Yeoman 2008; Kasim 2011; Poulaki 2018).

Several prerequisites must be met to ensure the efficient organisation and planning of the pilgrimage visit. The selection of the pilgrimage is crucial to ensure that the itinerary is not arduous and challenging for the pilgrims. The temporary duration of the trips is also significant. During times of abstaining from food and engaging in prayer, the individual should demonstrate exceptional endurance. The pilgrim is in search of a sanctuary where he might find solace and experience a heightened sense of spirituality.

In addition to studying ecclesiastical legacy, the pilgrim directs their attention towards religious experience by visiting places of worship, engaging in sacraments and religious rites, and showing reverence for icons and holy relics. In the context of pilgrimage, the term "experience" refers to the whole trip itself, including activities like fasting, making vows, and engaging in prayer, all leading towards the sacred destination. On the other hand, in religious travel, the term "experience" is limited to the destination alone, which may include places like churches, monasteries, or monuments. Religious regions sometimes possess a dual nature, including both religious and secular elements, since they attract pilgrims, travellers, religious devotees, and cultural tourists (Poulaki 2018).

Religious tourism exhibits a considerable degree of seasonality, even though religious sites are accessible to visitors year-round. Furthermore, agricultural labour often impacts pilgrimage journeys. For instance, only a small number of pilgrims from the vineyards of France visit Lourdes during the harvest season. The timing of religious tourism, which revolves on Church conferences and feasts, is governed by the Church administration, the holiday season, and weather circumstances. Women are well-represented in Lourdes. The pilgrims' motives are mostly driven by spiritual factors. Indeed, several contemporary pilgrimage journeys have striking similarities to secular excursions in terms of their organisation. They are often structured in a similar manner to mass tourist packages.

The pilgrimage is influenced not just by religious beliefs but also by several prevailing conditions, including political, economic, and social influences. Historical considerations, such as the presence of the holy monastery of Agia Lavra near Kalavryta, have a vital role. This monastery not only has religious significance but also has historical importance as it served as a symbol for the Greek revolution of 1821 (Poulaki 2018; Shinde 2012).

In addition to the significant Christian holy cities and reference sites, it is worth mentioning the religious centres of regional or local significance. These locations are often associated with notable religious occurrences such as miracles or apparitions of the Virgin Mary. We often refer to Mount Athos, which has significant political autonomy and serves as a major cultural hub. Other notable sites are the Panagia of Tinos, the Cave of the Apocalypse on Patmos, and the Monasteries of Meteora. Religious centres of comparable significance. Specifically for Roman Catholics, notable pilgrimage sites include the Vatican, Our Lady of Lourdes, Fatima in Portugal, Medjugorje in Bosnia and Herzegovina, the Black Virgin at Czestochowa in Poland, and Santo Nino in the Philippines. The religious-pilgrimage path to the church of St. James in Compostela, NW Spain (El Camino), is renowned and considered one of the oldest in the Christian world. In addition, religious rituals and events of significant importance worldwide captivate the attention of both believers and non-believers. These include the portrayal of the Divine Drama and the preservation of ancient customs, such as torch relays, parades of masked individuals, and theatrical performances. Notable examples include Semana Santa in Spain, the Philippines, and Latin America, the Passion Week in the Holy Land, and the portrayal of the Divine Nativity in Oberammergau, Bavaria (Raptis 2011; Lopez 2013; Havard 2017; Poulaki 2018).

4 Presentation of Samos Island: Sightseeing

Samos, an island in the eastern part of Greece, is situated on the coast of Asia Minor. It is considered the princess of Ionia among the Greek islands. The island's advantageous geographical location facilitated its growth and transformation into a significant hub of the Ionian civilisation. The extant structures, the dispersed archaeological sites, the discoveries, and the museum displays emphasise the historical significance of Samos. They provide evidence of the existence of an ancient city that had a prominent and well-developed civilisation. The settlement had a high level of organisation, with sturdy walls, paved streets, spacious squares, well-constructed dwellings, thriving shops, elegant villas adorned with exquisite mosaic floors, temples, a bustling market, public buildings, productive factories, necropolises, expansive and efficiently managed sports facilities, spas, and a remarkable water supply and sewage network.

Samos has been the birthplace or ancestral home of several distinguished individuals who have achieved greatness in the fields of science, politics, and the arts. Typically, we refer to notable individuals such as the mathematician Pythagoras, the philosopher Melissos, and the astronomer Aristarchus. Samos preserves its cultural legacy and is renowned for its many historical and religious landmarks. Specifically, we list the most notable ones. Additionally, the island of Samos has many and notable ancient structures, including temples dedicated to goddesses Athena and Demeter, as well as gods Poseidon, Apollo, and Hermes, among others.

The ancient city is situated on the southeastern periphery of the island, occupying the present-day Pythagorion, which is named in honour of the renowned mathematician and philosopher Pythagoras. It is regarded as one of the most significant cities from ancient times. The Tunnel of Eupalinos is widely regarded as one of the most remarkable architectural achievements of ancient times. The aqueduct of the ancient city is located deep into the mountain Kastro or Kastri (Poulaki et al. 2015). Herodotus refers to it as the "amphistomon trench". The tunnel is 1036 metres in length and begins on the northern side of the mountain, concluding on the southern side. The construction's exceptional quality and adherence to contemporary technological standards have garnered global appreciation, drawing in thousands of visitors.

The walls of Polycrates are located above Pythagoreion, namely in the NE and NW sections. Measuring 6 km in length. The partially demolished towers, gates, and battlements that remain are remnants of the defences that were destroyed in 439 BC during the 9-month siege of Samos by the Athenians led by Pericles. The theatre is constructed on the incline of the Monastery of Panagia Spiliani at Pythagorion, Samos. Only a small number of artefacts were discovered in the theatre. The structure was built during the Roman era and seems to have included three main parts: the cavea, the orchestra, and the stage.

In the capital of Samos, near the sports facilities, are located Thermes. The floors are covered with mosaic tiles and the walls are made of marble. The archaeological site of Therma is located west of the city, within the site of the sports facilities of ancient Samos. These are the ruins of Therma dating back to the second half of the 2nd century AD. On the southwest side of the port of Pythagorion (ancient Samos), there is a low hill on which parts of the Byzantine Castle of the city are preserved until today (Ministry of Culture 2015; Poulaki 2018).

The monument of Pythagoras located in the port of Pythagorion is special and impressive, as is the monument of Aristarchus in Karlovasi. Also worth mentioning are the Museums of Samos, such as: the Archaeological Museum of Samos, the Archaeological Museum of Pythagorion, the Natural History Museum of the Aegean/Paleontological Museum of Mytilinia, the Museum of Samian Wine, the Historical/Folklore Museum of Karlovasi, the Tannery Museum in Karlovasi and the Logothetis Tower (tower and museum) in Pythagorion (Poulaki 2018). Characteristics are the traditional settlements, towns, and picturesque villages of Samos, such as: Ano Vathi, Pythagorion, Heraion, Vourliotes, Karlovasi, Kastania,

Kokkari, Lekka, Manolates, Marathokampos, Myloi, Pagondas, Spatharaioi, Chora, Mytilinioi, Pyrgos, Ambelos, Koumaradaioi, Kastania, Lekka (Poulaki et al. 2015; Poulaki 2018). The capital of Samos is called Samos or Vathy and is located in the northeastern part of the island, it is built amphitheatrically around the bay. Beautiful neoclassical buildings house public services.

The caves are found as natural or artificial and are characterised as: simple cave, cave chasm, cave church, underground river, cave trench, mine, well, cave spring, and cave residence. Speleologists have recorded 84 caves scattered throughout Samos. Particularly noteworthy are the following caves:

- Kandyli Cave on Mount Kerkis
- Cave of Agia Triada on Mount Kerkis
- Cave "Window of the Virgin Mary"
- Cave of Kasolis
- The cave of Pythagoras is located on the slopes of Kerkis, and according to tradition, the philosopher found refuge here when he was chased by the local tyrant Polycrates, although historical sources have not yet ruled on the authenticity of the legend.
- Cave "Hole of the Jetze" in Kosmadeoi
- Cave "Iron Door" in Kiourka of Marathokampos
- Cave "Tsakalotripa" in Pythagorion (Poulaki 2018).

Many caves of Samos hide small churches, such as:

- Panagia Spiliani,
- Agios Antonios in Karlovasi
- Panagia Sarantaskaliotissa
- The Distant Virgin Mary on Mount Kerkis built in 1765
- Panagia Kakoperatiani on Mount Kerkis is called "Kakoperatiani" because of the poor condition of the road
- Agios Isidoros in the village of Kallithea (Poulaki 2018).

The Speleological Association of Samos "Efpalinos", founded in 2005, has taken important actions to promote and preserve this wealth. It is worth mentioning the at least 10 underwater caves in Samos, mainly in the area of Marathokampos, Kallithea, and Potami.

Samos has many remarkable hiking trails and is ideal for a variety of nature excursions in their gorges, streams, and caves. Kerkis and Karvounis, the two large mountains of Samos, are rich in trails and hiking routes. The Greek Mountaineering Hiking Club of Samos organises excursions to the island and at the same time gives useful information for hiking. Also, an interesting route starts from Kokkari and crosses the villages of Mount Karvouni, passing through Vourliotes. An old path leads to the village of Manolates, then to Stavrinides, from where the descent to the sea begins, ending in Agios Konstantinos. Another route starts from Votsalakia, from the side of Marathokampos reaches the monastery of Evangelistria of the 11th century, and from there, through a network of paths, it ends at Vigla, the peak of Kerkis, at an altitude of 1400 metres. An organised climbing field is located in

Potami, just outside Karlovasi, while very interesting is the field near Marathokampos, on a route called "the Pythagorean Theorem", as nearby is the cave of the ancient philosopher. Another climbing area is located near the village of Drakaioi, at the western end of Samos, on a slope with the possibility of opening two and three rope routes. A total of 60 climbing routes have been mapped in Samos, while the more adventurous can try their hand at canyoning, in one of the 27 small gorges of the island. Samos has sea, many beaches and coasts for water sports. There are many and impressive beaches of Samos with crystal clear waters, such as the following: Avlakia, Kokkari, Lemonakia, Megalo Seitani, Mikro Seitani, Potami, Tsamadou, Tsampou, Gagou, Licorice, Kerveli, Mykali, Panagia Eleousa, Psili Ammos, Votsalakia, Limnionas, Balos, Potokaki, and Golden Sand.

There are many interesting customs in Samos. Typical are the masquerades on Halloween, also known as "fratzolia" or "old men" or "bells". The "fragrant" used to tease passers-by with strange dances and songs. It is worth mentioning the various festivals that take place on the island and attract many visitors, such as Heraia, Karlovasia, the cultural week "Yiannis Ritsos", and many more.

5 Elements that Constitute Religious Tourism in Samos

Samos has amazing religious sites, monasteries and churches, which are a natural part of the national legacy and draw visitors who might help to boost religious tourism. The devotion and religious life of the people living on Pythagoras distinguishes the island. The many parish churches, monasteries, chapels, and private churches clearly show this. Samos is a perfect place for thousands of pilgrims because of the abundance of religious sites, the legends that endure throughout the years, and the many marvels that appeal in the lovely surroundings (Poulaki 2018).

Tourists are drawn to numerous religious sites in Samos, which are an essential component of our national heritage. Indicatively, the following churches are mentioned: Panagia tou Potami in Karlovasi, Agios Nikolaos in Potami, Agios Charalambos in Kallithea, Dormition of the Virgin Mary Butterfly, Saint Christodoulos in Heraion, Agios Ioannis in Pnaka Vourlioton, Agios Ioannis in Mytilinion, Dormition of the Virgin Mary Vourlioton, Panagia tou Vronta in Skoureika, Catholic of the Holy Monastery of Agia Zoni, Agios Taxiarchis of Megali Panagia, Evangelistria in Pano Vathi, Agioi Taxiarches in Malagari, Dormition of the Virgin Mary in Platanos, Agios Ioannis in Pano Vathi, Panagia Evangelistria in Ano Vathi, Transfiguration of Christ the Savior in Pythagorion, etc. (Poulaki 2018).

Speleologists have recorded 84 caves scattered on the island. Many caves of Samos hide small churches. Indicatively, the most important caves are mentioned: Panagia Spiliani, Panagia Sarantaskaliotissa, Agios Antonios in Karlovasi, Panagia Makrini, Agios Isidoros in Kallithea, etc. In addition, particularly noteworthy are the following holy monasteries of Samos: Vrontiani Monastery, Megali Panagia Monastery, etc. (Poulaki 2018).

Samos has many annual events. Varvounis (1995) states that Samians traditionally distribute a unique meal known as "Giorti" or "Kiskek" at celebrations and weddings. This tradition, originating from Asia Minor, is one of the most ancient practices that are annually renewed on the night of August 22nd. It is specifically devoted to the Virgin Mary. The goat meat is cooked with wheat and onions overnight in large cauldrons by very skilled individuals, mostly males. On August 23rd, the "Festival" is provided to all participants. A similar occurrence takes place during the celebration of other Saints in different regions of the island.

The customary burning of Judas, as well as Klidonas, occurs at several locations around the island. On the evening before Agios Ioannis Prodromos (June 24th), bonfires are ignited in most villages using the wreaths from May Day. The young people in the area then leap over these bonfires. On New Year's Eve, children present their parents with the "provenda", a meal consisting of vasilopita and other sweets, while also offering them the "boupistrina", which refers to both money and more sweets. The island hosts numerous festivals dedicated to various saints, including Agia Marina, Agia Paraskevi, Agios Panteleimonas, Transfiguration of Christ the Saviour, Agios Georgios, Agios Konstantinos, and Eleni, among others. These festivals attract a significant number of tourists and locals in all the villages of the island (Poulaki 2018).

Samos has the potential to attract religious tourists due to its mix of pilgrimages, monasteries, and museums. The Ecclesiastical-Byzantine Museum is specifically listed as a repository for items, utensils, and religious relics acquired from the monasteries on the island. The Byzantine-ecclesiastical collection showcases a variety of artistic pieces including silversmithing and gold embroidered works, icons and manuscripts, gospels, holy utensils, crosses, and vestments (Poulaki et al. 2015).

6 Method of the Study

It was conducted qualitatively with interviews and judgement sampling. For the sampling of the research, after planning the interviews, the stakeholders were approached, i.e. representatives of organisations and entrepreneurs who belong to the tourism industry and are involved either directly or indirectly with religious tourism. This was followed by interviews that took place in 2023 and involved stakeholders from the North Aegean Region, the Samos Regional Unit, the Samos Municipalities, the Samos Chamber of Commerce, the Samos Chamber of Commerce, the Port Authority, the Airport, Customs, the Holy Metropolis of Samos, owners travel agencies, hotels, rooms for rent, car rental offices, tourist shops and others.

7 Conclusions

The conclusions drawn from the qualitative research are as follows:

- Absence of a religious tourism development plan with specific strategic components, such as: (1) Contribution to local development by creating new favourable conditions for the lifting of the tourist isolation of the Northern Aegean islands. (2) Expanding tourist demand and lengthening the tourist season, (3) Creating an enriched and diversified tourist product, (4) Exploiting the rich history, culture and ecclesiastical history of the island of Samos. (5) Develop infrastructure and offer suitable facilities for tourists, and (6) Consolidate a development program aimed at tourism development.
- Lack of specific policy and appropriate planning for the development of religious tourism.
- There have been only fragmented actions, both at the local and regional level with limited cooperation between the agencies and the tourism businesses.
- Lack of highlighting the points that can make the island of Samos strong, such as: its geographical position in the Mediterranean Sea, the great sunshine throughout the year, the architectural interest of many settlements and buildings, the rich cultural heritage, the existence of many touristic untapped areas, the feeling of security and hospitality of the tourists, the preservation of "identity" and authenticity of the destination to a large extent, the rich natural and cultural background, the good quality and wide variety of local products.
- Failure to record the arrivals of religious tourists in Samos, as well as in the other islands of the North Aegean, so that it is not possible to cite relevant data. The specific complete lack of recording of the elements of the phenomenon on the islands of the North Aegean Region is an irrefutable witness to the sloppiness and the absence of an essential strategy for the development of religious tourism and other forms of tourism.
- Absence of development incentives and collaborations required at local, regional, and national level, with the consequence that the creation of an enriched and diversified tourist product is absent, to strengthen investment activity, to improve infrastructure and create a strong local identity.
- Lack of effective organisation, administration, and management of tourist development. Hesitation to open new markets and cooperation with new tour operators, for staff training, but also for the utilisation of the know-how of the University of the Aegean.
- Inefficient promotion of special and alternative forms of tourism in the context of sustainable development. There is an absence of organised actions to encourage them.
- Lack of information, awareness, and public actions regarding the development of religious tourism, which is achieved with information at the local and regional level, but also with the creation of a database of religious monuments.

- Ineffective tourist promotion actions of the islands of the Region to increase tourist traffic. The island's cultural wealth has not yet been properly exploited and showcased.
- Transportation problems of the islands of the Northern Aegean Region with the rest of the country and with Europe. As a result the entire operation depends exclusively on charters and tour operators.
- Short tourist season for most tourist companies. The size of the hotel units is small, because of this viability and competitiveness are limited.
- Lack of specific programs for the utilisation of religious monuments and the development of religious tourism.
- Inability to combine religious tourism with other special and alternative forms of tourism.
- Lack of infrastructure in monastic areas and the absence of affordable hotel accommodation.
- Insufficient promotion of the island's religious monuments and religious tourism.
- Limited cooperation between Church and State in matters of tourism.

The findings of the research reflect the current state of development of religious tourism in Samos. It is feasible to implement a comprehensive and specialised tourist development program in a gradual manner, thereby making a substantial contribution to the tourism industry of Samos. Collaboration among all competent stakeholders is essential for the most effective utilisation of the natural and cultural resources present on the islands of the region. Religious tourism can be the driving force of the economy for the North Aegean Region with the appropriate planning and strategic components. All these measures presuppose a planning process that will consider what a region has to offer as a tourist experience and whether the product is representative of local values and the identity of the host place. The active participation of all agencies is required, as well as the cooperation of both the productive classes and the social groups of the islands to implement specific actions.

References

Cassar G, Munro D (2016) Malta: a differentiated approach to the pilgrim-tourist dichotomy. Int J Relig Tour Pilgr 4(6). https://arrow.dit.ie/ijrtp/vol4/iss4/6

Cerutti ST, Piva EL (2016) The role of tourists' feedback in the enhancement of religious tourism destinations. Int J Relig Tour Pilgr 4(4). https://arrow.dit.ie/ijrtp/vol4/iss3/4

Cohen E (2003) Tourism and religion: a case study-visiting students in Israeli universities. J Travel Res 42:36. https://doi.org/10.1177/0047287503254247

Cohen E, Hattab K (2010) Struggles at holy sites and their outcomes: the evolution of the Western Wall Plaza in Jerusalem. J Herit Tour 5(2):125–139. https://doi.org/10.1080/17438730903518411

Dillon M (1997) Pilgrims and pilgrimage in ancient Greece. Routledge

Dubish J (2000) The religious pilgrimage in modern Greece. An ethnographic approach. Alexandria, Athens

Fleisher A (2000) The tourist behind the pilgrim in the Holy Land. Int J Hosp Management, 19(3), 311–326. https://doi.org/10.1016/S0278-4319(00)00022-0

Gassiot A, Prats L, Coromina L (2015) Analysing accessible tourism in religious destinations: the case of Lourdes, France. Int J Relig Tour Pilgr 3(6). https://arrow.dit.ie/ijrtp/vol3/iss2/6
Griffin K, Raj R (2012) Reflecting on religious tourism and pilgrimage. ATLAS, Arnhem
Havard T (2017) Pilgrimage and tourism: Convergence and divergence. Int J Relig Tour Pilgr 5(2):1–14.
Kaelber L (2006) The sociology of medieval pilgrimage: contested views and shifting boundaries. In: Kaelber L (ed), From medieval pilgrimage to religious tourism: the social and cultural economics of piety. Praeger, pp 41–66
Kasim A (2011) Balancing tourism and religious experience: understanding devotees' perspectives on Thaipusam in Batu Caves, Selangor, Malaysia. J Hosp Market Manag 20(3–4):441–456. https://doi.org/10.1080/19368623.2011.562437
Kumar R (2016) Travelling through caste. Int J Relig Tour Pilgr 4(11). https://arrow.dit.ie/ijrtp/vol4/iss6/11
Lagos D, Poulaki P (2014) Religious monuments and tourism: the case of Chios. Sustain Reg Dev Cult Herit 1-A
Lopez L (2013) How long does the pilgrimage tourism experience to Santiago de Compostela last? Int J Relig Tour Pilgr 1(2). https://arrow.dit.ie/ijrtp/vol1/iss1/2
Lucarno G (2016) The Camino de Santiago de Compostela (Spain) and The Via Francigena (Italy): a comparison between two important historic pilgrimage routes in Europe. Int J Relig Tour Pilgr 4(7). https://arrow.dit.ie/ijrtp/vol4/iss7/7
Metreveli M, Timothy DJ (2010) Effects of the August 2008 war in Georgia on tourism and its resources. Tour Geogr 12(4):503–520. https://doi.org/10.1080/14616688.2010.520776
Ministry of Culture (2015) Annual report on cultural heritage and tourism. Ministry of Culture Publications.
Munro D (2017) Historical perspectives of shifting motives for faith-based travel. Int J Relig Tour Pilgr 5(5). https://arrow.dit.ie/ijrtp/vol5/iss2/5
Polyzos S (2010) Religious tourism in Greece: spatial analysis and contribution to the development of disadvantaged areas. Res Ser 16(9):203–222. University of Thessaly
Poulaki P (2013) Prospects for the development of religious tourism in the Prefecture of Chios. Master's thesis, University of the Aegean, Chios, Greece. http://hdl.handle.net/11610/11468
Poulaki P (2016) The Ottoman monuments of Chios. Proc "Philip Argenti" Int Working Grp Study Intercult Osmosis East Mediterr Black Sea Reg Mid East Forum, pp 95–121
Poulaki P (2018) The development of religious-pilgrimage tourism in the islands of the North Aegean. Doctoral dissertation, University of the Aegean, Chios, Greece. http://hdl.handle.net/10442/hedi/43317
Poulaki P, Lagos D, Balomenou C (2015) Religious tourism in Greece and regional development: the case of Samos Island. In: Proc 55th ERSA Congress, World Renaissance: Changing Roles for People Places, Lisbon, Portugal, 25–29 Aug 2015
Poulaki P, Kritikos A, Vasilakis N (2021a) Digital technology as a means of promoting cultural tourism in Greece – case study: Acropolis Museum. J Sustain Dev Cult Tradit 1a/2021. https://doi.org/10.26341/issn.2241-4002-2021-1a-1
Poulaki P, Kritikos A, Vasilakis N, Valeri M (2021b) The contribution of female creativity to the development of gastronomic tourism in Greece: the case of the island of Naxos in the South Aegean. New Gov Manag Tour Destin:13. https://doi.org/10.4018/978-1-6684-3889-3.ch015
Poulaki P, Lagos D, Lambrou P (2021c) The impact of the Covid-19 pandemic on the tourism industry. In: Rezaei N (ed) Covid-19. Springer, Tehran University of Medical Sciences, Iran
Poulaki P, Lagou M, Valeri M (2021d) The contribution of female entrepreneurship to the development of agritourism in Greece. In: Valeri M, Katsoni V (eds) Gender Tour Chall Oppor. Emerald Publ, Niccolo Cusano Univ, Rome, Italy, Univ West Attica, Greece
Poulaki P, Stavrakakis I, Tarazonas D, Vasilakis N, Valeri M (2021e) Sustainable development and cultural heritage in Greece. In: Valeri M, Scuttari A, Pechlaner H (eds) Resilience and sustainability: global dynamics and local actions. Giappichelli Publ

Poulaki P, Stavrakakis I, Tarazonas D, Vasilakis N, Valeri M (2022) Crisis management in tourism: Covid-19 pandemic in Greece. In: Valeri M (ed) Tour risk crisis recovery manag. Emerald Publ Ltd. ISBN: 9781801177092

Rachiotis T, Poulaki P (2022) The contribution of cultural routes to the enhancement of urban cultural tourism. J Hosp Tour 20(2):31–46. https://doi.org/10.5281/zenodo.8322203

Rachiotis T, Poulaki P (2024a) Cultural routes as a factor of sustainable management of cultural reserve and development of cultural tourism. Sustain Dev Cult Tradit 4(A):50–72. https://doi.org/10.26341/issn.2241-4002-2024-4a-4-T02072

Rachiotis T, Poulaki P (2024b) Exploring the sustainability and management of overtourism in globally recognized destinations. J Reg Socio-Econ Issues 14(2):37–45. https://doi.org/10.26215/heal.3n9m-8v15

Rachiotis T, Poulaki P (2024c) Innovating cultural tourism in Greece: the strategic role of tourist guides in heritage promotion. J Reg Socio-Econ Issues 14(2):14–28. https://doi.org/10.26215/heal.3n9m-8v15

Rachiotis T, Poulaki P (2024d) Management of cultural routes as the new status quo in urban cultural tourism. In: Cultural tourism in urban areas. Taylor Francis, pp 165–181. https://doi.org/10.4324/9781032633374-14

Raj R, Bozonelos D (2015) Pilgrimage experience and consumption of travel to the city of Makkah for Hajj ritual. Int J Relig Tour Pilgr 3(6). https://arrow.dit.ie/ijrtp/vol3/iss1/6

Raj R, Morpeth N (2007a) Religious– tourism and pilgrimage management: an international perspective. CABI Publ, Wallingford, UK

Raj R, Morpeth N (2007b) Introduction: establishing linkages between religious travel and tourism. In: Raj R, Morpeth N (eds) Religious tourism and pilgrimage management: an international perspective. CABI Publ, Wallingford, UK, pp 1–14

Raptis N (2011) Religious festivals and tourism development in Northern Greece. Tourismos: Int Multidiscip J Tour 6(2):339–357

Rinschede G (1992) Forms of religious tourism. Ann Tour Res 19(1):51–67. https://doi.org/10.1016/0160-7383(92)90106-Y

Ron AS (2009) Managing sacred sites: service provision and visitor experiences. Tour Manag 30(5):703–712. https://doi.org/10.1016/j.tourman.2008.10.024

Santos XM (2002) Pilgrimage and tourism at Santiago de Compostela. Tour Recreat Res 27(2):41–50. https://doi.org/10.1080/02508281.2002.11081219

Shinde K (2012) Policy, planning, and management for religious tourism in Indian pilgrimage sites. Coll Archit Bharati Vidyapeeth Deemed Univ Pune India

Sfakianakis G (2000) Religious tourism in Greece: the case of Tinos. Anatolia 11(1):23–34. https://doi.org/10.1080/13032917.2000.9686960

Stausberg M (2011) Religion and tourism: crossroads, destinations, and encounters. Routledge, London/New York

Timothy DJ, Olsen DH (2006) Tourism, religion and spiritual journeys. Routledge. https://doi.org/10.4324/9780203001073

Varvounis MG (1995) Greek religious customs and traditions. Ekdotiki Athinon.

Vukonic B (2002) Religion, tourism and economics: a convenient symbiosis. Tour Recreat Res 27(2):59–64. https://doi.org/10.1080/02508281.2002.11081221

Yeoman I (2008) Tomorrow's tourist: scenarios & trends. Elsevier

Sustainable Development Practices in Tourism and Hospitality Industry: A Case Study of Garhwal Himalayan Region of Uttarakhand, India

Madhu Kumari, Sandeep Guleria, and Suneel Kumar

Abstract Uttarakhand's Garhwal Himalayan area, known for its peaks, vistas, and cultural history, has emerged as a vital tourism and hospitality engine, propelling economic and job growth. However, the region has severe problems, including environmental deterioration, pollution, poor infrastructure, and resource depletion, which are exacerbated by climate change and natural catastrophes. To maintain long-term tourism development, it is critical to implement and improve corporate social responsibility (CSR) strategies that prioritize environmental conservation, resource management, and community engagement. The purpose of this study is to assess the effectiveness of these practices and their role in achieving sustainable, inclusive development that is consistent with the Sustainable Development Goals (SDGs).

1 Introduction

"Nature-based tourism" is a tourism that is "mainly focused on the direct experience of some relatively undisturbed natural occurrence." (Valentine 1992). It is a type of tourism that is reliant on the outdoors and natural environments (Hall and Boyd 2005), and it includes pursuits like either avitourism or birdwatching, which are

M. Kumari
Department of Tourism and Hospitality Management, Chandigarh University, Mohali, Punjab, India

S. Guleria
Department of Tourism, Chandigarh University, Mohali, Punjab, India
e-mail: hod.uithm.tm@cumail.in

S. Kumar (✉)
Department of Commerce, Shaheed Bhagat Singh College, University of Delhi, New Delhi, India

M. Valeri (ed.), *Contemporary Religious Tourism*, Tourism, Hospitality & Event Management, https://doi.org/10.1007/978-3-031-79074-4_12

crucial to the overall economic development of nations, particularly rural ones. Avitourism, as outlined by Nicolaides (2013), is a form of economic activity designed to support a country's long-term economic and social growth while also promoting environmental protection and preserving cultural heritage. It contributes to the country's economic and social growth as well as the protection of the environment and its cultural legacy. This approach emphasizes the active involvement of local communities and other stakeholders. Activities such as kayaking and hiking further enhance this model by offering opportunities for individuals to connect with nature, relieve the stresses of everyday life, and engage with natural landscapes and wildlife (Blamey 2001; Deb et al. 2022). Participating in such activities has grown to be a significant part of both local and international tourism (TIA 2002; TIES 2003). Both adventure and nature tourism have emerged as two of the tourism industry's fastest-growing subsectors during the past 20 years. The sector that appears to be developing the fastest is nature tourism, with a growth rate of 10% to 30% every year. Currently, it holds a 20% market share for international travel (Christ et al. 2003; UNWTO 2005).

Many different flora and wildlife species have called mountains home. Mountains are frequently regarded as the lifeblood of the regions through which perennial rivers run because they are the principal sources of many of these rivers (Whiteman 1988). The Himalayan mountain range is one of the most prominent mountain systems in the world. It is recognized for being the world's highest, youngest, and most environmentally sensitive mountain range. 20% of the world's population, or 1.2 billion people, live in mountains, which occupy one-fifth of the planet's surface. There is a lot of worry about how ecosystem services can be extracted and distributed. In comparison, the population in mountainous places is less advanced and developed. Due to shoddy construction in the alpine areas, sustainability challenges in these places have recently attracted increased attention (Sati 2005).

In November 2000, Uttar Pradesh was divided into Uttarakhand, India's 27th state, commonly known as "DevBhoomi" or the "Land of Gods" and created from inside it. Its fortunate location at the base of the Himalayan mountain ranges and its hilly topography make it a country of potential. A total of 53,483 square kilometers make up Uttarakhand, of which about 86% are mountainous and 70% are covered in forest. It shares a border with Nepal and Tibet and makes up part of the country's northern border. It is bordered to the north by Tibet, to the west and north-west by Himachal Pradesh, to the south by the Gangetic plains of Uttar Pradesh, and to the east by Nepal. It extends from the southern foothills to the northern, snow-covered mountains. The state is rich in natural resources, with plenty of water and forest reserves, as well as several glaciers, rivers, and lush woods, all situated in a peaceful and unpolluted setting.

Since the state has many potential for leisure, natural, and religious/spiritual tourism, as well as limitless opportunities in the tourism-related services, it is apparent that the state's economic plan is focused on the development of the tourist industry. The majestic mountains are the sacred home of the Char-dhams, which include the four highly revered Hindu temples: Badrinath, Kedarnath, Gangotri, and Yamunotri. Uttarakhand is a popular tourist destination due to the availability of numerous hill resorts, wildlife reserves, places of worship, and trekking routes. A total of 105,880 tourists visited the state in 2015–16, up from 101,970 in 2014–15. A state in the Himalayas, Uttarakhand, welcomes millions of visitors during the peak travel season. However, due to its limited industries, a considerable amount of the money spent by tourists contributes to the region's economic growth and opens up opportunities for development. Various businesses word to meet the various demands of travelers. The topography of the state gives nature tourists the chance to engage in a variety of nature-focused tourist activities.

Objectives in the present study are as follows:

1. To study the recent development in the tourism and hospitality industry for establishing green practices for sustainable development of the hotel industry in Himalaya Garhwal.
2. To address the issues of adoption of green and sustainable practices in the tourism and hospitality.
3. To determine the influencing factors toward the implementation of sustainable development practices in tourism and hospitality industry.

2 Review of Literature

Uttarakhand, especially the Garhwal area, is very popular with tourists from a wide range of backgrounds, including history, art, and culture, trekking, climbing, pilgrimage, and outdoor enthusiasts. In addition, the area provides conference and convention tourism, adventure, ecotourism, wildlife, historical, and rural tourism. There is room for expansion for travel agencies in Haridwar, Rishikesh, and Dehradun since they are essential in organizing both leisure and business travel. But the tourist sector needs skilled workers and creative product creation. India's economy depends heavily on tourism, thus Garhwal needs to compete and get visitors. But expansion is hampered by problems including poor pay, frequent employee turnover, informal employment, and a lack of skilled workers. The global tourist industry is competitive and dynamic, necessitating flexibility in response to changing circumstances.

Author & Year	Findings
Liu et al. (1987)	In a study on how locals perceived the effects of tourism, it was found that locals experienced these effects on an equal basis regardless of where they lived geographically or how developed the industry. Their findings revealed that residents of areas with highly developed tourism industries are very conscious of both the good and negative environmental effects of tourism
Johnson et al. (1994)	The author claims that locals had great hopes for the socioeconomic advancement that would come from more tourism.
Nicolaides (2008)	Emphasized how important providing top-notch service is to keeping tourism alive.

Author & Year	Findings
Ernoul (2009)	Observed the integration of locals' worries and apprehensions toward the new type of tourism development in their area.
Singh et al. (2009)	In the Bhyundar Valley of Uttarakhand, they investigated how the locals felt about mountain tourism.
Nicolaides and Sumbana (2011)	According to researcher without effective leadership, any initiative is bound to failure.
Sdrali et al. (2015)	They emphasized the crucial role of local communities in fostering tourism and highlighted the opportunities for residents to actively engage in the development process.
Nicolaides (2015)	All stakeholders must be properly taken into account and, to the greatest extent feasible, their needs in terms of participation met. The friendliness, expansion, and sustainability of tourism are influenced by the views of the local population.
Nicolaides (2015)	In their study they recommend including locals in tourism-related decisions, and Nicolaides (2012) highlights the need of providing good service in regional eateries to draw in customers.
Bagri and Kala (2016)	Examined locals' perceptions of the effects of tourism in the Koti-Kanasar, Indroli, and Pattyur circuits in Uttarakhand, as well as the influence of demographic factors on those perceptions. They found that the locals were aware of the whole range of the economic, social, and environmental effects of tourism in their hilly region.
Grobler and Nicolaides (2016)	The study found that ecodevelopment committees were used to involve the local population in promoting ecotourism and ensuring regional sustainability in order to combat the detrimental effects of tourism. In a similar vein, it is crucial to take into account rules of behavior in tourism operations so that stakeholders can be controlled.
Gupta and Rout (2016)	In order to emphasize the community's potential and the significance of their participation in the development process, the researchers in their study used a value chain approach. In order to secure ethical tourism growth in mountainous natural areas, they emphasized the significance of involving stakeholders and the necessity of taking their opinions into consideration.
Nicolaides and Tassiopoulos (2017)	In order to promote community-based ecotourism in the Uttarakhand region of Jaunsar Bawar.
Ashish Nag (2018)	A study on poverty and environmental issues highlights the challenges faced by poor villages in LDCs/DCs, leading to deforestation and overlooking environmental protection. Ecotourism, a method promoting environmental protection and money production, is gaining popularity among environmentalists.
Joveriya and Mariya (2019)	The study explores tourism as a vital service sector, generating national income and job opportunities, and promoting cultural, economic, and social growth. It highlights its development potential, especially in developing countries like India, which boasts natural beauty and historical sites.

Source: Author's own compilation

The study by Liu et al. (1987) found that locals experienced the effects of tourism equally regardless of their geographical location or industry development. Residents in well-developed tourism locations are aware of both beneficial and negative environmental effects. Johnson et al. (1994) discovered that there were high expectations for the socioeconomic benefits of tourism development. Similarly, concluded that locals' contentment was influenced by their perceptions of tourism's effects, which provides useful insights for future growth planning. Nicolaides (2008) emphasized the importance of high-quality service for sustaining tourism. Ernoul (2009) observed locals' concerns and apprehensions toward tourism development. Recommended including locals in tourism-related decisions and providing good service in regional eateries. Discovered that inhabitants were highly aware of the effects of tourism on their hilly terrain. Grobler and Nicolaides (2016) found that ecodevelopment committees effectively engaged local communities in promoting ecotourism and guaranteeing regional sustainability. Gupta and Rout (2016) used a value chain method to highlight the community's potential and the crucial importance of their participation in the development process.

3 Research Methodology

The research aims to explore how local stakeholders, including inhabitants and tourist service providers, perceive the benefits of sustainable development in tourism and hospitality in the study region. In Uttarakhand's Bhilangana Valley, the research collected primary data from communities and tourism service providers. There are numerous river systems in the Garhwal Himalaya. The Bhilangana valley, located in Tehri Garhwal's Uttarakhand district, is a beautiful and uninhabited area that drains the Gangotri river.

The Ghuttu, Ghansali, Akhori tourism circuit in the Bhilangana Valley was the site of this qualitative study. The researcher used the qualitative technique in order to get more in-depth information from the participants on the connection between sustainable development practices and its function in the area's social change. Qualitative approaches are useful for examining the contextual component of complicated issues in real-life situations using a variety of data sources and when the target group is easily accessible in their home. A set of interview questions has been created as a result of the researcher's several visits to the study area, rounds of conversations, and meetings with the local tourist stakeholders. Purposive and snowball sampling were used to choose research participants. The researcher first spoke with one villager, after which he or she suggested the next, and so on. To capture a variety of perspectives on the growth of Tourism and Hospitality in the area, an attempt has been made to engage participants of all ages, genders, professions, and economic levels. Based on the intended goal and information gathered from in-depth interviews, the researcher is able to identify the different variables in the Garhwal region's of Uttarakhand social change in total: 41 in-depth from the Tourist and Hospitality potential of the area.

4 Discussion of the Study

The study evaluated the adoption of sustainable development in the tourism and hospitality sectors of the Himalayan Garhwal region by examining local stakeholders' perspectives on nature tourism and its consequences. It identified problems such as the replacement of traditional eco-friendly architecture with dangerous, poorly designed structures, insufficient road infrastructure, insufficient waste management, air pollution, watershed degradation, and the depletion of natural resources, biodiversity, and ecosystem services. The study also looked at essential features of sustainable tourism, which arose from the notion of Sustainable Development, which was presented some 15 years earlier. Sustainable tourism seeks to integrate the tourism industry with broader sustainable development goals by providing high-quality, low-impact experiences built on four pillars: economic, ecological, cultural, and local sustainability.

4.1 The Region's Attractions, Which Have Enormous Potential for Tourism

When asked whether they believed that the area had enormous potential for tourism, the respondents said they did. One responder from Ghana stated that the majority of respondents agreed with this assertion. The researchers were informed by a resident of Akhori village that during the 2013 Kedarnath disaster, some people chose to take an alternate mountainous route in order to survive the floods and reach Akhori village. The valley is home to several significant tourism-specific locations, including the Buddha Kedar, the Panwali Kantha, the Khatling Glacier, and more. These locations have the potential to draw large numbers of tourists interested in nature, adventure, and pilgrimage tourism. He recommended that a different path from the Akhori hamlet to the Kedarnath shrine be investigated in order to enhance the number of tourists visiting the valley (Table 1).

4.2 The Distinct Garhwali Culture Prevalent in the Area

There is nothing else in the nation like the distinctive Garhwali culture found in this region. When asked about the distinctiveness of their culture, an elderly respondent from Akhori village said that the region's food, music, dress styles, and dance forms

Table 1 SWOT analysis of the research areas of Garhwal Himalayan region of Uttarakhand in the Tourism and Hospitality Industry

Strengths	Weakness
• The existence of snow-capped summits, rivers, meadows, mountains, and glaciers	• Absence of knowledgeable and experienced tour guides with formal training
• Ideal locations to fully appreciate the splendor and grandeur of the Himalayan mountain ranges	• Advertising in the vicinity as a surprise
• There is an abundance of pathways available for hiking	• Low-quality design in the construction of tourism facilities
• An ideal environment for nature tourism is provided by the existence of natural landscapes	• Inadequate agency-to-agency cooperation
• Natural Beauty of Garhwal	• Lack of effective public transportation
• Potential for pilgrimage and health tourism	• Insufficient community participation in the tourist industry
• Situated amidst the majestic Himalayas	• Unequal allocation of tourism advantages to the surrounding community

Opportunities	Threats
• Possibility of promoting natural tourism in the research area at different locations	• Inclined to natural calamities, breaching of carrying capacity
• Possibility of wildlife tourism because of the profusion of flora and fauna	• The persistence of inadequate road upkeep and insufficient visitor services deters the entry of quality tourists
• Possibilities for adventure tourism because the area has a large number of hiking destinations	• Haphazard growth of infrastructure
• It is possible to use the unique culture of the area to draw tourists from rural and village areas	• "Environmental Pollution" refers to the harmful effects of human activities on the environment
• Opportunities for second homes and holiday houses to be built in different parts of the region	• Economic disparity

Source: Author's own compilation

are all unique and can only be experienced by traveling to the valley and engaging with the locals. However, over time, the culture of the region is gradually disappearing due to people's disinterest and migration from the area to larger cities like Delhi and Dehradun in search of a better quality of life. The majority of respondents said that they would like to get involved in reviving the lost culture of the region through the promotion and development of sustainability, especially after learning from the researchers that their unique culture may be in demand with the growth of sustainable development practices in the area.

4.3 Enhancement of the Local Population's Standard of Living

During their research travels to the valley, researchers discovered that women, schoolchildren, and the elderly (mainly retired) make up the majority of the village's population. There were fewer adult males in the village since many had left their hometowns to seek jobs in larger cities. One responder from Ghuttu Village stated, "After completing their education, most people move to big cities in search of jobs and later settle there," referring to the dearth of employment possibilities in the area. Another participant noted that the growth of tourism may potentially create job possibilities within the area. When no one leaves their community and jobs from tourism improves their standard of living, the issue of outmigration will be resolved. Pastoralism and agriculture were the region's primary sources of revenue. For the locals, tourism may be an additional source of revenue.

4.4 Enhancement of the Development of Infrastructure

The study area is a rural location with modest services and infrastructure accessible to both locals and visitors. According to a primary survey, most of the valley's villages have access to basic medical care, a primary school, and roads. While there were not many hotels or dining establishments in the villages of Ghuttu and Akhori, Ghansali Town had enough of both. One of the responders from Ghansali highlighted the need for improved shopping facilities in the area. The majority of respondents said that as sustainable development practices are developed in the area, an increase in tourism would bring in more visitors, which would raise the need for greater amenities and ultimately result in an upgrade of the region's current infrastructure. For example, improved transportation, lodging, and dining options enhanced security and safety, round-the-clock electricity and water supplies, and superior medical facilities.

4.5 A Way for the Locals to Generate Revenue

Currently, pastoralism and agriculture are the region's primary sources of revenue. The tourism industry is attracting young people in the study region who would like to work as guides and hiking guides for visitors visiting the valley. One responder from Ghuttu village said that most visitors from Delhi and Dehradun arrive with large travel agencies and hardly engage with the locals. Additionally, locals lack any trekking-related training that would be helpful to visitors. The majority of occupations that locals do are menial ones like giving tourists meals or lugging their bags. However, I have observed the changes in the region when traveling to Kedarnath. The area as a whole will gain if some of the visitors that come to Kedarnath also come here, as tourism would provide new opportunities for employment and infrastructure. The government should take steps to promote the valley's distinctive culture and tourist attractions, which would encourage travelers to visit the area and the Char Dham.

4.6 Raising Awareness of the Environment

The research region is home to a diverse range of plants and animals, exhibiting remarkable biodiversity. One of the hotel owners in Ghuttu stated that 'The meadows of Panwali Kantha during the rainy season are filled with colorful flowers and plants species which cannot be seen anywhere in the world'. Also people from outside who do not have much knowledge about the place set up camps at high altitudes and leave behind litter which harms the natural environment. The involvement of local people in tourism development and decision making would ensure that the effects of tourism do not harm the natural environment, and the money earned from tourism would be used for its preservation and conservation. Since valley is their home, local people will never encourage tourism at the cost of harming the environment.

4.7 Approaches for Sustainable Tourism

Both the government and non-governmental organizations have recognized the importance of promoting environmentally friendly tourism given the volume of adventure sports and tourist activity in the area. It is indeed exceedingly difficult to change the popular, inefficient, and exploitative form of tourism. It is not entirely doable, though. The government tries to promote environmentally friendly tourism by disseminating information through its authorities. Locals' individual attempts to go green are encouraged and have had great success. To promote regional development, the Uttar Pradesh government founded the Garhwal and Kumaon Mandal Vikas Nigams (GMVNs), which are currently overseen by the Uttaranchal government. Experts advise that GMVN and KMVN should think about privatization and more institutionalization in order to actually foster regional growth. With this change, they would be able to stop concentrating on tourism-related businesses that are motivated by profit and instead monitor and improve tourism. The Department of Tourism launched a project in 1996–1997 with the goal of "Developing and Beautifying" the pilgrimage path between Badrinath and Kedarnath in order to draw in more tourists. Now this project has come to an end, reforms that were created in partnership with the community and local institutions and were motivated by local needs be put into place. Three methods for safeguarding ecosystems along climbing paths were suggested by Colonel Balwant Sandhu. In order to identify acceptable forms of mobility, such as bicycling, hiking, and driving, one tactic is to develop a national approach to parks, sanctuaries, mountain resorts, and recreational places. Limiting the size of wilderness areas to sustainable levels is another tactic. Recommended a two-pronged approach to tourism sector development, consisting of creating infrastructure in a variety of tourist locations and supporting traditional crafts and regionally significant economic activity. This should highlight the use of family labor, especially that of women, and make use of local manufacturing technologies. In order to support tourist-related activities while empowering women, tourism planning should also take into account job possibilities, local production capabilities, marketing potential for locally produced items, access to healthcare and education, financial support systems, essential technologies, and credit facilities. Local inhabitants have been marginalized throughout time as a result of the exploitation of their lands and the wildlife that surrounds them, which is sometimes seen as a barrier to development.

Globally, there is recognition of the connection between tourism and the environment. In light of the latest evaluation conducted by the UNEP/IE Industry and Environment Office, the environment is the resource most essential to the growth of tourism. Tourism is a sensitive business to the environment, and as a result, its growth is reliant on the state of the environment. As soon as adverse environmental effects make traveling less enjoyable, tourism will stop growing. "Ecotourism" is the term used to describe tourism that protects the natural and social environment. Even within the travel and tourism sector itself, the exact definition of this phrase is still unclear. Ecotourism differs from regular tourism, though, in that it seeks to lessen the adverse impacts of tourism on the environment while simultaneously promoting the interests of local populations and the environment.

Mass tourism's negative impacts on the local economy, ecology, and society are becoming more widely recognized nowadays. With this growing knowledge, the ideas of sustainable tourism gained acceptance:

4.8 Approaches for Sustainable Trends in Hospitality

The Tourism Industry is booming, bringing in large numbers of tourists and making good money. However, this is making it more difficult for us to exploit non-renewable resources and straining the ecology. Providing for both the present and the future is a prerequisite for sustainable growth. The hotel sector started adopting and pursuing environmentally friendly practices in 1990 as a response to the unstable economy and the sector's emphasis on customer satisfaction and service. This is the main reason eco-friendly techniques are becoming more popular, which also explains why visitors want hotels to implement them. Through proactive measures to mitigate resource depletion, the hotel sector has the potential to yield substantial environmental benefits. The number of hotels implementing these strategies has increased recently, and both hotel staff and guests are eager to fully embrace them. The contemporary generation is fully aware of the harm that regular hotel and business activities cause to the environment. The trend of hotels implementing eco-friendly practices is rather noticeable. It is said that the natural world can be successfully conserved and preserved through ecologically appropriate practices. Every year, millions of tourists travel to the Dehradun District (Rishikesh, Mussoorie), most of them for vacation. For this reason, it's typical to witness people cutting down trees, constructing hotels, and installing HVAC systems. Human activity has an impact on the environment worldwide. Along with destroying the environment, they also contribute to contamination of the air, water, soil, and noise. According to estimates from the Indian Ministry of Environment and Forests, a 150-room hotel consumes the same amount of energy in a week as 100 typical houses do in a year. The greatest method to decrease the negative effects of tourism is to adopt eco-friendly practices including reducing, reusing, and recycling as well as solid waste management, community awareness, and energy and water conservation. In the Dehradun District, though, very few hotels have completed this change as of yet.

India is a fantastic travel destination because of its diverse range of landscapes, extensive cultural heritage, and specialized tourism offers. Amazing India has been a major contributor to the growth of the labor force and the tourism sector. It is anticipated that the current 0.4 million Indians who cruise would treble to 4 million. By 2020, the economic impact of the cruise sector is expected to increase from its current $110 million to $5.5 billion. The construction of multi-use structures, such as hotels, resorts, and restaurants, as well as transportation infrastructure like highways, railroads, and shipping ports, is frequently spurred by the influx of tourists. In terms of foreign visitor arrivals, Tourism Trade Conference (WTTC) is ranked twelfth in the world (2020). In the Travel & Tourism Development Index published by the World Economic Forum, India is ranked number 54 as of 2021. By 2030, it is anticipated that business travel from India will rank among the top five worldwide. Under the automated procedure, 100% FDI is allowed in the tourism sector. Investment from foreign direct companies is welcome to be used entirely in the development of tourism infrastructure, such as hotels and theme parks.

4.9 Green Hotel Practices and Environmental Sustainability in the Hospitality Industry

The hotel industry is one of several tourism-related factors that seriously endanger the environment. Hotels produce a lot of solid and raw waste due to their high use of non-durable commodities, energy, and water. If these resources are mishandled, both the environment and a company's financial performance will suffer. A number of factors, including the growing environmental consciousness of visitors, the need to reduce the industry's negative environmental impact, the desire to project a positive image, and the possibility of financial reward, have pushed the hotel industry to become more environmentally friendly in recent years. Thus, in an effort to reduce their environmental impact and address customers' concerns about the condition of the world, a lot of hotels have begun implementing innovative ways to improve the "greenness" of their operations. Environmental concerns with visitors are growing. In order to draw in this expanding customer base, hotels are modifying their operations to become greener. In order for an activity in the hospitality industry to be considered ecologically sustainable, it must achieve a delicate equilibrium with the surrounding environment, allowing for interaction without resulting in harm to either party. The hospitality industry should prioritize addressing environmental issues related to water and energy efficiency and preservation, climate change and greenhouse gas emissions, recycling and waste management, biodiversity and the preservation of natural resources, reducing environmental impacts, environmentally responsible sourcing, and developing green construction standards for new hotel construction. How to manage these issues has become a priority in recent years as environmental consciousness has increased across society, and among visitors in particular. Numerous studies have focused on the strong relationship between environmental sustainability and holiday experiences. Indicates that visitors who believe a hotel is ecologically conscious are more likely to have a positive experience, recommend it to others, and even pay more for it. Nowadays, customers anticipate that companies will contribute to the environment. Hotels that use this approach would, have a competitive edge. According to research by environmental sustainability positively and directly affects visitors' satisfaction with their stay and, consequently, the overall quality of their stay.

"Green practices" can refer to a wide range of activities and are frequently associated with different approaches in the hotel industry. It as "a value-added business approach that benefits a hospitality organization that invests in environmental preservation measures." defined "green" as "environmentally friendly, that is, operating business in a manner that avoids waste, conserves energy, and generally supports environmental health." state that "green practices" aim to reduce their impact on the environment by using renewable resources and energy sources, as well as minimizing waste. Among the "green hotel practices" are recycling and eco-friendly purchasing, to name just two. "Green hotels" are defined similarly: they are "eco-friendly establishments whose management is prepared to launch initiatives that conserve water, save electricity, and reduce solid waste—while saving money to help safeguard our only planet." "It is a less environmentally damaging property that has committed to several ecologically sound activities including conserving water, saving electricity, and minimizing solid waste." Although green practices in the hotel industry have been extensively researched, the majority of the studies identify three primary motivators: stakeholder interactions, financial incentives, and customer happiness. However, discovered that the main justifications for putting eco-label requirements into practice were their positive effects on the environment and the hotel's reputation. The implementation of environmentally friendly practices has primarily driven by cost savings. However, in recent times, the focus has switched to growing market share via fortifying relationships with important stakeholders. The adoption rate of green techniques in the hotel sector may be impacted by external business variables such as stakeholder pressure and environmental limitations.

The biggest hotel chains in the world are prioritizing sustainability and have included it into their strategy plans. As an example of this commitment, the Intercontinental Hotel Group states that "our rules and standards spell forth our stance on social, environmental, and ethical concerns." As our business grows, we understand that we must contribute to the achievement of the Sustainable Development Goals set out by the United Nations. We are certain that we can make the most difference on seven of these objectives by utilizing our scale and global reach. "Sustainability is a core part of our business strategy, and it shows in every aspect of Marriott International," the company stated. "At the same time, Hilton Worldwide explains the 2030 value chain goal: By 2030, we are determined to increase our investment in social impact and lower our environmental footprint to half via responsible hospitality throughout our value chain." "By creating and managing sustainable hotels in collaboration with our coworkers, hotel owners, franchisees, brands, suppliers, business partners, customers, and guests, we proactively reduce our environmental impact and potential business risks."

4.10 Green Tourism

One of the most important issues facing the travel industry in recent years has been sustainable tourism. A growing number of developing countries are utilizing tourism as a tool for economic development after discovering the potential financial benefits of the industry. When development happens too fast and the surrounding ecology is not given enough consideration, issues arise. Because of this, sustainability studies try to educate builders about issues to be aware of, especially those pertaining to the preservation of natural resources in the tourism industry. Any type of travel that is concerned with or supports sound environmental management, or "green," practices is referred to as "green tourism." This includes, but is not limited to, travel that highlights a place's natural environment and cultural heritage. Any type of travel that is concerned with or supports green or excellent environmental management practices is referred to as "green tourism," including but not limited to itineraries that highlight the region's natural resources and cultural legacy.

4.11 Development of Sustainable Tourism Based on a Four Pillar Concept

In actuality, there are four pillars supporting the rise of sustainable tourism: the community, ecological, cultural, and economic. Numerous suggestions have been presented for each pillar's expansion. Sustainable economy: lucrative in the long run as well as the short term. Form alliances with companies at every stage of the supply chain, from small, regional companies to large, international enterprises.

Make use of internationally acknowledged and evaluated training and certification standards. Urge clients to behave sustainably and morally. By developing a wide range of tourism activities, you may diversify the products. A percentage of the profits should be allocated to training, ethical marketing, and product development. Offer financial incentives to organizations that implement sustainability practices. Ecological sustainability is the preservation of fundamental ecological processes, biological variety, and biological resources while allowing for development. All levels of tourism should have codes of conduct developed. Establishing guidelines for the operation of the tourism industry, impact assessment, and monitoring of cumulative effects; creating national, regional, and local tourism policies and development strategies that are in keeping with the overarching goals of sustainable development; Establish baseline studies for environmental impact assessments. Ensure that sustainability concepts are included into facility design, planning, development, and operation. Ensure that tourism in protected areas, such as national parks, is incorporated into and subject to solid management plans. Keep track of and carry out research on the actual effects of tourism. Identify appropriate tourist conduct. Encourage responsible tourist practices. Cultural sustainability strengthens community identity, gives people more control over their lives, and is compatible with the cultures and beliefs of those affected. Establish education and training programs to improve and manage heritage and natural resources. Preserve cultural diversity, Respect traditional inhabitants' rights to their land and property. Ensure the protection of nature, local and indigenous cultures, and especially traditional knowledge. Actively collaborate with indigenous leaders and minority groups to ensure that indigenous peoples are respected. Foster, strengthen, and promote the community's capacity to use and uphold traditional skills. Informing travelers about appropriate and desirable conduct; Inform the tourism sector of the proper and proper conduct. That is intended to assist local communities and create or maintain income in such communities is known as local or community sustainability. Tourism should create high-quality employment for locals. The community should keep influence over tourism development. Provide financial incentives for local firms to enter the tourism industry. Encourage enterprises to limit their negative effects on local communities and make positive contributions to them. Ensure an equal distribution of financial advantages throughout the whole supply chain.

5 Findings and Conclusion

The Garhwal Himalayan region's Bhilangana Valley is blessed with the abundance of nature. Some of the country's most amazing natural tourism attractions are found in this region, making it a veritable gold mine of natural attractions. This area is home to uncommon indigenous plants and wildlife, making it a hotspot for biodiversity and further demonstrating the environmental significance of the area. The state and the participants in the tourism business were unable to properly use the benefits of the industry, despite the region's plethora of tourist attractions. The policy and decision makers' poor planning and development plans were to blame for this, which increased host population anger. Additionally compounding their troubles, the hilly area also faced issues with illiteracy, emigration, unemployment, and deforestation. The current study was carried out in an effort to identify long-term remedies for the issues facing the area. Data analysis conducted for the study revealed that the benefits of nature tourism in the valley are positively viewed by local stakeholders. As a sought-after and sustainable location for nature tourism, they expected the tourism policy and planners to actively develop and market the area. They understood that the area offered abundant opportunities for environmental tourism, including forests of oak trees, waterfalls, and verdant meadows. One of the most important parts of the Himalayan Mountain System, the Uttaranchal Himalaya is known for its charming and beautiful scenery. Its three principal vertical zones are the Greater Himalaya, Middle Himalaya, and Shivalik Ranges, or Lesser Himalaya. Tourism has been drawn to the sweeping and gorgeous areas for generations. The four dhams are the main draw, which helps to further promote tourism in this country. In summary, we can say that there is a lot of room for tourism growth in the Uttaranchal Himalaya. It has been noted that tourism has both positively and negatively impacted the state's economy and society because Uttaranchal's mainland is more susceptible to natural disasters and environmental degradation. This is because tourism activity has intensified these problems. The Uttaranchal Himalaya has a long history of tourism. But the introduction of modern tourism, brought about by the profound penetration of modern transportation, has had negative effects on the local economy, ecology, and society.

References

Bagri SC, Kala D (2016) Residents' attitudes toward tourism development and impacts in Koti-Kanasar, Indroli, Pattyur tourism circuit of Uttarakhand State, India

Christ C et al (2003) Tourism and biodiversity – mapping tourism's global footprint. Conservation International: Washington 2003. Available at http://www.unep.org/PDF/Tourism_and_biodiversity_report.pdf

Deb SK, Nafi S, Valeri M (2022) Promoting tourism business through digital marketing in the new normal era: a sustainable approach. Eur J Innov Manag. https://doi.org/10.1108/EJIM-04-2022-0218

Ernoul L (2009) Residents' perception of tourist development and the environment: a study from Morocco. Int J Sustain Dev World Ecol 16(4):228–233

Grobler S, Nicolaides A (2016) Sustainable tourism development for Clarens in the Free State Province of South Africa. Afr J Hosp Tour Leis 5(2):1–18

Gupta SK, Rout PC (2016) The value chain approach in community based ecotourism: a conceptual framework on sustainable mountain development in The Jaunsar Bawar Region of Uttarakhand. Amity Res J Tour Aviat Hosp 1(01)

Hall CM, Boyd S (2005) Nature-based tourism in peripheral areas. Development or disaster. Channel View Publications, Clevedon, p 280

Johnson JD, Snepenger DJ, Akis S (1994) Residents' perceptions of tourism development. Ann Tour Res 21(3):629–642

Joveriya, Mariya (2019) Problems and prospects of tourism industry in Uttarakhand. Int J Geogr Geol Environ 1(1):10–16
Ko DW, Stewart WP (2002) A structural equation model of residents' attitudes for tourism development. Tour Manag 23(5):521–530
Liu JC, Sheldon PJ, Var T (1987) Resident perception of the environmental impacts of tourism. Ann Tour Res 14(1):17–37
Nag A (2018) A study of tourism industry of Himachal Pradesh with special reference to ecotourism. Sumedha J Manag 7(1):85–106
Nicolaides A (2008) Service quality, empowerment and ethics in the South African hospitality and tourism industry and the road ahead using ISO9000/1. Unpublished PhD theses, University of Zululand, KwaZulu-Natal
Nicolaides A (2012) Quality provision in upmarket restaurants: a survey of diners in three restaurants in a Gauteng casino complex. Afr J Hosp Tour Leis 2(2)
Nicolaides A (2013) Promoting Avitourism as a special niche area of Ecotourism in South Africa. Afr J Hosp Tour Leis 2(3)
Nicolaides A (2015) Tourism Stakeholder Theory in practice: instrumental business grounds, fundamental normative demands or a descriptive application? Afr J Hosp Tour Leis 4(2):1–26. Available from http://www.ajhtl.com/archive.html
Nicolaides A, Sumbana H (2011) Ethical business practice in the South African Tourism Industry. Afr J Hosp Tour Leis 1(4)
Nicolaides A, Tassiopoulos D (2017) Effective codes of ethics in the events industry. Afr J Hosp Tour Leis 6(2)
Sati VP (2005) Natural resource conditions and economic development in the Uttaranchal Himalaya, India. J Mountain Sci 2(4):336–350
Sdrali D, Goussia-Rizou M, Kiourtidou P (2015) Residents' perception of tourism development as a vital step for participatory tourism plan: a research in a Greek protected area. Environ Dev Sustain 17(4):923–939
Singh RB, Mal S, Kala CP (2009) Community responses to mountain tourism: a case in Bhyundar Valley, Indian Himalaya. J Mountain Sci 6(4):394–404
TIA (2002) Travel Scope. Travel Industry Association of America
TIES (2003) Ecotourism Statistical Fact Sheet. The International Ecotourism Society
UNWTO: Tourism Highlights – (2005 Edition). Summary Information Brochure. UNWTO, Madrid 2005
Valentine P (1992) Review: nature-based tourism. In: Weiler T, Hall C (eds) Special interest tourism. Belhaven Press, London, pp 105–128
Whiteman PT (1988) Mountain agronomy in Ethiopia, Nepal and Pakistan. In: Allan NJR, Knapp GW, Stadel C (eds) Human impact on mountains, pp 57–82

A Heavenly Trail: The *Cammino Celeste* as a Non-place of Personal Religion for a Self-Healing Religious Walk Experience

Jean-Gabriel Bela and Peter Lah

Abstract Those who set out on pilgrimage, whether believers or not, embark on a personal spiritual quest. Basilicas, churches, and shrines they encounter along the way are not always the (spiritual) goal of their journey; they are as much a part of the landscape as nature, trails, and inns. Their journey is an expression of a personal religion decoupled from ecclesiastical and theological systems of meaning. In fact, the personal experience of pilgrimage, alone or in a group, separates one from all the institutions that structures one's life. One wants to find within oneself and in spiritual communion with nature the path to spiritual well-being. Religion no longer necessarily refers to a transcendent entity above the celestial vault. Henceforth, one of the characteristics of the religious tourism is therefore the bracketing of the Christian referent and the emergence of a subject who transcendentally communicates with the world around him.

1 Introduction

1.1 The Non-place of Religion in the Sphere of Tourism

When you open the website of the World Tourism Organization (UN Tourism), the term "religious tourism" is a non-place in the sense that it is non-existent. The spiritual component of tourism appears as a part of the great ensemble of cultural tourism, one of the goals of which, in this regard, is only the preservation of places of worship that are part of the landscape and of cultural heritage. It follows that the motivations of the religious tourist can be described in terms of:

> To learn, discover, experience, and consume the tangible and intangible cultural attractions/products in a tourism destination. These attractions/products relate to a set of distinctive material, intellectual, spiritual, and emotional features of a society that encompasses arts

J.-G. Bela (✉) · P. Lah
Faculty of Social Sciences, Pontifical Gregorian University, Rome, Italy
e-mail: jg.bela@unigre.it; lah@unigre.it

M. Valeri (ed.), *Contemporary Religious Tourism*, Tourism, Hospitality & Event Management, https://doi.org/10.1007/978-3-031-79074-4_13

> and architecture, historical and cultural heritage, culinary heritage, literature, music, creative industries, and the living cultures with their lifestyles, value systems, beliefs, and traditions (World Tourism Organization (UNWTO) 2012).

Religious tourism is therefore first and foremost cultural tourism. On the booking sites of hotels, asked to indicate the reason for the trip, one's choice is limited to "Leisure" or "Business." The religious and the spiritual would then fit into the "leisure" box. The problem with this understanding of tourism, namely presenting the religious and the spiritual only as consumer products, is that it negatively sanctions individuals' quest for meaning. This is therefore the first non-place of religious tourism. It is reduced to leisure as consumer product, in the lexicon of what Marc Augé calls *supermodernity*.

1.2 Non-places: Marc Augé and Michel De Certeau

Marc-Augé borrows the concept of "non-place" from jurisprudence. It refers to a court decision which, after investigations have been concluded, orders the cessation of proceedings. It is a kind of a legal non-being. This is the meaning which the concept assumes in Augé's anthropology. In his exposition of the non-place, he evokes at length Michel De Certeau (1980) who, before him, spoke summarily of non-places to designate the various modes of passage to which anthropological spaces are subject (Certeau and Giard 2010, pp. 158–159). The crossing of spaces by its practitioners, the proper names given to places, or specific instructions, transform a place into a non-place, into a place of passage, or even into "an absence of the place to itself" (Augé 1992, p. 108). There is therefore in Michel De Certeau a superposition of these two realities that are not opposed. The non-place is the place subverted or disguised by the wanderings of those who cross it.

Augé will broaden the concept and finally oppose place to non-place, by "introducing three constitutive data of anthropological reasoning: time, space, and the subject" (Abélès 1994, p. 194). In the current epoch that Augé calls *supermodernity*, these three categories are broken down, respectively, into the overabundance of events characterized by the acceleration of history, the spatial overabundance characterized by the shrinking of the planet, and the oversizing of the ego and of the individual characterized by the individualization of approaches and the revaluation of individual histories (Augé 1992, pp. 35–52). Succinctly, because of *supermodernity*, for Marc Augé, a non-place is a geographically identifiable space where nothing consistent happens and which does not make history. In fact, it is "a space that cannot be defined as identity, relational, or historical" (Augé 1992, p. 100). Unlike the anthropological place, which creates the organic social, the non-place creates solitary contractuality (Augé 1992, p. 119) and by this same fact becomes the place of the tourist (Debarbieux 1993, p. 90). A kind of liquefied and uneventful place where men and women who frequent them are passers-by rather than inhabitants. In doing so, Augé positioned himself as a precursor of the liquid modernity of Zygmunt

Bauman, who noted the entry of modernity into a liquid phase in which all social forms were less durable because of their ephemeral nature (Bauman 2007, p. 7). Bauman's modernity is the Augé's *supermodernity* which "imposes on individual consciousness very new experiences and trials of solitude, directly linked to the appearance and proliferation of non-places" (Augé 1992, p. 117): airports, highways, supermarkets, in short, these places of transit characterized by solitary contractuality.

This work consists of questioning religious tourism by opposing it to the religious pilgrimage, made on foot or by bike. The hypothesis is that, in a context where travelers move away from the conveniences of modernity in order to take a break from the frenzy that it produces or to heal themselves, they withdrew from the non-places to find other non-places. The aim here is to extend Marc Augé's anthropology of non-places to other spaces that are not necessarily produced by *supermodernity* but are redefined and acquire new meanings because the travelers who frequent them are products of this *supermodernity*. These spaces are the itineraries of religious pilgrimages, whose final objectives are often a place of worship or a sanctuary. Though Augé does not consider the latter as non-places, since they are historical spaces (Augé 1992, p. 76), De Certeau considers them as non-places, since they are crossed. It is therefore a question here of rethinking the space of the religious itinerary as a non-place (Augé) that leads the pilgrim to another non-place (De Certeau).

To do this, the *Cammino Celeste* route served as a field of study. The *Cammino Celeste*, officially inaugurated in 2006, is the longest route dedicated to the pilgrimage walks in the north-east of Italy, more precisely in the Friuli-Venezia-Giulia region "whose story can be heard in the silence of walking and in spaces between lines" (Bellavite et al. 2011, p. 12). Spanning from the Adriatic coast to the heart of the Julian Alps, the *Cammino Celeste* is a trail that connects two great sanctuaries of the region, the one of the island of Barbana and the one of Mount Lussari. The trail is marked by other Marian sanctuaries and Christian places of worship. Most of them are dedicated to the cult of Mary, Mother of Jesus. Over the course of ten days or so, the pilgrim (or hiker) covers 222 km between the sea and the mountain top, making stops at places of worship imbued with historical and religious significance such as the ancient basilica of Aquileia. In addition to its religious character, the guide of *Cammino Celeste* mentions at the end of each stage other things to visit (*cosa c'è da vedere?*), such as: "We warmly suggest that you visit the National Archaeological Museum" (Bellavite et al. 2011, p. 59).

2 The Methodological Approach

This study is essentially qualitative. In the choice of the people who participated to the survey, and in the means used to collect the data, the anthropological dimension was privileged. Lack of other data on the subject matter of study—the *Cammino Celeste*—and the circumstances of this study could only allow for a method that relies on a sample of volunteers (Paugam 2017, p. 84). These volunteers had to meet two important criteria: the first was that of having completed the *Cammino Celeste*. Pilgrims at Monte Lussari asked for the stamp marking the completion of the pilgrimage were approached with a request for a semi-structured interview. The second criterium consisted in the requirement that they must have walked for a minimum of seven days.

Another important element is that the group interviews during this research followed the rhythm of the arrivals of pilgrims who came in couples or in small groups. With few exceptions who were in a hurry, when approached with the request for interview, they graciously welcomed the idea of talking about their experience at the end of their journey. This was an opportunity for them to make a first point of the experience they had just lived. So, the data collected here are the very first impressions of pilgrims at the end of their journey. They were first collected between June and August 2023, in the period when the route sees the largest number of pilgrims. The initial analysis data has left several question marks in relation to the theme of religious tourism which required further interviews in June and July 2024, including the founders and promoters of the *Cammino Celeste*, namely the president of the association *"Iter Aquileiense,"* the manager of the *Cammino Celeste* trail, the president of the *"Circo Culturale Navarca,"* and the person responsible for the route across the topography.

This research consists of a total of 14 interviews. The following are the individuals and groups of individuals who participated in the interviews:

Pilgrims	Dates	Number of days of walking
A woman	13/06/2023	10 days
A man	18/06/2023	8 days
A man	18/06/2023	11 days
A woman	22/06/2023	11 days
A man + a woman (couple)	24/06/2023	10 days
A man	07/07/2023	10 days
Two women	10/07/2023	11 days
Five women + one man	21/07/2023	15 days
Two women	08/08/2023	9 days
A man + a woman (couple)	14/08/2023	13 days
A man	18/08/2023	7 days
A man + a woman (couple)	15/06/2024	10 days
Six men (Mountain Bike Friuli, MBF)	23/06/2024	7 days
Three women	12/07/2024	8 days
Three men	11/07/2024	Founders and promoters of the *Cammino Celeste*

30 pilgrims participated in this survey, including 17 women and 13 men, women having a greater tendency to do so in groups compared to men. Then, at the end of the investigation, the founders of the association *Cammino Celeste* were added. A

total of 33 people were interviewed, with ten interviews taking place in 2023, and four in 2024.

Apart from the semi-structured interviews, participant observation was also a valuable source of information because of the researchers' immersion in the life of the Lussari sanctuary.

3 An Anthropology of the Pilgrimage Walk

3.1 The Non-place and the Pilgrimage Walk, the "Delinquencies" That Produce a Personal Religion

Two elements intersect in the religious experience of the pilgrimage walk: the geographical space of the itinerary and the agents of the itinerary (both travelers and local residents). A continuous interaction of these elements determines the degree of religiosity of this path. The topography of the *Cammino celeste* follows the most important religious buildings and places of worship along the trail. Thus, the *religious* aspect marks the road that goes from the sea to the mountains of the Alps, independently of who travels it or how, the trail invariably passes by a church or a sanctuary. Whether one is a believer or not, and independently of personal motivations, the route obliges the traveler to be in contact with the religious element. It is the very idea of the topographical route to make it a journey of faith. The founders of this route are quite explicit about their initial intentions. They let it be understood that:

> It is a journey of faith born in 2006 in the lower Friuli. An idea that we liked to develop after our return from the Camino of Santiago in Spain. We said to ourselves "it can't end like this"… And it is a journey of faith because it connects the most important sanctuaries of the region (Barbana, Aquilea, Castel Monte, Prato di Resia, Rosa mistica di Cormons, Lussari) … It is called celestial because it goes from the sea to the mountains. We rise towards the sky. (Interview of 11/07/2024).

The spatial organization of the trail is therefore already a set of places that have religious character regardless of the functions they will fulfill for the pilgrim. They were not built for the itinerary, it is the itinerary that followed them, like landmarks, to mark the traveler's way. They are places of worship that belong to dioceses, and therefore to institutions that are geographically located and indefinitely limited in time. Pilgrims therefore have no right to it and are not even given priority. The double relationship of place, that is to say its origin, and of non-place (its modality of place of passage) is already present in the very discursive elaboration of the project (Certeau and Giard 2010, p. 158). It is therefore not surprising that most of these pilgrims find these churches closed during their various passages, even though the guide of the *Cammino Celeste* guaranteed them that they would be able to visit them, pray there, and/or enjoy their beauty.

The fact that the practitioners of the itinerary are unable to visit the churches that dot this a priori religious itinerary leads to disappointment, disillusionment, and occasionally indifference. "I was able to pray," says a pilgrim, "but unfortunately, more often than not, you find the churches closed" (interview of 13/06/2023). Another pilgrim says: "If I found a chapel in the early morning, I prayed there, but during the day… We must admit that there is a big problem, the churches are still closed" (interview of 18/06/2023). A year later, three female pilgrims recounted the same experience: "We are very disappointed to find the churches closed throughout the route" (interview of 12/07/2024). In fact, one of the disillusionments of the latter has roots in a belief in Marian presence in institutionally approved places. If they cannot meet Her in places dedicated to her worship and presence, they resign themselves to a Marian imagination as this pilgrim recounts: "I knew that it was an itinerary dedicated to the Virgin Mary… When I was little, I didn't feel it, but this time I felt it. She is a figure who accompanies you all along the road" (interview of 13/06/2023).

The other disillusionment is the impossibility for them to experience a moment of prayer, an experience of their religion far from their community, in places where they are known to no one. Not always being fervent Christian practitioners in their territorially defined community, or not having the possibility to exercise religion freely in all its practices, sometimes for reasons of sacramental discipline, the churches along the way are an opportunity for religious or sacramental practice far from the eyes of those who know and judge them. The president of the association *"Iter Aquileiense"* recounts an episode of one of his encounters while they were on their way to Santiago: "This is the case of this divorced gentleman, who now has another partner, and who could not have access to a certain number of 'sacramental services' in the Catholic Church, since the latter was still very reluctant to the idea of divorce." The church of the trail is therefore the means of freeing oneself from the constraints of the practices of one's community of origin, just as the path itself is the space that frees one from the rules of the church-as-institution. It becomes, through its geometric spatiality, the anthropological place that invents the non-place (Certeau and Giard 2010, p. 159), the place where the pilgrim can allow him- or herself to transgress, walking being by nature the act of "crossing" and therefore also of "transgressing"; in the wandering movement, there is always a certain "delinquency" (Certeau and Giard 2010, pp. 153–190). The disciplinary prohibitions imposed elsewhere are therefore no longer relevant. One can appropriate religion and tailor it to one's needs. Gone is the heavy specter of the institution and the constraining authority, the pilgrim experiences a personalization of one's religion. This is precisely what William James says: "personal religion is something more fundamental than theological systems and ecclesiastical institutions" (James 2001, p. 70).

In these sanctuaries and places of worship along the way, and in this case that of Mount Lussari, the final destination on the *Cammino Celeste*, those who attend services in the church and the ministers of religion who exercise functions there are unknown to each other. Both categories of participants operate in the depersonalized space of religious practices. The community present at the religious services is not territorially located and the participants do not know each other. They are passers-by. Identity liquefaction is instead moving to non-identity. The sanctuary lives and survives from the absence of identities. The church-institution liquefies in these places and loses its rigidity. Sacramental inclusion then becomes possible. One can allow oneself to make a confession or take communion, something one would not dare to do in the community of origin. The self-healing process that began on the way in the ephemeral but expiatory encounters with other pilgrims or with local residents is celebrated in every church or shrine that offers the opportunity to do so.

The closed churches on the trail thus are not good news for pilgrims. But the fact that these are closed also means that they exist not for passers-by but primarily for the inhabitants of the places where they are located. Here, the solid reality of a territory delimited and under the control of its identifiable inhabitants intersects with the liquid reality of the evanescence of the pilgrim who is not part of the human herd registered in this territory. "Having been victims of theft and acts of vandalism in the past, due to their geographical isolation in the countryside," says the president of the *Circolo Culturale Navarca,* "the Church has preferred to keep its buildings closed and only open them at the time of the religious functions that are held there." In doing so, the Church also participates in the postmodern transformations of religion, because it accelerates the autonomy of the religious sphere and gives people the right to choose their own religious context and the way in which they want to adhere to it (Rosito 2018a, p. 26). As Augé said about the singular character of Catholic practice in *supermodernity*: "Practitioners tend to practice in their own way" (1992, p. 51). Moreover, according to the statements of the socio-founders, "85% of pilgrims of the *Cammino Celeste* are not practicing it for religious reasons, only 15% of pilgrims venture on this route for reasons of faith." The following statements confirm this:

> I'm a believer in my own way, says a woman. I don't know if God exists…, Jesus, yes, he existed… in any case, it suits me well to believe that there is something else, I don't know if it's God (interview of 22/06/2023).

> Yes, I can say that my pilgrimage is a bit of a matter of faith in my own way, says another pilgrim, a way of getting in touch with God, but which is not Sunday Mass. Rather, it is to look at and admire creation. Immersing oneself in the beauty of nature is a different way of praying (interview of 07/07/2023).

> I'm very unreligious, "says the only man in a group of six. I believe in man, in the goodness of man… I believe that God is nature (interview of 21/07/2023).

> Even though I went to a church to pray during this pilgrimage, I don't go to church. I did it just because it's part of the itinerary (interview of 18/08/2023).

For the remaining 15% of believing and practicing pilgrims who cannot visit a church on their journey because of its unavailability, their inner space is their only religious space. The practice of the itinerary thus becomes a movement of alternation of two spaces: the reflective moment when the pilgrim is alone with his thoughts, refocusing on oneself, and immersed in personal reflections and meditations, and the extroverted moment when he can admire a church from the outside, or have his naturalistic experience of God. Religion is then relocated in the private and personal sphere, transforming the pilgrim into a kind of "argonaut of the spirit" who is now forced to navigate away from traditional religious institutions (Rosito 2018b, p. 27). This is a far cry from the collective and social eminence of the Durkheimian religion (1912), which is organized around a system of beliefs and practices relating to sacred things, and of which the perfect example is the Church (Durkheim 2014, pp. 95–96). Furthermore, when we consider the motivations of the pilgrims—such as those of the *Mountain Bike Friuli* who pray together an "Our Father" every evening to strengthen the bonds of their friendship, as one of the members stated (interview of 23/06/2024)—or the motivation of a couple who said, "We have been looking for a child for a long time, without success," their journey takes on a deeply personal dimension. "In fact, added the couple, this baby is adopted, and he arrived after our first pilgrimage. We were tired of the monotony of everyday life" (interview of 15/06/2024). Here, we are closer to the Weberian notion of religion (1922) which privileges the subjective experience of people, their goals and quests for meaning, but all always turned toward the here below (Weber and Kalinowski 2013, pp. 79–80).

In the end, whether for believers or for those who no longer have the Church as an institutional referent of their faith, the *Cammino Celeste* is an opportunity to individualize and personalize ways of believing in God or in something else transcendent, and also a means of healing. The laconic creeds in man or in nature operate in a resizing of God and religion and the institutions that assume them.

3.2 The Non-place as a Place of Therapeutic Encounter

Three elements are to be considered here: the place of the encounter, the actors of the encounter, and the type of encounter. The interweaving of these three factors makes pilgrim encounters cathartic events from which the actors take abundant advantage. As for the place of the encounter, the passage through the small and abandoned towns of the *Cammino Celeste* makes this route a particularly unattractive route for a tourism of leisure. It is an accumulation of solitudes that cover this itinerary: a solitude of places, a spatial solitude, a human solitude. "We find ourselves in the middle of the groves," says a pilgrim, "because it is an itinerary where we can go for days without meeting anyone; except for the one who is supposed to host you; and sometimes he even leaves you the key at the entrance of the house while recommending that you leave it in the same place when you leave." (interview of 18/06/2023). The pilgrims note with great sadness this absence of life in these places. A solitude of the place that also has its rough and wild side as described by this pilgrim, "It's a kind of Far West. It would be nice if we thought of activities that can make these abandoned places 'bloom again'. I am really disappointed, because in these conditions it is difficult to see the panorama, there are trees everywhere, only vegetation … However, in the past, we took good care of the groves" (interview of 13/06/2023). The pilgrim does not cross these places in a detached manner; rather, he questions the desolation of the place. He tries to make sense of this sore state of the space: it is not untouched wilderness; it is a place that once was cultivated and is now abandoned.

Such, then, is the spatial framework that hosts the pilgrim encounters. A space that, through its desolation, slows down time and reduces the rhythm of life. Outside of *supermodernity*, we travel slowly, on foot or by bike; we are not bombarded by the flow of information made possible by the new information and communication technologies. The direct consequence of this aloneness is that the residents are so isolated from each other that they are not able to give clear directions to these pilgrims who ask for them. "One thing about this route," says one of the ladies in a group of pilgrims, "is that no one knows how to give you directions, such as where to find food, drink or a bar." (interview of 23/08/2023). A clear description of the non-place: "it exists, and it does not house any organic society" (Augé 1992, p. 140).

At first glance, the sadness of this description would preclude any positive action in such a space. No! Beyond the desolation of the place, the encounters turn out to be cathartic experiences in every way. People on the trail, both pilgrims and local residents, experience a liberating and regenerating moment. All of this is facilitated by anonymity. Not an anonymity of the erasure or annihilation of the other, as is the case in cities where "more is built to isolate and protect than to connect and integrate" (François 2013, p. 75), but an anonymity of differential inclusion (Rosito 2018a, p. 75). The pilgrim then gives to local residents an opportunity to talk and to come out of their isolation. And the local resident, on his return, in the act of welcoming, gives the pilgrim the opportunity to meet a humanity other than his own in these deserted places. The pilgrim is thus this stranger who is not of the place and who is not concerned by the vicissitudes of the place. With him, we take the opportunity to exchange. We fulfill the natural need to communicate with a fellow human. Again, anonymity guarantees openness. Meeting people on this route is therefore a joy for pilgrims and local residents. The encounter here is naked. The power and difference of the encounters in the city are due to the vulnerability of the individual. On the trail, the precariousness of the place reminds the strangers who meet of their common vulnerability so that they are obliged to trust each other. There is no need for pretense, from the moment their paths will probably separate forever. Their different untraceable subjectivities can then allow themselves the indelicacy that they would not dare in a metropolis. The reality of the everyday sense of insecurity in postmodernity (Bauman and Cupellaro 2017, p. 26), with its social fears and securities obsession (Bauman 2007) and sometimes fed by politicians is fading. The barriers of daily life in the cities are broken down in the experiences of this type of pilgrimage. All the distancing mechanisms and devices that are used in large, populated cities disappear in the face of the lived experience of shared difficulties. When a word is spoken, it transforms the solitary experience. It opens the possibility of the cathartic encounter with the other where one can open oneself in front of a perfect stranger and receive from him without fear that what was said will be disclosed to the world. A young 22-year-old pilgrim recounts, "I met two Slovenian ladies, in their sixties, beautiful and energetic. We walked a part of the way together and spent the night in the same refuge; The next day we left together. They told me about their lives; And then with one of them, I talked a little more about what I was going through in my life, and she gave me advice." (interview of 18/06/2023). And then this other pilgrim who says: "I feel richer: I have spoken with new people who have told me about their lives and experiences in these places" (interview of 07/07/2023).

The unknown—its space and the unknown traveler who crosses it from one side to the other—helps to evince what the known world cannot do. One no longer hides from the gaze of the other, on the contrary he looks for that face, that physical presence that is appreciated or accepted as a companion on the road without inner resistance (François 2013, pp. 88–91). In this space, the unknown does not stress or frighten; rather, it liberates. It is exorcizing. One wants to take advantage of this presence to nourish his daily life in a different way. Solitudes meet to form a punctual and furtive community, as the president of the *Circolo Culturale Navarca*: "When local residents see the pilgrims, they come out of their houses because they want to talk; look, unfortunately nowadays, people are walled up in their houses" (interview of 11/07/2024).

Whether between pilgrims or with local residents comes out is a human relationship free of the mistrust that reigns in cities. These meetings are not places of competition, but spaces-moments where people can live again without being judged, criticized, where they do not need to perform in expected ways, and where people do not judge themselves. They just need to listen and welcome the other. Whether they are intergenerational encounters as just seen, age being one of those variables that Sen designates as focal (2012, pp. 20–21), to show its classificatory and categorizing potential, these encounters reinvent the human relationship and often give it a punctual intensity, bordering on therapeutic, whose power lies in a paradoxical fact that it is ephemeral. A pilgrim encounter has an early expiration date, a miniscule life expectancy. This temporal emptiness favors the fullness of the encounter that cannot survive outside this space-time. This moment is the one that William James also referred to as the "specious present" as opposed to the "obvious past," while distinguishing it from the "real present" (James 1890, p. 609). It is a kind of present inhabited by the past and the anticipation of the near future (Leclerq and Galetic 2012, p. 232). Pilgrims know in advance that these encounters have no future and will remain prisoners of the present that made them possible. This is so true that this therapeutic trust for pilgrims comes to a dead stop at the threshold of overcrowded shrines. The pilgrim then dissolves into the crowd of a thousand faces who visit or pray. We no longer distinguish between the tourist and the pilgrim, because the sanctuary brings them all together. In this respect, the sanctuary is no different from a shopping mall or an airport. It is a place of *"spiritual transit"* where people cross paths without meeting. In short, a type of place and non-place at the same time. The encounters made on the way remain an emotion and a memory and there is no reason for pilgrims to continue these relationships after the pilgrimage experience. "In fact," says a pilgrim, "it's difficult to maintain a relationship. Of my many experiences, yes at the beginning a few messages, but the memory remains just that of the pilgrim walk… I never had a friendship from these experiences. It's very difficult" (interview of 22/06/2023).

3.3 *The Hospitality of Poverty, the Value of Things and the Spirit of Conviviality, Antidotes to the Consumer Society*

Through its 3 visible poles, which are the pilgrim, the local resident, and the promoter of the trail, the *Cammino Celeste* practices an economy of conviviality. The frugality and sobriety of the economic exchanges that are practiced expose the pilgrim to the reality of economic de-growth (Latouche 2019). On his path as a traveler, most often solitary (even if they are in a group, the fatigue of walking and the slowing down of time reorganize the position of each other, puts them in line, one after the other and sends them each back to their thoughts—interview of 21/07/2023), the pilgrim learns to organize his days in the scarcity of means. He only has a 10, or a 15 kilograms backpack containing everything he needs to survive for several days. He resigns himself to the Marxian truth (1867) which established that things must always be considered from the dual point of view, that of quality and that of quantity (Marx and Rubel 2008, p. 110). He does not have a lot and consequently cannot afford the profligacy of the consumer society. He rediscovers the use value of things. "We learn to appreciate the little things again," admits a pilgrim. (interview (22/06/2023). Moreover, the pilgrim's path is an abandoned and rough area and the spaces where shops are not always available. And if in the middle of the day the inns are closed, as the pilgrims point out, this suggests that the economy here does not follow the logic of the exorcizing rite of the compulsive shopping of the consumerist society (Bauman and Minucci 2016, p. 85). The pilgrim's itinerary teaches him how much he can learn to live with just what is necessary and that the society from which he comes is a society that has industrialized needs and prevents any satisfaction of them (Illich 2014, p. 41). A pilgrim says it so well: "You realize that you have far too many things that are absolutely useless to you" (interview of 14/08/2023).

The pilgrim route of the *Cammino Celeste* is therefore a kenotic journey to which those who live in the overabundance of postmodernity (Bauman) or *supermodernity* (Augé) do not have immediate access. We are in an economic model that has a soul and where the pilgrims are welcomed as family. Hosts and guests don't just talk about their lease agreement, but much more about their families and their lives too. The residents are not in the logic of gain and profit, they share with the pilgrims what they have in their homes. This is confirmed by a pilgrim who remained amazed by the humanity in service: "The lady who hosted me last night, she served me a cake for breakfast, of course I paid, but she did it with such kindness of heart that I came to thank her" (interview of 22/06/2023). "I was stunned," says another pilgrim, "by the welcome in these localities. Even when someone is not able to help you directly, they activate a support network to satisfy your requests. We never felt alone" (interview of 21/07/2023). Solidarity survives here because neoliberal theory and practice have not become part of the way of life and exchange (Bauman 2000, p. 37). The economics of large hotel chains don't work here; a person is not a slave to the tool (Illich 2014, p. 27); He is not forced to interact with a website to find where to sleep, it only takes a phone call to book and even sometimes with or without a reservation, he can knock on a door and have it opened for a night. An economy that is not focused on money but on the person. The cost-benefit logic fails here, and with it, the social classes and discrimination. It is the economy of life, the economy of the dignity of persons. This is why the promoters of the *Cammino Celeste* have brought together all the economic or exchange practices related to the reception of pilgrims under one concept: hospitality of the poor. "The hospitality of poverty," says one of the socio-founders, "is the concept that designates the spirit of the contract between the reception structures and the pilgrims, with free offering as its driving force" (interview of 11/07/2024).

In the end, the pilgrim routes are routes of the gift attitude, routes of humanity where the pilgrims have discovered that "before any production and acquisition made by man, there is a gift made by the nature" (Berthoud 2005, p. 53). Residents do not hesitate to give free of charge to pilgrims who always have good memories of it. The kindness experienced along the journey between travelers or with hosts indicates another economic model that mixes feelings, compassion, trust, and reciprocity as in the good old days of the Neapolitan School of Economics (1754) (Genovesi and Dal Degan 2013:XI, pp. 135–142).

However, the advent of the Covid-19 event has put the economy of the hospitality of poverty under strain. The pandemic hit Italy hard and resulted in the closing these reception facilities where the cost never exceeded 20 euros for a night. In doing so, travelers who were once just pilgrims have turned into customers for B&Bs who have found a new market. The technology of *supermodernity* has intruded into this sector, complicating the experience of travelers who are forced to spend time on booking sites. One of them explains why it's difficult: "We can't manage anything in the emergency mode anymore. Before, you could call, and you were expected. Now you call, they tell you to go to the website. It may turn out that the cellphone battery is dead, or that the area where we are finding ourselves is not covered by the internet" (interview of 18/08/2023).

If Ivan Illich (1973) contrasts conviviality to industrialized society and Serge Latouche contrasts de-growth to economic growth, there is a reason, and that is that of the recentralization of humans in the economy and not their exclusion in favor of gain and profit or the tool that could produce these benefits. And this is precisely the characteristic of the economic exchanges along the *Cammino Celeste*. People are at the center and economic issues are secondary. To understand how this itinerary tries to keep the human being at the center and not primarily the economic dimension, we must turn to the political.

The political management of pilgrimage routes is a factor to be taken into consideration. Although the *Cammino Celeste* is the longest pilgrimage route in the Friuli Venezia Giulia region, it does not receive as much funding as other smaller routes. For a simple reason: the *Cammino Celeste* is a private initiative. And because of its legal status, its promoters feel it is their duty to keep it alive through personal sacrifices. Unlike the other routes in the region, which are administered by the municipalities and therefore also through public financing, the *Cammino Celeste* is the result of the sacrifices of the promoters. The direct consequence is that it is not that easily eligible for public funding, which can amount up to 60,000 euros. To receive it, one must contribute from its own resources 20% of the amount requested. The small association of *"Iter Aquileaense,"* which is only a small group of enthusiasts, is therefore not able to ask for large amounts of funding. It can only resort to very small financing below 15,000 euros, considering that 20% would come from their small contributions. "As we are not able to provide the 20% share," explains the president of the *Circolo Culturale Navarca*, "we have to finance the various activities useful for the route ourselves, such as the most important monuments on the route, the marking of the route with information plaques, etc., and this costs us money" (interview of 11/07/2024).

The convivial spirit that is manifested by the different pilgrim solidarities is therefore the consequence of a model of political management of the routes. The pilgrim would believe that the attentions experienced in the field are all spontaneous. They are the result of an educational process undertaken by the members of the founding association of the *Cammino Celeste*. It is by dint of conferences, held by this association in the 20 municipalities through which this route crosses, that the local population has been awakened to this attitude of conviviality and solidarity toward travelers. "At some meetings, we had up to 150 participants, at other 35, or 30. And there, these participants talked to other people too, and so on." (interview of 11/0/2024). In doing so, these promoters have understood that it is not a question of "increasing the services produced by institutions, but rather of providing an institutional framework where, without being constrained, people can constantly educate themselves in action, participation, and the possibility of acting themselves" (Illich and Durand 2015, p. 110).

4 Conclusion

During one of the group interviews, a woman expressed her sadness for the end of the pilgrimage and the return to ordinary life, identifying it with the circuit of the washing machine that runs non-stop (interview of 21/0772023). The possibility of a return to the frenzy of the rhythm of life in the metropolis terrifies her. This fear is a common experience for most of the participants in the survey, and that was not their first pilgrimage on foot. The spirit of the pilgrims on the trail can be summed up in these words: "Now I'm going home, but next week I'll pack my bag again, or next month, for another pilgrimage" (interview of 10/07/2023). Once they have started walking, they continue to walk and to frequent anthropological places. But they are the makers of the non-places. Products of *supermodernity* or reconfigured by it, non-places and its practitioners, in this case pilgrims on foot or by bicycle, redraw the face of religious tourism through transgressive practices that aim above all to cure them, either from the rigidities of their religion, or from the acceleration of time and history, which produces social malaise in the metropolis. Where the Church and modern society are a kind of washing machine with rhythms that no one can stop, the non-place then becomes the place of pause and transgression. Transitory or ephemeral as it may be, making history or not, it is the "specious present" that de-institutionalizes the Church and its religion.

In religion, the non-place is not the void where nothing happens; rather, it is the parenthesis of modernity, a creative nothingness of sorts. Just like airports, shopping malls or highways, places par excellence of non-place, where the traveler only passes for the sake of passing, the sanctuaries and church-museums of pilgrimage routes are places of passage too and therefore also to a certain extent non-places in the sense of Michel De Certeau. These sanctuaries and churches become places where believers regain their spiritual health without the advice of the religious authority, and places that non-believers cross with their feet as a shopping mall. Thus, as non-places, they become anthropological spaces where the resizing of religious authority and curative spiritual subversion are played out; and also the spaces where the quest for meaning and the spiritualities of believers and non-believers are assimilated to what Diotallevi calls unchurched spirituality (Diotallevi 2017, p. 29).

References

Abélès M (1994) M. Augé, Non-lieux. Introduction à une anthropologie de la surmodernité. Revue française d'anthropologie Tome 34(N°129):193–194. https://www.persee.fr/doc/spgeo_0046-2497_1993_num_22_1_3142. Accessed 23 July 2024

Augé M (1992) Non-lieux: introduction à une anthropologie de la surmodernité. Seuil, Paris

Bauman Z (2000) La solitudine del cittadino globale. Feltrinelli Editore, Milano

Bauman Z (2007) Le Présent liquide. Peurs sociales et obsession sécuritaire. Le Seuil, Paris

Bauman Z, Cupellaro M (2017) Stranieri alle porte. Gius Laterza & Figli Spa, Bari-Roma

Bauman Z, Minucci S (2016) Modernità liquida. Laterza, Bari-Roma

Bellavite A, Bregant M, Perini T (2011) Il Cammino celeste. A piedi da Aquileia al Monte Lussari. Ediciclo, Portogruaro

Berthoud A (2005) Une philosophie de la consommation: Agent économique et sujet moral. P U septentrion, Villeneuve d'Ascq

Certeau M, Giard L (2010) L'invenzione del quotidiano. Lavoro, Roma

Debarbieux B (1993) Non-lieux. Tome 22(n°1):90–91. https://www.persee.fr/doc/spgeo_0046-2497_1993_num_22_1_3142. Accessed 20 July 2024

Diotallevi L (2017) Fine corsa. La crisi del Cristianesimo come religione confessionale. EDB, Bologna

Durkheim E (2014) Les formes élémentaires de la vie religieuse: le système totémique en Australie. CNRS Éd, Paris

François (2013) La Joie de l'Évangile: Exhortation apostolique Evangelii gaudium. Artège Edition, Perpignan

Genovesi A, Dal Degan F (2013) Lezioni di economia civile. V&P Vita e Pensiero, Milano

Illich I (2014) La convivialit*é*. Éd. Points, Paris

Illich I, Durand G (2015) Une société sans école. Points, Paris

James W (1890) The principles of psychology, vol 1. Henry Holt and Company, New York

James W (2001) Les formes multiples de l'expérience religieuse: essai de psychologie descriptive. Exergue, Chambéry

Latouche S (2019) La décroissance implique de sortir de la modernité. Marianne. https://www.marianne.net/debattons/entretiens/serge-latouche-la-decroissance-implique-de-sortir-de-la-modernite. Accessed 4 May 2020

Leclerq B, Galetic S (2012) James et Husserl: Perception des formes et polarisation des flux de conscience. Revue internationale de philosophie 260(2):229–250. https://doi.org/10.3917/rip.260.0229
Marx K, Rubel M (2008) Le capital. Gallimard, Paris
Paugam S (2017) L'enquête sociologique. Presses Universitaires de France, Paris
Rosito V (2018a) Dio delle città: cristianesimo e vita urbana. EDB, Bologna
Rosito V (2018b) Postsecolarismo: Passaggi e provocazioni del religioso nel mondo contemporaneo. EDB - Edizioni Dehoniane, Bologna
Sen A (2012) Repenser l'inégalité. Points, Paris
Weber M, Kalinowski I (2013) Sociologie de la religion: Economie et société. Flammarion, Paris
World Tourism Organization (2012) Tourism and Intangible Cultural Heritage. https://www.e-unwto.org/doi/book/10.18111/9789284414796. Accessed 25 July 2024

Determinants of the Development of Religious Tourism in Greece

Paraskevas F. Tsorvas, Maria Lagou, and Anthi Gousiou

Abstract This study examines the factors influencing the growth of religious tourism in Greece, with the objective of facilitating its further development. An empirical study utilizing factor analysis with principal components analysis was conducted to identify the key determinants of religious tourism. The multiple regression analysis identified independent variables demonstrating correlations that can be utilized to create mathematical models for forecasting tourism trends. Additionally, a framework for a tourism policy focused on religious tourism is proposed, including selected actions for planning, management, and administration. This research is notable for integrating both factor and regression analyses to forecast trends and tailor tourism policies to the needs of religious tourism in Greece.

1 Introduction

Religious tourism represents one of the earliest and most enduring forms of tourism, garnering the attention from researchers (Rinschede 1992). This distinctive phenomenon encompasses visits to locations of religious significance for religious reasons. It includes not only active religious sites, such as churches and monasteries, but also the exploration of places associated with extinct religions that have left behind a rich legacy of artistic and cultural monuments.

In the contemporary era, religious tourism can be defined as a distinct form of travel reflecting an individual's aspiration to leave their place of permanent residence and visit a location they regard as sacred, such as a monastery, temple, or the tomb of a saint. These journeys are often timed to coincide with specific occasions, such as Holy Week, Christmas, August 15th, or the celebration of a saint's feast day.

P. F. Tsorvas · A. Gousiou (✉)
Department of Business Administration, University of the Aegean, Mytilene, Greece

M. Lagou
Department of Accounting and Finance Administration, University of West Attica, Athens, Greece

M. Valeri (ed.), *Contemporary Religious Tourism*, Tourism, Hospitality & Event Management, https://doi.org/10.1007/978-3-031-79074-4_14

During these pilgrimages, individuals engage in activities aimed at communicating with the divine, making offerings, and expressing gratitude.

The World Tourism Organization (WTO) estimates that 300 million visits are made annually to countries with a well-developed religious tourism sector, resulting in a total of 600 million national and international religious trips across the globe. In this total, 40% occurs in Europe, representing a significant portion of the global total. Most European countries are significant tourist destinations, possessing a considerable number of religious and cultural heritage monuments (Valeri 2022). This has led to the development of tourism and, subsequently, economic growth. Furthermore, these monuments offer significant incentives for tourists, facilitating the vital contributions of respect, tolerance, and mutual understanding between different cultures (Griffin and Raj 2017; Shekhar and Valeri 2023).

Religious holidays, regardless of whether they are associated with pilgrimage sites, serve as genuine opportunities to attract tourists. Holidays are essentially time-bound attractions, with all the prospects and challenges associated with any special event (Greenwood 1989).

In Greece, there are many regions (e.g., Mount Athos, Meteora, Tinos, and Patmos) with numerous religious sites, such as churches and monasteries, which are an integral part of the national heritage and attract a significant number of tourists. Furthermore, numerous tourist attractions in Greece have been designated as UNESCO World Heritage Sites (UNESCO 2024). Among these are Mount Athos, Meteora, the church of Saint Demetrios in Thessaloniki, the monasteries of Daphne and Saint Luke, and the New Monastery in Chios. Additionally, this list includes the Monastery of St. John the Baptist and the Cave of the Apocalypse in Patmos. Of particular significance are the monuments of Greek Orthodoxy, which constitute an essential component of the country's national heritage and represent a primary attraction for visitors.

The plethora of historical churches, both Byzantine and post-Byzantine, situated in towns and villages across the country, in conjunction with the cathedrals and picturesque chapels, along with the pilgrimages in rural areas and the monasteries, metochia and hermitages, illustrate a profound and centuries-old relationship between artistic creation and religious practice. This is evident in the churches' rich decoration with hagiographies, icons, and mosaics. Religious tourism in Greece utilizes the cultural heritage of the past, which, when viewed from a different perspective, serves as a reference point for cultural tourists, religious tourists, and especially pilgrims, as a daily expression of living religion (Griffin and Raj 2017). Furthermore, the contribution of religious tourism to the economy should not be underestimated. The benefits to local, regional, and national economies are considerable, with revenues generated in areas that host religious tourists.

One significant benefit of religious tourism is its potential for development outside the high season, particularly during the spring and around Easter, when festivals are held. This helps to extend the tourist season and increase occupancy (Cerutti and Piva 2016: 8; Terzidou et al. 2018).

The total demand for religious tourism is contingent upon the availability of suitable destination locations. Thus, as pilgrimage destinations develop and gain prominence, the relative demand for them will correspondingly increase (Olsen 2010: 848; Hill-Smith 2011: 236–246). In Greece, where cultural tourism and recreation are combined with religious or pilgrimage monuments, demand is increasing annually.

Taking the aforementioned considerations into account, this article aims to investigate the factors influencing religious tourism in Greece. To achieve this objective, the development of religious tourism in Greece is initially described, followed by the presentation of the research's methodological framework and results. In conclusion, the findings are presented, along with recommendations for tourism policy.

2 The Development of Religious Tourism in Greece

The historical monuments of the Greek Orthodox tradition constitute an essential component of the national cultural heritage and serve as significant points of interest for visitors from all over the world. A multitude of early Christian, Byzantine and post-Byzantine churches, chapels, monasteries, metochia, and hermitages, adorned with mosaics, hagiographies, and icons, are scattered across cities and villages and combined with Christian holidays and festivals of great importance, spiritually connecting religious worship with art, architecture, tradition, and culture (Tzovolos 2016).

Historically, the movement did not possess the characteristics of a tourist excursion and was not an attempt to escape from everyday life; rather, it was undertaken for specific reasons (commercial, military, colonization). Pilgrims commonly visited holy sites situated at a distance from their residences. They climbed the mountain of Zeus, Olympus, and other mountain peaks, including the highest mountain of Aegina, known as Zeus the Panhellenic (Hellenic). Structures were built on these mountain peaks to accommodate arriving pilgrims, where they offered vows and donations, a practice that persists today. In this context, it is also important to consider the visits to the oracles of the period, which were regarded as sacred and to which people turned to receive the guidance they sought (Rodosthenous 2012; Kumar et al. 2022).

The advent of Christianity transformed the utilization of existing places of worship. Christian churches were constructed on the grounds of former Greek or Hellenistic temples. The oldest extant churches in the Greek area date from the late Roman period (fourth to sixth century). During this same period, several important ancient Greek temples (Parthenon, Erechtheum, and Thisium in Athens) were converted into churches. Additionally, Christian temples were constructed near to or on the sites of pagan sanctuaries that were significant centers of religious pilgrimage (Daphni, Eleusis, Delphi, Epidaurus, Delos) (Vouronikou 2017: 34). This evidence demonstrates the continuity of ancient Greek piety and its evolution into Christian faith.

The transition from the Hellenistic and Roman periods to the Byzantine era was characterized by the absence of any violent interruption to religious worship in

Greece. The same religious practices were observed in each location. During the nineteenth century, European scholars traveling through Greece noted the ancient customs of visiting holy places and paying homage to saints or God. Scholars of the European Enlightenment struggled to properly evaluate such practices when they encountered them.

In the present era, other significant categories of religious tourism destinations are associated with religious symbols, natural sites from which holy events are believed to flow, locations where great religious figures lived and worked, such as the monastery of Agia Lavra, and places where important religious events or festivals are held periodically, as well as extensive religious conferences and "religious routes." The route "In the footsteps of the Apostle Paul" exemplifies this, covering the journeys of the Apostle Paul through Greek territory and the areas he visited and taught. Consequently, tourists may choose to combine their visits to religious and archeological sites with the exploration of the historical, cultural, and architectural traditions of the surrounding areas. Today, these routes are followed by many individuals. In many of these events, which are also pilgrimages, religious ceremonies are held, such as in Veria, where the Pauline Events are celebrated in honor of the Apostle Paul. The proceedings conclude with an open-air vespers service within the Orthodox tradition at the Apostle Paul's Step, where it is believed he stood and preached the Gospel to the local community (Lagos and Christogianni 2006: 473; Moira 2009: 60).

In the context of an analytical categorization (see Table 1), we find active religious buildings and monuments in Greece that maintain their religious function even during periods when visitors are present.

It is notable that a significant number of religious tourism centers in Greece are buildings or monuments of worship associated with a diverse array of doctrines and cults (Poulaki and Lagos 2016). For instance, the islands of the Cyclades (e.g., Syros and Tinos) host a considerable number of churches and chapels belonging to the Roman Catholic denomination. Additionally, numerous historical Muslim monuments (mosques) can be found in regions of Greece that were liberated from Ottoman rule, including Macedonia, Thessaly, Crete, and the Dodecanese. These sites could potentially be incorporated into a national initiative to promote religious tourism.

Each Regional Unit (Prefecture) of Greece (see Table 2) contains religious monuments (some of which are places of pilgrimage), contributing to the value of the Greek capital as a center for religious tourism.

The majority of religious tourists or pilgrims in Greece are Greek Orthodox Christians, primarily interested in visiting pilgrimage site associated with the Orthodox Church. There is considerable demand for religious pilgrimage tourism from tourists of Cypriot, Russian, Bulgarian, Serbian, and Romanian (mainly Orthodox) nationality, in addition to Greek expatriates abroad. For example, the island of Patmos attracts the greatest number of distinct tourist categories (target groups) during the Easter period, with visitors exceeding 3000 individuals from Greece and Cyprus (Poulaki 2018: 128).

Table 1 Greek monuments of world cultural heritage

Monument, Location, and Inclusion Date	
• Temple of Apollo Epicurius at Bassae	1986
• Archaeological Site of Acropolis	1987
• Archaeological Site of Delphi	1987
• Sanctuary of Asclepios at Epidaurus	1988
• Mount Athos—Athos	1988
• Meteora	1988
• Paleochristian and Byzantine Monuments of Thessaloniki	1988
• *Church of Acheiropoietos* • *Rotonda* • *Church of St. Demetrios*	
• *Latomou Monastery* • Church of St. Sophia	
• *Church of Panagia Chalkeon* • *Church of St. Panteleimon*	
• *Church of the Holy Apostles* • *Church of St. Nicholas Orphanos*	
• *Church of St. Catherine* • *Church of Christ Saviour*	
• *Blatades Monastery* • *Church of Prophet Helias*	
• *Byzantine Bath* • *City Walls of Thessaloniki*	
• Medieval City of Rhodes	1988
• Archaeological Site of Olympia	1989
• Archaeological Site of Mystras	1989
• Archaeological Site of Delos	1990
• Monastery of Daphni, Monastery of Osios Loukas, New Monastery of Chios	1990
• Archaeological Site of Herai Samos	1992
• Archaeological Site of Aigon (Vergina)	1996
• Archaeological Sites of Mycenae and Tiryns	1999
• Historical Center (Town) with the Monastery of Saint John the Theologian and the Cave of the Apocalypse in Patmos	1999
• Old Town of Corfu	2007

Source: UNESCO (2024)

Until the 1960s, the monastic state of Mount Athos received approximately 2500 visitors per year. Since then, the number of visitors has exceeded 55,000, including Orthodox individuals of all nationalities, with a small percentage (approximately 4%) representing heterodox groups. The demand for entry is even greater, as the procedures for obtaining an entry permit for heterodox visitors are strict and act as a significant deterrent (Sidiropoulos 1998: 76). Following the fall of the former communist regimes in Eastern Europe and the Balkans, Northern Greece and its Orthodox monuments have become key points of interest for Orthodox people.

Additionally, the Meteora region is a significant contributor to the growth of religious tourism. The distinctive geomorphological characteristics of the area result in the site attracting a specific type of tourist on each visit. Similarly, religious monuments, such as those found in Meteora, tend to attract tourists and pilgrims from other countries, including China, whose interest in the area is notable (Albani 2014: 100–101).

It is anticipated that emerging trends influencing the broader landscape of tourism will also impact the religious tourism sector. Consequently, the formulation of

Table 2 Distribution and number of pilgrimages per regional unit (Prefecture)

Prefecture	Total pilgrimages	Prefecture	Total pilgrimages	Prefecture	Total pilgrimages
Etoloakarnania	25	Thesprotia	11	Magnisia	25
Argolida	14	Thessaloniki	25	Messinia	20
Arcadia	33	Ioannina	17	Xanthi	4
Arta	17	Kavala	12	Pella	8
Attica	69	Karditsa	8	Pieria	6
Achaia	28	Kastoria	5	Preveza	5
Boeotia	26	Corfu	24	Rethymno	13
Grevena	2	Kefalonia	14	Rhodope	1
Drama	2	Kilkis	3	Samos	30
Dodecanese	67	Kozani	11	Serres	17
Evros	7	Corinthia	43	Trikala	14
Evia	18	Cyclades	41	Fthiotida	19
Evrytania	8	Laconia	16	Florina	5
Zakynthos	9	Larissa	19	Fokis	7
Ilia	29	Lasithi	10	Halkidiki	8
Imathia	13	Lesvos	14	Chania	11
Heraklion	27	Lefkada	7	Chios	22

Source: Greek National Tourism Organization (2005)

an integrated strategic approach and policy can exploit Greece's comparative advantage within this sector, establishing new tourist sites that can attract tourists with special interests and spiritual pursuits. In addition to the economic benefits that will result, this approach is expected to contribute to the mitigation of tourist seasonality and the expansion of multicultural identity.

3 The Methodological Framework of the Research

The research employs a factor analysis methodology, based on a set of selected questions that serve as variables, expressing the degree of satisfaction or desire on a 5-point scale. To ensure the effectiveness and reliability of the factor analysis, it is necessary to cross-check the results with those obtained from Principal Components Analysis (PCA). For the final solution to determine the factor components, the rotated component matrix was chosen. This matrix shows the two factors with their respective loadings on the variable after orthogonal rotation. The rotation aims to increase the large loadings and correspondingly reduce the small ones to facilitate result interpretation.

In order to ascertain the mathematical model that describes, predicts, and interprets the values of the selected variables in relation to those of other variables that affect them, a linear regression analysis was conducted.

To apply the aforementioned methodological framework, primary field research conducted at the doctoral thesis level (Tsorvas 2024) was employed within the context of non-probability sampling, utilizing an appropriately structured questionnaire. A factor analysis was conducted on two questions pertaining to the factors that both facilitate and impede the growth of the religious tourism in Greece.

4 Results of the Empirical Research

The steps of the methodological framework of the factor analysis per question (variable) are presented below (Tsorvas 2024):

A. Factor analysis of the facilitating factors for the growth of religious tourism in Greece.

Table 3 presents the Kaiser–Meyer–Olkin and Bartlett's Test statistics.

As indicated in Table 3, the Kaiser–Meyer–Olkin index provides evidence of the adequacy in the size of the sample (0.898 > 0.50), thus allowing for the use of factor analysis. Additionally, the Bartlett's Test of Sphericity index, which assesses the correlation between variables/factors and their suitability for factor analysis, yielded a value of 0.00015 ($p < 0.05$), further supporting the application of factor analysis. Subsequently, the total variance of the eigenvalues of the factors is calculated using the PCA method (see Table 4).

Table 4 indicates that factors 1, 2, 3, and 4, namely, (a) attracting tourists/pilgrims from traditional and emerging religious tourism markets, (b) implementing targeted promotion and advertising of Greece's "religious image," (c) developing appropriate tourism planning and formation of customer management strategy, and (d) strengthening of bodies related to religious tourism, are perceived by respondents as the most significant contributors to the development of religious tourism in Greece. More specifically, the most significant, as identified by 25.2% of the participants, are the attractions offered to tourists and pilgrims from traditional and emerging markets of religious tourism. This is followed by the importance of implementing specialized and targeted promotion and advertising of Greece's "religious image" (24.4%), the development of appropriate tourism planning and formation of customer management strategy (23.1%), and the strengthening of bodies related to religious tourism (9.3%).

Table 3 Correlation measures on the factors that facilitate the growth of religious tourism through the Kaiser–Meyer–Olkin and Bartlett's Test indicators

KMO and Bartlett's test		
Kaiser–Meyer–Olkin measure of sampling adequacy		0.898
Barlett's test of sphericity	Approx. Chi-Square	991.001
	df	101
	Sig.	0.00015

Table 4 Total variance of the eigenvalues of the factors that facilitate the growth of religious tourism

Total variance explained									
	Initial Eigenvalues			Extraction sums of squared loadings					
	Total	% of Variance	Cumulative %	Total	% of Variance	Cumulative	Total	% of Variance	Cumulative
1	9.187	25.200	25.200	9.187	25.200	25.200	9.180	25.100	25.100
2	9.093	24.400	49.600	9.093	24.400	49.600	9.100	24.500	49.600
3	8.796	23.100	72.700	8.796	23.100	72.700	6.796	18.100	67.700
4	1.947	9.300	82.000	1.947	9.300	82.000	3.947	14.300	82.000
5	0.872	8.100	90.100						
6	0.421	7.205	97.305						
7	0.332	2.102	99.407						
8	0.108	0.461	99.868						
9	0.101	0.118	99.986						
10	0.100	0.014	100.00						

Extraction Method: PCA

Table 5 illustrates the adjustment coefficients (communalities) of each factor that facilitates the growth of religious tourism in Greece, using the PCA method.

The analysis of the adjustment coefficients (communalities) demonstrates the relative importance of each factor. This indicates that the initial value of 1 for the factor attracting tourists/pilgrims from traditional and emerging markets of religious tourism, following factor analysis, became 0.729 (72.9%). Similarly, the initial value of 1 for the factor of implementing specialized and targeted promotion and advertising of the "religious image" of Greece, after factor analysis, became 0.712 (71.2%). The initial value of 1 for the factor of suitable tourism planning and formation of customer management strategy became 0.624 (62.4%), while the initial value of 1 for the strengthening factor of agencies related to religious tourism became 0.613 (61.3%).

Figure 1 shows the scree plot of the eigenvalue of each factor, highlighting the most important factors. The factors that appear to be important are the attraction of tourists/pilgrims from traditional and emerging religious tourism markets, the implementation of specialized and targeted promotion and advertising of the "religious image" of Greece, the development of appropriate tourism planning and formation of customer management strategy, and the strengthening of bodies related to religious tourism.

Subsequently, the multiple linear model

$$Y = b_0 + b_1X_1 + b_2X_2 + b_3X_3 + b_4X_4 + \varepsilon$$

is applied, with the dependent variable representing the factors that enhance religious tourism and the independent variables encompassing the following factors:

Table 5 Adjustment coefficients of each factor that facilitates the growth of religious tourism

Communalities		
	Initial	Extraction
Attract tourists/pilgrims from traditional and emerging religious tourism markets	1.000	0.729
Implementing specialized and targeted promotion and advertising of the "religious image" of Greece	1.000	0.712
Develop appropriate tourism planning and formation of customer management strategy	1.000	0.624
Strengthening of bodies related to religious tourism	1.000	0.613
Encourage cooperation among stakeholders (e.g., the church, the tourism industry, local government organizations) involved in religious tourism	1.000	0.447
Upgrading the food and accommodation facilities in the monastery complexes for pilgrims	1.000	0.420
Establish an organized network for the provision of information and the organization of guided tours of religious sites	1.000	0.368
Foster collaboration with Higher Education Institutions	1.000	0.328
Invest in tourist facilities associated with religious tourism	1.000	0.201
Enhance the transport network and accessibility to religious monuments	1.000	0.129

Extraction Method: PCA

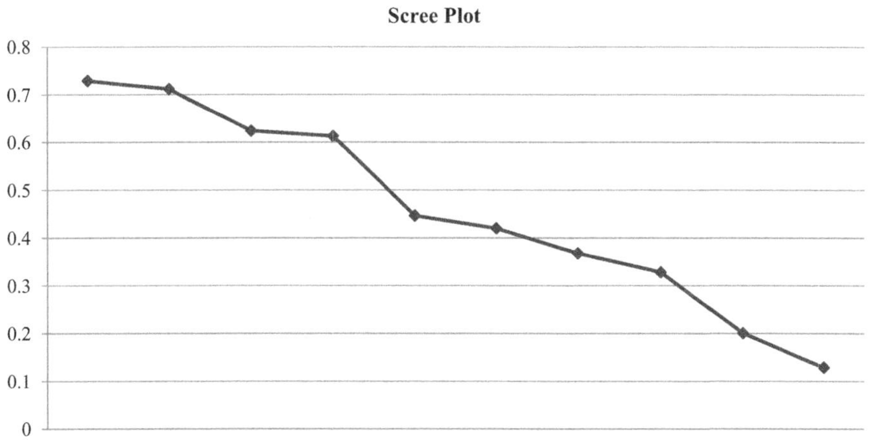

Fig. 1 Scree plot on the facilitating factors for the development of religious tourism

the attraction of tourists/visitors from traditional and emerging religious tourism markets; the implementation of targeted promotion and advertising of Greece's "religious image"; the development of appropriate tourism planning and formation of customer management strategy; and the strengthening of bodies related to religious tourism.

The regression analysis, based on the determination index $R^2 = 0.88$ (see Table 6), indicates that the percentage of the dependent variable explained by the independent variable is 88%.

Table 6 Multiple correlation coefficients (index of determination, corrected index of determination, and standard error of estimate) of the regression model on the facilitating factors for the development of religious tourism

Model Summary				
Model	*R*	*R* Square	Adjusted	Std. error
1	0.96	0.88	0.698	11.307

Table 7 Coefficients of the multiple regression model for the dependent variable "factors that facilitate the development of religious tourism"

Coefficients							
	B	Std error	Beta	*t*	Sig.	Lower	Upper
(Intercept)	520,627.10	1614.10	−0.90	322,322.10	0.200	517,461.10	523,794.10
Attract tourists/pilgrims from traditional and emerging religious tourism markets	9186.10	1459.10	0.09	6290.10	0.002	6323.10	12,049.10
Implement specialized and targeted promotion and advertising of the "religious image" of Greece	7244.00	1476.00	0.070	4908.00	0.002	4351.00	10,138.00
Develop appropriate tourism planning and formation of customer management strategy	10,976.00	1477.00	0.105	7429.00	0.001	8080.00	13,873.00
Strengthening of bodies related to religious tourism	6244.00	1476.00	0.070	4908.00	0.000	4351.00	10,138.00

Table 7 presents the framework of the results of the multiple regression model, which determines which of the independent variables contribute to or influence the prediction of the dependent variable, namely "factors that facilitate the development of religious tourism."

The beta values provide the contribution of each variable to the prediction of the dependent variable. The *t*-tests presented in Table 7 correspond to the following hypotheses: H0: b0 = 0, H0: b1 = 0, H0: b2 = 0, H0: b3 = 0, H0: b4 = 0. The *p*-values (0.002, 0.002, 0.001 and 0.000) indicate that the respective hypotheses H0: b1 = 0 and H0: b2 = 0, H0: b3 = 0, H0: b4 = 0 can be rejected (at the 5% significance level). Therefore, in accordance with the above, the attraction of tourists/pilgrims from traditional and emerging markets of religious tourism, the implementation of specialized and targeted promotion and advertising of the "religious image" of Greece, the development of appropriate tourism planning and formation of strategic customer management, and the strengthening of bodies related to religious tourism facilitate the growth of religious tourism in Greece.

B. Factor analysis of the factors impeding the growth of religious tourism in Greece.

Table 8 presents the Kaiser–Meyer–Olkin and Bartlett's Test statistics.

The data presented in Table 8 indicate that the Kaiser–Meyer–Olkin index assesses the adequacy of the sample (0.958 > 0.50) and can be used in the factor analysis, while the Bartlett's Test of Sphericity index, which evaluates whether the correlations between the variables/factors permit the application of factor analysis, is 0.00028 ($p < 0.05$). Next, the method of PCA is used to calculate the total variance of the eigenvalues of the factors (see Table 9).

Table 9 shows that factors 1 and 2, namely, (a) the inadequate promotion of religious sites, and (b) the lack of cooperation among stakeholders (e.g., the church, the tourism industry, local government organizations) involved in religious tourism, are

Table 8 Correlation measures on the factors that impede the growth of religious tourism through the Kaiser–Meyer–Olkin and Bartlett's Test indicators

KMO and Bartlett's test		
Kaiser–Meyer–Olkin measure of sampling adequacy		0.958
Bartlett's test of sphericity	Approx. Chi-Square	901.001
	df	101
	Sig.	0.00028

Table 9 Total variance of the factors impeding the development of religious tourism

Total variance explained									
	Initial Eigenvalues			Extraction sums of squared loadings					
	Total	% of Variance	Cumulative %	Total	% of Variance	Cumulative	Total	% of Variance	Cumulative %
1	10.187	42.200	42.200	11.187	42.200	42.200	10.677	41.900	41.900
2	9.993	41.400	83.600	9.993	41.400	83.600	10.503	41.700	83.600
3	0.997	13.100	96.700						
4	0.947	1.300	98.000						
5	0.872	1.100	99.100						
6	0.721	0.205	99.305						
7	0.432	0.102	99.407						
8	0.108	0.191	99.598						
9	0.101	0.188	99.786						
10	0.100	0.138	99.924						
11	0.098	0.065	99.989						
12	0.076	0.01	99.999						
13	0.0076	0.001	100.00						

Extraction Method: PCA

the primary obstacles hindering the development of religious tourism, as perceived by respondents. Specifically, the inadequate promotion of religious sites is identified as a factor impeding the development of religious tourism by 42.2% of respondents, while the inability to cooperate between the bodies (e.g., church, tourist industry, local authorities) involved in the development of religious tourism is identified as a factor by 41.4%.

Table 10 presents the coefficients of adjustment (communalities) for each factor that impedes the growth of religious tourism in Greece, using the PCA method.

The analysis of the adjustment coefficients (communalities) shows the relative importance of each factor. This indicates that the initial value of 1 for the factor "Inadequate promotion of religious sites" after the factor analysis became 0.888 (88.8%). Similarly, the initial value of 1 for the factor "lack of cooperation among stakeholders (e.g., church, tourism industry, local government organizations) involved in religious tourism" became 0.827 (82.7%) after the factor analysis.

Figure 2 shows the scree plot of the eigenvalue of each factor, highlighting the most important factors. The factors that seem to be important are the inadequate promotion of religious sites and the lack of cooperation among stakeholders (e.g., church, tourism industry, local government organizations) involved in religious tourism.

Subsequently, the multiple linear model

$$Y = b_0 + b_1X_1 + b_2X_2 + b_3X_3 + b_4X_4 + \varepsilon$$

Table 10 Coefficients of adjustment of each factor impeding the growth of religious tourism

Communalities		
	Initial	Extraction
Inadequate promotion of religious sites	1.000	0.888
Lack of cooperation among stakeholders (e.g., the church, the tourism industry, local government organizations) involved in religious tourism	1.000	0.827
Lack of an organized network for the provision of information and the organization of guided tours of religious sites	1.000	0.496
There are no tour operators willing to attract religious tourists	1.000	0.469
Inability to approach the religious monuments	1.000	0.397
Lack of cooperation with specialized religious tourism agencies at both national and international level	1.000	0.380
Absence of strategic plan for the development of religious tourism	1.000	0.368
Non-existent cooperation between the Church and the State	1.000	0.328
Absence of a central organization promoting religious tourism	1.000	0.301
Non-existent guided tours of religious sites	1.000	0.229
Negative attitude of the Church authorities	1.000	0.222
The Church does not wish to take a more active role in religious tourism	1.000	0.211
Inability to combine religious tourism with other special and alternative forms of tourism	1.000	0.201

Extraction Method: PCA

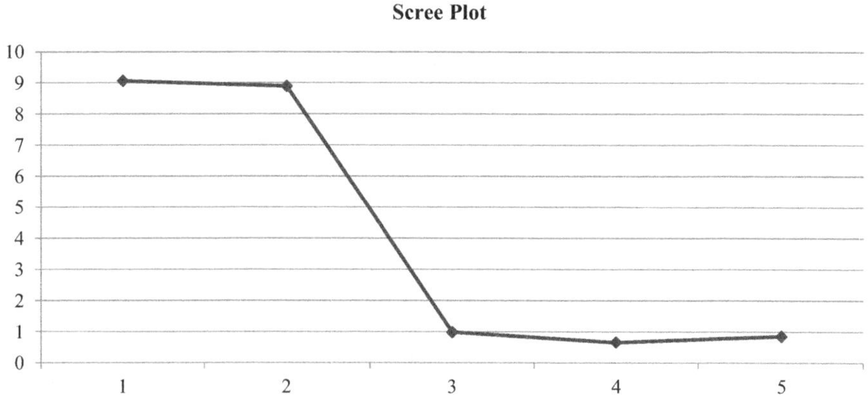

Fig. 2 Scree plot on the factors impeding the growth of religious tourism

Table 11 Multiple correlation coefficients (index of determination, corrected index of determination, and standard error of estimate) of the regression model on the factors that impede the growth of religious tourism

Model summary				
Model	*R*	*R* Square	Adjusted	Std. error
1	09.2	0.81	0.698	11.307

is applied, with the dependent variable representing the factors that impede the growth of religious tourism and the independent variables encompassing the following factors: "the inadequate promotion of religious sites," "inability to approach the religious monuments," "the lack of an organized network for the provision of information and the organization of guided tours of religious sites," etc.

The regression analysis, based on the determination index $R^2 = 0.81$ (see Table 11), indicates that the percentage of the dependent variable explained by the independent variable is 81%.

Table 12 presents the framework of the results of the multiple regression model for the dependent variable "factors impeding the development of religious tourism."

The beta index values demonstrate the extent to which each variable contributes to the prediction of the dependent variable. The *t*-tests presented in Table 12 correspond to the following null hypotheses: H0: b0 = 0, H0: b1 = 0, H0: b2 = 0. The *p*-values (0.04 and 0.031) indicate that the hypotheses H0: b1 = 0, H0: b2 = 0 can be rejected with a 5% level of significance. Consequently, based on the above, only the variables of inadequate promotion of religious sites and lack of cooperation among stakeholders (e.g., church, tourism industry, local government organizations) involved in religious tourism seem to hinder the growth of religious tourism.

Subsequently, Table 13 presents the Component Matrix for the variables of the two questions. This table presents the two factors (Components 1 and 2) and the corresponding loadings of the variable on these two factors. Their values range from −1 to 1.

Table 14 presents the Rotated Component Matrix, which shows the two factors (Components 1 and 2) with the respective loadings of the variable after orthogonal rotation. The rotation is intended to increase the large loadings and correspondingly decrease the small ones, thus facilitating the interpretation of the results.

Table 14 reveals that variables 11–23 are associated with the first factor, which may be designated as "Factors that impede the development of religious tourism in Greece." In contrast, the remaining variables 1–10 are linked to the second factor,

Table 12 Coefficients of the multiple regression model for the dependent variable "factors impeding the development of religious tourism"

	B	Std error	Beta	*t*	Sig.	Lower	Upper
(Intercept)	−28.9	19.32		−1.565	0.155	−67.877	11.69
Inadequate promotion of religious sites	3.912	1.24	0.512	3.133	0.04	1.346	6.78
Lack of cooperation among stakeholders (e.g., church, tourism industry, local government organizations) involved in religious tourism	19.67	8.62	0.367	2.28	0.031	1.933	37.408

Table 13 Component matrix for the variables of the questions

A/N	Variables	Component 1	Component 2
1.	Attract tourists/pilgrims from traditional and emerging religious tourism markets	0.659	0.29
2.	Implement specialized and targeted promotion and advertising of the "religious image" of Greece	0.743	0.214
3.	Develop appropriate tourism planning and formation of customer management strategy	0.791	0.123
4.	Strengthening of bodies related to religious tourism	0.813	0.2
5.	Cooperation among stakeholders (e.g., the church, the tourism industry, local government organizations) involved in religious tourism	0.677	0.108
6.	Upgrading the food and accommodation facilities in the monastery complexes for pilgrims	0.765	0.107
7.	Establish an organized network for the provision of information and the organization of guided tours of religious sites	0.651	0.401
8.	Foster collaboration with Higher Education Institutions	0.751	0.305
9.	Invest in tourist facilities associated with religious tourism	0.83	0.28
10.	Enhance the transport network and accessibility to religious monuments	0.678	0.384
11.	Inadequate promotion of religious sites	0.301	0.871
12.	Lack of cooperation among stakeholders (e.g., the church, the tourism industry, local government organizations) involved in religious tourism	0.398	0.765

(continued)

Table 13 (continued)

A/N	Variables	Component 1	Component 2
13.	Lack of an organized network for the provision of information and the organization of guided tours of religious sites	0.143	0.666
14.	There are no tour operators willing to attract religious tourists	0.29	0.61
15.	Inability to approach the religious monuments	−0.153	0.83
16.	Lack of cooperation with specialized religious tourism agencies at both national and international level	0.133	0.671
17.	Absence of strategic plan for the development of religious tourism	0.2	0.751
18.	Non-existent cooperation between the Church and the State	0.23	0.83
19.	Absence of a central organization promoting religious tourism	0.312	0.495
20.	Non-existent guided tours of religious sites	0.401	0.567
21.	Negative attitude of the Church authorities	0.305	0.659
22.	The Church does not wish to take a more active role in religious tourism	0.199	0.439
23.	Inability to combine religious tourism with other special and alternative forms of tourism	0.299	0.81

Table 14 Rotated component matrix for the variables of the questions

A/N	Variables	Component 1	Component 2
1.	Attract tourists/pilgrims from traditional and emerging religious tourism markets	0.71172	0.2813
2.	Implement specialized and targeted promotion and advertising of the "religious image" of Greece	0.80244	0.20758
3.	Develop appropriate tourism planning and formation of customer management strategy	0.85428	0.11931
4.	Strengthening of bodies related to religious tourism	0.87804	0.194
5.	Encourage cooperation among stakeholders (e.g., the church, tourism industry, local government organizations) involved in religious tourism	0.73116	0.10476
6.	Upgrading the food and accommodation facilities in the monastery complexes for pilgrims	0.8262	0.10379
7.	Establish an organized network for the provision of information and the organization of guided tours of religious sites	0.70308	0.38897
8.	Foster collaboration with Higher Education Institutions	0.81108	0.29585
9.	Invest in tourist facilities associated with religious tourism	0.8964	0.2716
10.	Enhance the transport network and accessibility to religious monuments	0.73224	0.37248
11.	Inadequate promotion of religious sites	0.281	0.891
12.	Lack of cooperation among stakeholders (e.g., the church, the tourism industry, local government organizations) involved in religious tourism	0.378	0.785

(continued)

Table 14 (continued)

A/N	Variables	Component 1	Component 2
13.	Lack of an organized network for the provision of information and the organization of guided tours at religious sites	0.123	0.686
14.	There are no tour operators willing to attract religious tourists	0.27	0.63
15.	Inability to approach the religious monuments	−0.173	0.85
16.	Lack of cooperation with specialized religious tourism agencies at both national and international level	0.113	0.691
17.	Absence of strategic plan for the development of religious tourism	0.18	0.771
18.	Non-existent cooperation between the Church and the State	0.21	0.85
19.	Absence of a central organization promoting religious tourism	0.292	0.515
20.	Non-existent guided tours of religious sites	0.381	0.587
21.	Negative attitude of the Church authorities	0.285	0.679
22.	The Church does not wish to take a more active role in religious tourism	0.179	0.459
23.	Inability to combine religious tourism with other special and alternative forms of tourism	0.279	0.83

which could be labeled "Factors that facilitate the development of religious tourism in Greece."

5 Conclusion and Policy Recommendations

The findings of the research indicate that some variables function as both facilitators and impediments to the growth of religious tourism in Greece. The factors facilitating this growth include attracting pilgrims from traditional and emerging religious tourism markets, implementing specialized and targeted promotion and advertising of the "religious image," developing appropriate tourism planning and formation of customer management strategy, and strengthening of bodies related to religious tourism. Conversely, the factors hindering development include insufficient promotion of religious sites, lack of cooperation among stakeholders (such as the church, the tourist industry, local government organizations) involved in religious tourism, the absence of an organized network for the provision of information and the organization of guided tours of religious sites, and the lack of strategic planning for the development of religious tourism.

Additionally, the multiple regression analysis conducted revealed independent variables expressing functional relationships, which can be utilized to create mathematical models for tourism predictions. Specifically, the dependent variable "factors that improve religious tourism" is influenced by several independent variables, including the attraction of tourists/pilgrims from traditional and emerging religious tourism markets, the implementation of targeted and specialized promotion and advertising of the "religious image," the development of appropriate tourism planning and the formation of strategic customer management, and the strengthening of bodies related to religious tourism. Furthermore, the dependent variable "factors hindering the development of religious tourism" is influenced by independent variables, such as the inadequate promotion of religious sites and the lack of cooperation between the bodies (e.g., church, tourist industry, local government organizations) involved in the development of religious tourism.

In light of these factors shaping the growth of religious tourism, an integrated framework for strategic planning and policy for local tourism development can be formulated to capitalize the Greece's religious tourism potential. To achieve this objective, the following four key parameters should be taken into consideration:

1. Status of Existing Religious Sites: This includes the preservation and protection of religious heritage and initiatives to promote monuments of religious interest in a sustainable manner.
2. Specialization and Cooperation Among Stakeholders: This focuses on enhancing collaboration among the various bodies involved in the growth of religious tourism in Greece.
3. Promotion and Advertising Strategies: This encompasses a range of strategies for promoting and advertising religious tourism, both domestically and internationally.
4. Local Activity Grid: This pertains to establishing a network of activities at the local level, with the religious motivation of tourists serving as a primary pillar. This must be achieved in alignment with the fundamental principles of sustainable tourism development and linked to the specific and multi-faceted nature of tourism demand in Greece.

By implementing these policy recommendations and building on the strengths identified in the research, Greece can enhance its appeal as a religious tourism destination, fostering economic growth while promoting cultural exchange and mutual understanding between different faiths and cultures.

References

Albani X (2014) Inbound religious tourism in Greece: A case study of Thessaloniki. [Master's Thesis, Aristotle University of Thessaloniki]. Institutional Repository of Scientific Works of the Aristotle University of Thessaloniki. https://doi.org/10.26262/heal.auth.ir.135267 (in Greek)

Cerutti S, Piva E (2016) The role of tourists' feedback in the enhancement of religious tourism destinations. Int J Religious Tour Pilgrimage 4(3):4

Greek National Tourism Organisation (2005) Religious monuments in Greece. Athens. (in Greek)

Greenwood D (1989) Culture by the pound: An anthropological perspective on tourism as cultural commoditization. In: Gmelch SB (ed) Hosts and guests: the anthropology of tourism, 2nd edn. Pennsylvania Press, Philadelphia, pp 157–169

Griffin K, Raj R (2017) The importance of religious tourism and pilgrimage: reflecting on definitions, motives and data. Int J Religious Tour Pilgrimage 5(3):1–9. ISSN 2009-7379. https://arrow.dit.ie/ijrtp/vol5/iss3/2/

Hill-Smith C (2011) Cyberpilgrimage: The (virtual) reality of online pilgrimage experience. Relig Compass 5(6):236–246

Kumar S, Shekhar, Valeri M, Sageena G (2022) Spiritual tourism transition: An ethos of hope during COVID-19 crisis from Indian perspectives. In: Valeri M (ed) Tourism risk. Crisis and recovery management. Emerald Publishing

Lagos D, Christogianni P (2006) The development of religious tourism in Greece. In: Vogiatzis A (ed) Maria Negreponti-Delivani Honorary. University of Macedonia, Thessaloniki, pp 465–478. (in Greek)

Moira P (2009) Religious tourism. Interbooks Publications, Athens (in Greek)

Olsen DH (2010) Pilgrims, Tourists and max Weber's 'Ideal Types. Ann Tour Res 37(3):848–851

Poulaki P (2018) The development of the religious - pilgrimage tourism in the North Aegean islands [Doctoral dissertation, University of the Aegean]. National Archive of PhD Theses. http://hdl.handle.net/10442/hedi/43317 (in Greek)

Poulaki P, Lagos D (2016) Ottoman monuments of chios and cultural tourism. Lambert Academic Publishing

Rinschede G (1992) Forms of Religious Tourism. Ann Tour Res 19(1):51–67

Rodosthenous N (2012) Religious and pilgrimage tourism in Cyprus: pastoral approach. [Doctoral dissertation, Aristotle University of Thessaloniki]. Repository of Scientific Works of the Aristotle University of Thessaloniki. https://doi.org/10.26262/heal.auth.ir.132307 (in Greek)

Shekhar, Valeri M (2023) Modelling the effect of spiritual tourism motivators on spiritual tourism consumption. Tourism 71(3):634–646. https://doi.org/10.37741/t.71.3.13

Sidiropoulos G (1998) The Athos Peninsula as a communication and cultural pole of tourist attraction. Soc Res Rev 95:69–85. (in Greek)

Terzidou M, Scarles C, Saunders MN (2018). The complexities of religious tourism motivations: Sacred places, vows and visions. Annals of Tourism Research, 70, 54–65

Tsorvas P (2024) The religious pilgrimage as a tourist experience: The case of Greece [Unpublished doctoral dissertation]. University of the Aegean. (in Greek)

Tzovolos A (2016) The Religious - Pilgrimage tourism in Greece and its contribution to local development. The case study of the Island of Aegina. [Master's Thesis, Hellenic Open University]. Apothesis Repository of Hellenic Open University. https://apothesis.eap.gr/archive/item/80178 (in Greek)

UNESCO (2024, July) World Heritage List. UNESCO, World Heritage Convention. https://whc.unesco.org/en/list/

Valeri M (2022) New governance and management in tourist destinations. IGI Global Publishing, Hershey, PA

Vouronikou D (2017) Religious tourism in Meteora. [Master's Thesis, University of Thessaly]. Library and Information Centre of the University of Thessaly. https://ir.lib.uth.gr/xmlui/handle/11615/41186?locale-attribute=it (in Greek)

Religion and Tourism: Management of Cognitive, Emotional, and Spiritual Experiences in Destination Bulgaria

Mariya Stankova and Svetoslav Kaleychev

Abstract Religious tourism should be considered a complex cultural phenomenon that goes beyond religion as a sign of determining followers and visitors to a given destination. And since religion is among the oldest drivers of human travel, its potential for impact as a motive is well known in tourism. Various products and services are offered in connection with it so that the need for a person to achieve peace of mind is transformed into an opportunity for development and economic benefits. This wide scope and interpenetration suggest that religious tourism should be approached holistically. Moreover, when practicing it, a rendezvous occurs between people with different origins, backgrounds, and cultural perceptions. Tourist destinations, which have the potential for the development of religious tourism, should take all this into account, as well as to consider the fragile nature of the resources used. One worthy destination for religious tourism is Bulgaria. As location for tourism, for its 1343-year history, the country guards the treasures of previous civilizations from the Balkan region, as well as witnesses the progress of Christianity in Europe and the traditions of Islam and some fine examples, related to Judaism, offering interesting experience on the tourist paths.

1 Introduction

Religious tourism is an important component of the global tourism industry, transforming religious sites, artifacts, events, and phenomena into experience, that attract the interest of millions of people. And since religion is among the oldest drivers of

M. Stankova
Faculty of Economics, South-West University, Blagoevgrad, Bulgaria
e-mail: mstankova@swu.bg

S. Kaleychev (✉)
Faculty of Economics of Infrastructure, University of National and World Economy, Sofia, Bulgaria
e-mail: skaleychev@unwe.bg

M. Valeri (ed.), *Contemporary Religious Tourism*, Tourism, Hospitality & Event Management, https://doi.org/10.1007/978-3-031-79074-4_15

human travel, its potential for impact as a motive is well known in tourism. Various products and services are offered in connection with it so that the need for a person to achieve peace of mind is transformed into an opportunity for development and economic benefits.

The main thesis, the realization of which is sought in this study, is directly related to the possibilities of satisfying the cognitive and/or spiritual needs of people achieved through religious tourism when considering the practice in the destination Bulgaria. It is accepted that religion concerning tourism should be considered a complex cultural phenomenon that goes beyond religion as a sign of determining followers and visitors to a given destination. Religion also offers a philosophical system for understanding the surrounding world, but it is also related to achieving certain suggestions and mental states that can provoke behavioral patterns. The inter-religious differences, recognized when accessing objects of different origins and religious affiliations, contribute to ethnic tolerance. In economic terms, the flow of tourists to these destinations contributes to competitiveness, being able to enliven not-so-popular sites, alongside those enjoying national or international attention.

In the given focus, religious tourism should be approached holistically. On the one hand, one should consider the location and significance of religious tourism sites and events—monasteries, churches, sanctuaries, religious festivals, and rites that are or could be presented as desirable places to visit. On the other hand, bearing in mind the words of the UNWTO Secretary-General, Taleb Rifai (UNWO 2016), when he said “We suffer from deficits of different kind, being the lack of tolerance and understanding, a major one. Religious tourism is one of the driving forces to bring people from varied backgrounds together under a common cause: the admiration and protection of heritage of tangible and intangible nature.”

The object of the study is bound to places and events related to religious tourism in the Rila-Pirin tourist area of the destination Bulgaria. It is one of a total of nine tourist areas in the country that are large enough to be clearly recognizable on the tourist map and small enough to be managed effectively. Each region has a territorial scope that corresponds to the real dimensions of tourism development (including the existing organizational structures) and/or to the revealed tourist potential in terms of features such as attractiveness, infrastructural provision, homogeneity, territorial integrity, and tolerance (Concept of tourist zoning of Bulgaria 2018).

On this basis, principal (two types of tourism, which together define the uniqueness of the regions) and extended (complementary types of tourism) specialization is offered for each tourist area. In the case of the Rila-Pirin region, its nationally defined specialization is related to mountain (all types—skiing, hiking, recreational) and religious (and pilgrimage) tourism (Table 4)—p. 30—Tourism.Goverment 2018—Concept for touristic zoning of Bulgaria.

In the conditions of lack of information about the real tourist demand in the area regarding religious tourism and the impossibility of carrying out a detailed study within the framework of the present study, the scope was determined, taking into account the “internal” provision of inherent resources of the territory by municipalities. It was established: on the territory of a total of 12 municipalities, there are 100 objects of the Orthodox faith, 40 with pagan roots, 10 of Islam, and 1 related to Judaism:

Samokov Municipality—15 sites of the Orthodox faith; 1 object from the pagan spectrum; 1 of Islam.
Kostenets municipality—1 object of the Orthodox faith; 15 objects from the pagan spectrum.
Dolna Banya Municipality—1 object of the Orthodox faith; 22 objects from the pagan spectrum.
Municipality of Sapareva Banya—9 sites of the Orthodox faith; 2 objects from the pagan spectrum.
Dupnitsa Municipality—19 sites of the Orthodox faith; 1 object from the pagan spectrum; 1 of Islam; 1 of other religions—Judaism.
Municipality of Kocherinovo—1 object of the Orthodox faith; 1 object of the pagan spectrum.
Rila Municipality—7 sites of the Orthodox faith; 1 object of the pagan spectrum.
Blagoevgrad Municipality—11 objects of the Orthodox faith; 1 object from the pagan spectrum; 1 of Islam.
Razlog Municipality—10 objects of the Orthodox faith; 1 object of the pagan spectrum.
Belitsa municipality—2 objects of the Orthodox faith; 1 object from the pagan spectrum; 4 of Islam.
Yakoruda Municipality—1 object of the Orthodox faith; 1 object from the pagan spectrum; 3 of Islam.
Belovo Municipality—10 sites of the Orthodox faith; 2 objects from the pagan spectrum.

The subject of the study is the systematization of information about the religious sites of the Rila-Pirin tourist area, to study the potential of selected religious sites and practices and, on this basis, to reach specific recommendations concerning the management of cognitive and spiritual experiences in destination Bulgaria.

2 Religious Tourism in Focus

In the course of the time horizon of human civilization, travel has played a vitally important role in the formation of a worldview. The discovery of new and unknown lands, the trading of goods and services, and the exchange of cultures and knowledge would not be possible without making various journeys. Some of the first forms of travel and tourism arose from religious, health, and commercial motives. People have traveled and traveled, whether motivated by their own religious beliefs or out of curiosity about the religion practiced by others, to visit holy places such as Jerusalem, Lourdes, Vatican City, Angkor Wat, Mecca, Medina, etc.

Jusserand (1920) talks about how religious travel, as a type of movement (and tourism), has its origins in pilgrimage as early as Antiquity, and later in the Middle

Ages, defining pilgrimage as non-sequential. It has as many forms as there are pilgrims themselves—a vow given in case of illness or impending danger, atonement for sins that find expression in fasting, begging, self-flagellation, or walking barefoot. Again, interesting examples are found at Jusserand (1920)—in the fourteenth century, Rome, Jerusalem, and Galicia were the most frequently visited places and pilgrimages to them were extremely challenging for believers who faced difficulties of a different nature. Often attacked and robbed, they traveled under constant threat. In an attempt to control this type of travel, in early 1388 King Richard II issued an order requiring English pilgrims to possess and carry permits, the prototypes of modern passports. Eleven years later, the number of ports from which pilgrims could sail to overseas destinations was again limited by royal order to two, Dover and Plymouth, unless others received special permission from him. Fees for the transport of passengers and cargo, mandatory for all "carriers" (The Examination of Master William Thorpe, Arber's Engl. Garner 1407), were also established. According to Jusserand (1920), pilgrims were offered special services—for a fee, they could enjoy the company of performers of religious songs or, upon reaching the goal of their journey, have their arrival announced by blowing horns.

Apparently, religion and spirituality provoke tourists' interest in tourist destinations, and many of them have taken advantage of their potential as a travel motivation factor. This position is substantiated in one of the often-cited works on tourism and religion—that of Timothy and Olsen (2006). In this publication, as well as in the large part of existing research on the subject, definitions of religious tourism are proposed, its manifestations are sought and the effects it has on a spiritual, economic, and socio-cultural nature are tracked (Rinschede 1992; Rashid 2018; Stausberg 2014; Raj et al. 2015). What makes an impression, however, is that Religious tourism is highly difficult to define. There is no unique definition. Broadly, Timothy and Olsen (2006) states that religious tourism includes pilgrimages and visits to specific religious events, as well as visits made to churches, cathedrals, and shrines primarily for their cultural or historical (as opposed to religious) significance. According to Shackley (2004), religious tourism also includes staying at retreat houses. Much of the analysis of religious travel has concentrated on pilgrimage, including classifications of sites, discussion about the travel patterns of religious tourists, and differentiation between pilgrim groups (Timothy and Olsen 2006).

Religious tourism is now well accepted as a niche market in tourism "within the segment of cultural tourism, comprising four specialist niches: pilgrimages, visiting sacred. sites, church, mosque, and temple tourism, and travel for the pon or worship" according to the European Market Potential for Religious Tourism (2010). Against this background, however, there is a discussion on where pilgrimage tourism should be positioned. McConnell (1973) makes a statement that much academic discussion has also centered on whether pilgrimage constitutes a legitimate branch of tourism with much of the literature on pilgrimage concentrating on the definition of the term. In this context, religious tourism is identified as visiting religious sites and places, primarily motivated by the satisfaction of spiritual needs. Pilgrimage tourism, in addition to cult tourism, implies travel for worship before holy places of faith or participation in religious rituals, regardless of the cultural or architectural value of the sites or events. This extension is provided by a definition of the pilgrim as "… one who strives to obtain salvation on of his/her soul through a physical journey in which caritas Caritasfor God, and not cupiditas, love for material things, drives them" (Shackley and Welton 2008).

Experts also speak of Christian tourism, a subcategory of religious tourism. In this variant of species decomposition, Christian and religious tourism overlap with visiting sites such as shrines and churches, which, however, except for spiritual enlightenment, do not make special efforts to attract visitors to the destinations where they are located, according to Ron and Timothy (2019). However, Benea and Petroman (2007) said that given the large number of believers of other religions and because it is not correct to limit in any way the possibility of developing different tourism products, it would be terminologically correct to use the term religious tourism.

Already at the beginning of the new millennium, it is accepted that religious tourism on a global scale pilgrimage probably attracts 250 million people/year. Ten years later, Tarlow (2010) notes that "it is estimated that in the US alone some 25 percent of the traveling public is interested in faith-based tourism… and on a global scale, the projection of religious tourism reaches up to USD 18 billion and 300 million travelers." Thus, religious tourism to this day is being imposed as one of the fastest-growing segments and has recently attracted considerable interest from tourism planners and marketers who have realized their economic potential. According to Global Religious Tourism Market (2021), on a global market scale, religious tourism is expected to grow at a CAGR of 7.5% from 2022 to 2030. Additionally, as López (2023) states the world market value of tourism, based on faith is anticipated to reach 15 billion U.S. dollars, with a compounded annual growth rate of around 10.5% from 2023 to 2033.

Within these trends, at the regional level, one of the most significant destinations for religion-related tourism is Europe. This statement is not based so much on the number of visitors as on the number of sites and their religious, spiritual, and cultural value. Inside Europe, religious tourism focused on urban centers, churches, cathedrals, and monasteries as well as shrines and pilgrimage routes (Nolan and Nolan 1989). According to Russell (1999), tourists may also be moved by religious emotions as well as pilgrims, but these may not necessarily be "religious tourists." According to Religious Tourism In The European Union: Resources And Trends (2014), "Much of the recent interest in religious travel has come from the (belated) realization that such activities are of major economic importance, although the undoubted relationship between religion and tourism has also received attention from articles including Bywater (1994), Vuconic (1996), Rinschede (1992), Coleman and Elsner (1995), Fernandes et al. (2003)." However, there are a sufficient number of sources and authors that provide descriptions of the development and practices of religious tourism, including Nolan and Nolan (1989, 1992), Enschede (1992), Coleman and Elsner (1995), Vukonic (1996), Timothy and Olsen (2006), Russell (1999), Jackowski (2000a, b), Fernandes et al. (2003), Iliev (2020), and more.

Concerning the tourist destinations of religious tourism, the research points to the presentation of the accumulated impressions and experience, where it is committed to offering a combined tourist product of religious and spiritual experiences and cultural elements demonstrating the traditions of the local communities. For example, Blackwell (2007) talks about the simultaneous satisfaction of tourists' religious and cultural interests, their curiosity, and the search for new meaning in life. For their part, Heydari Chianeh et al. (2018) draw attention to the possibilities of religious tourism to go beyond the experiences of religious rituals by engaging visitors with the cultural heritage, artistic values, and historical events of the destination. Also of interest is a study of Harghita country council (WP3—Training needs and population wider attitude in Romania, Projects 2014–2020.interregerope.eu 2014) introducing a SWOT analysis, in which four groups of elements were identified about religious tourism experiences. It is striking that they are bound with the interaction of the participants, as more specifically expressing themselves to: (a) Existential-religious elements; (b) Social-helpful elements; (c) Cultural-formative elements; (d) Emotional elements. Besides the spiritual pull, other motives than religious are "the beauty of places and the monastery," "curiosity "relaxation," "cultural motivation," and "place visit" (Training and eds and population wider attitude in Romania 2019). At the same time, according to Timothy and Boyd (2003) as a result of marketing programs and the high interest in cultural tourism, religious sites are more and more visited by rather curiosity-driven visitors as compared to genuine pilgrims and this is the reason why they are promoted, adapted, and included in products meant for the tourism market.

3 Description of Context

The etymology of the word "context" refers to the most general representation of "the parts of discourse that surround a word or passage and can throw light on its meaning" or "the interconnected conditions under which something exists or happens" (Merriam-Webster 2024). And it is precisely on the second case of use that the research focuses on concerning religious tourism in the destination Bulgaria. Accepting the context as a synonym for environment or setting, first of all, the specific objects of religious tourism and their inherent characteristics, such as attractiveness, accessibility, and possibilities of residence, are identified. Next, groups of visitors are defined, according to the most general categories: those traveling by religious affiliation to holy places, pilgrims in traditional pilgrimage centers, participants in religious holidays and rituals, travelers for educational purposes, travelers with scientific or educational purposes, and to believers in the healing and spiritual powers of religious objects.

Entering the scope of the Rilo-Pirin tourist area, the first conclusion to be drawn is that a relatively large territory is being considered, which includes municipalities heterogeneous in terms of religious affiliation, such as Blagoevgrad, Bansko, Belitsa, Garmen, Gotse Delchev, Dupnitsa, Kresna, Kostenets, Petrich, Razlog, Sandanski, Sapareva Banya, Satovcha, Hadjidimovo, and Yakoruda. The main religions practiced by the population are Orthodox Christianity and Islam; part of the population is of Pomak origin. For the region, the specific demographic figures are as follows: out of a total of 292,227 inhabitants, as of 2021 census, 196,841 identified themselves as belonging to the Christian faith; 52,958 are Muslims, 46 are Jews, and 3925 people do not profess a particular religion (NSI 2024).

Of the listed objects above in the text, within the territory of the Rila-Pirin region, a total of 111 are related to the three main religions, forming the profile of the population. Certainly, the main part of inhabitants belongs to Orthodox Christianity, which is also official for the Republic of Bulgaria. Bearing this in mind, without belittling the importance of the religious sites of the other two religions, the research focus is placed on the two key ones for Orthodoxy and the most important in religious terms, sites in the region—Rila and Rozhen monasteries. The first monastery, in fact, is the largest in Bulgaria and is included in the UNESCO World Heritage List. It is also one of the symbols of the country, created by its patron saint—Saint Ivan of Rila. Over the centuries, the monastery has established itself as a holy place in, as described by the patriarch of Bulgarian literature, Ivan Vazov:

> When a man comes here to repent,
> must forget sinful peace,
> to flee temptations and seek peace.

The Rila Monastery "St. Ivan of Rila" (Rila Monastery 2024) was founded in the tenth century. It operates as a men's holy monastery, with a temple holiday on October 19 and under the direct jurisdiction of the Bulgarian St. Synod. Most significant for pilgrimage visits by local and foreign visitors, however, is August 15, the day of the Ascension of the Mother of God, after which the monastery church is named.

The monastery is located above the town of Rila, near the Rila National Park (2024) and the Rila Monastery Nature Park (2024). It is 100 km from the capital Sofia and 20 km from the largest city in the region—Blagoevgrad. 27 cultural monuments are also available in the vicinity, among which the church "St. Archangel Michael" (twelfth to thirteenth century), and the church "St. Nicholas" (eighteenth century). The monastery "Orlitsa" also belongs to the monastery, with the church "St. Peter and Paul," which dates back to the fourteenth century. The famous Bulgarian film "The Goat's Horn" was filmed there. Next to the parks, an interesting natural landmark is the "Stobsky pyramids" (Btsbg.org 2024, Pyramidsmonastery), which has museums, a library, and a wing intended for the accommodation of guests.

The other important religious and pilgrimage site in the Rila-Pirin region—the Rozhen Monastery (Pravoslavieto. com 2024)—is actually the second largest holy abode here. There is registered data about him from the seventeenth century, and the first information is from the thirteenth century. He acts as a male, with a temple holiday of September 8. It has been preserved relatively intact since the Middle Ages, despite the complicated historical times it has passed through, like the Rila Monastery. The monastery church is notable for its frescoes from 1597. It also has an interesting wooden iconostasis and stained glass in the eastern part. Having established itself as an important religious and cultural center on the territory of Bulgaria, the monastery contributed to the interaction between the Balkan peoples.

This role is predetermined by its geographical location, near the border with Greece and today's North Macedonia. 6 km from it is the smallest town in Bulgaria, Melnik—a popular tourist destination associated with architectural landmarks, natural phenomena such as the Melnik sand pyramids, and famous wine cellars where movies are shot. To the southwest is the border mountain Slavyanka. In this monastery, there is also an organized exposition.

Along with the described two sites of religious and pilgrimage tourism, the Rila-Pirin region is also distinguished by another, a particularly special place that attracts visitors— Rupite (Bulgariatravel. org 2024) area and the ecclesiastical and spiritual monument—"St. Petka Bulgarska." This is the place where the prophetess Vanga lived and was buried in 1996. It attracts many pilgrims from different parts of the country, neighboring countries, and especially from Russia and former Soviet republics. Vanga, also known as Grandma Vanga, is known for her numerous predictions that have come true. Legend has it that as a child, in a storm, she lost her sight, but in this accident, she received a vision that gave rise to her unique abilities. In the last years of her life, Vanga lived in Rupite, as she considered the area to be an energy center. Built as a church in 1994, St. Petka Bulgarska" is a Christian temple, but it differs from the canon with its frescoes. They are the work of the famous Bulgarian artist Svetlin Rusev and impress with their realism. In the area of Rupite, which is also a protected area, there once existed a volcano—Kozhuh, the remains of which are today a natural landmark. The area is also known for its healing hot mineral springs. Another particularly significant object here is the ancient city located on the southern slopes of Kozhuh, which existed in the period of the fourth century BC to the sixth century AD—Heraclea Sintica.

The visit to each of the described sites is accompanied by lasting memories in the minds of the visitors and brings a spiritual experience. Therefore, regardless of the purposes and the motive of the tourists, the benefits of practicing religion-related tourism are undoubted. Looking at the significance of the research focus, for each of the three religious sites, when trying to determine the size tourist flows to them, it was found that there is no information about the realized visits or the information found is fragmentary.

4 Methodology

The chosen methodology is tailored to the research thesis, object, and subject. It is aimed at three selected places related to religious tourism in the Rila-Pirin tourist area of destination Bulgaria, to study their potential and to specify recommendations in relation to the management of cognitive and spiritual experiences in destination Bulgaria. Complex in nature, the methodological apparatus uses quantitative and qualitative methods. More specifically, it includes a review of specialized publications on the topic, as well as uses the methods of observation, expert evaluation, analysis, and synthesis. The survey method was applied in relation to the study of tourists' attitudes toward religious tourism in the selected sites—Rila Monastery,

Rozhen Monastery and the Rupite area, and the temple monument "St. Petka Bulgarska." In terms of information, data from the National Statistical Institute (NSI 2024) were used, as well as own results obtained on-site in the three investigated sites. The analysis of the results is carried out using the Excel software product—Pivot. The synthesis methodology is applied to the conclusions and summaries to outline some basic approaches for managing cognitive and spiritual experiences and enhancing tourist interest.

The survey method for researching tourist attitudes is preferred for several reasons, but mostly because it is a tool for relatively quick collection of heterogeneous data according to Riddick and Russell (1999), including demographic and additional data such as respondent attributes, experience facts, knowledge acquisition, attitudes, preferences, and beliefs. Bearing in mind that "the fatigue of the respondents occurs approximately in 15 to 20 minutes" (Lundegren and Farrell 1985), a questionnaire of 14 questions was systematized for the purposes of the specific survey. Some main points are taken into account, namely: during the initial stage, the main target topic of the study is formulated, working questions are developed (selecting the question type (Yes/No, multiple choice, importance scale). Next place, the working questions were edited, to achieve clarity and ambiguity, the elaboration of explanations, in case of necessity. In the third stage, the questions are positioned in a logical sequence, the attitudes and expectations of visitors to the three selected sites as practitioners or non-religious tourism.

5 Discussion

Going into detail, the survey questionnaire was multiplied into 700 copies and distributed among visitors in the three selected sites. A total of 500 responses were collected from respondents, as follows: 275 from Rila Monastery, 102 from Rozhen Monastery, and 123 from Rupite.

The summary profile of the respondents shows that of the 500 participants in the survey, 60% are women and 40% are men. Figure 1 shows the percentage distribution of respondents from the three sites.

Specifically, the detailed gender profile, derived by sites (Fig. 2), shows that on average per site, the highest proportion of women visited the Rozhen Monastery, followed by the Rupite. It is important to note that among the visitors of the Rila Monastery, the female/male ratio is the most balanced, and the logical explanation for this could be found in the general popularity of the site, as well as its presence on the UNESCO World Heritage List.

The age distribution of the participants in the survey is as follows (Fig. 3): in both monasteries, the highest share of visitors is in the age category between 50 and 59 years. At the same time, the share of tourists between the ages of 18–29 is not small. These are young people who in some cases accompany elderly members of their families or are people who like to travel and wish to get to know the rich cultural and religious heritage of the destination Bulgaria.

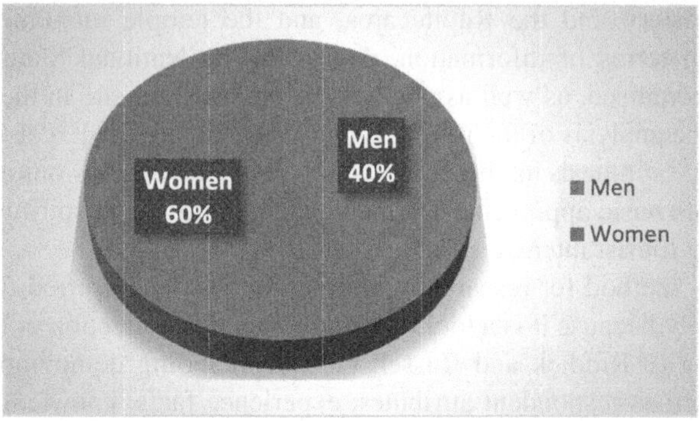

Fig. 1 Proportion of examined males and females in total for the three sites (Authors' research)

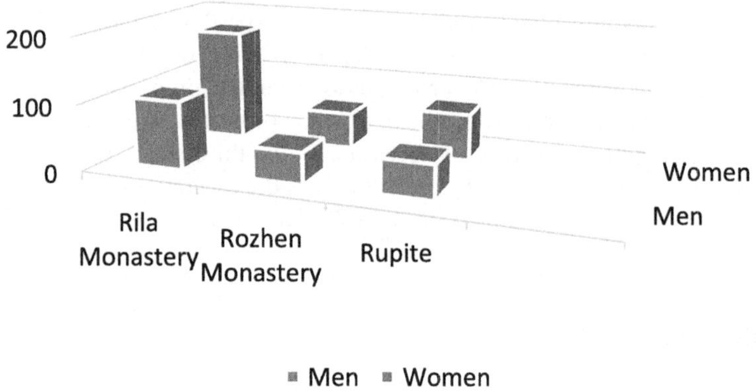

Fig. 2 Proportion of examined men and women in total for each of the sites (Authors' research)

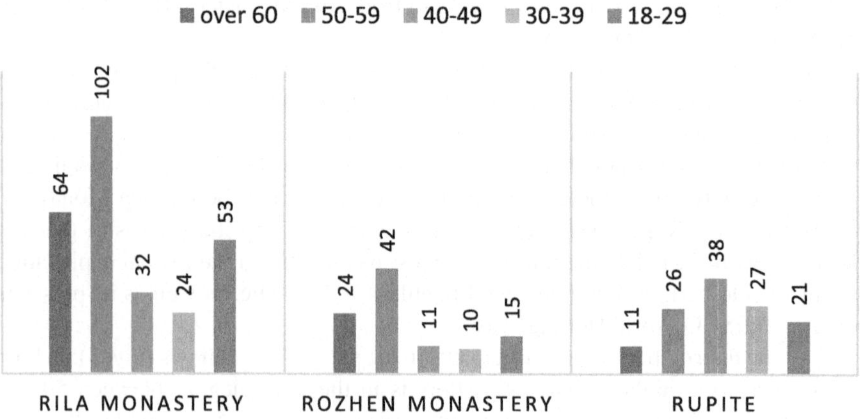

Fig. 3 Ratio of respondents by age distribution (Authors' research)

The research data show an interesting feature of the Rupite. There, the largest number of visitors is between the ages of 40–49, and even younger. This fact is explained by the mystical nature of the object, which unites culture, history, religion, and spiritual experiences. A large part of them also visit the mineral springs located in the immediate vicinity. Expectations regarding this site are for a significant change in the number of tourists, provoked by the recent unique finds in the ancient city of Heraclea Sintica.

In the three religious sites studied, the largest group of visitors with Bulgarian nationality is the largest in terms of numbers—339 people in total; in next place are the representatives of the Republic of Greece—63 respondents, and in third place are those from the Republic of North Macedonia—43 people. These are mainly Orthodox Christians. Most foreigners register at the Rila Monastery, which is also visited by visitors from Italy and Spain, although it can be assumed that these tourists belong to the Catholic faith. A total of 15 Muslims are also found among the respondents (Fig. 4).

The results obtained during the processing of the questionnaires show that, at all three sites, the largest number of tourists visiting them for religious purposes make at least one trip within 3 months on average. The majority of visitors to the Rila and Rozhen monasteries indicate that they make religious or pilgrimage trips every 3 or 6 months; however, visitors to Rupite note that they make such a trip at least once a month. This feature could be explained by the specific nature of the site, which combines religion, spiritual development, recreation, and relaxation, definitely having the role of factors generating additional motivation to travel.

In general, the respondents indicate religious motives as the main reasons for visiting the Rila and Rozhen monasteries, which are leading for all; secondly, their interest is provoked by participation in religious holidays, and thirdly, there are reasons of a personal nature. The experiences resulting from them are developed against the background of cultural enrichment and familiarization with the history of the objects. Differences are again found at the third investigated site—Rupite.

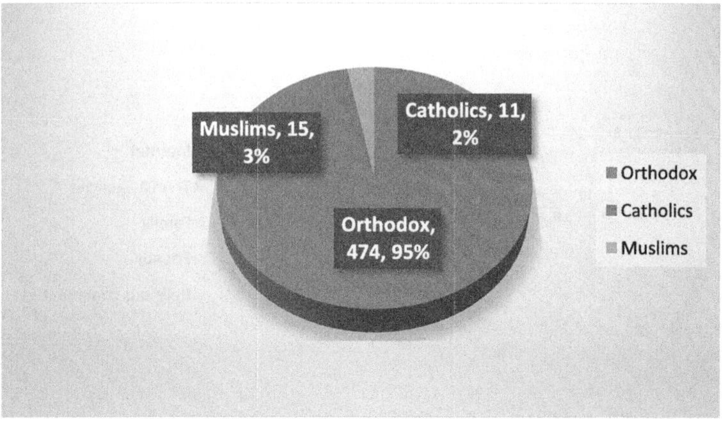

Fig. 4 Ratio of respondents by religious affiliation (Authors' research)

The analysis of the responses indicates that the site attracts tourists in a specific way, and the leading reason for visiting according to the respondents is its image as an energy center. Next, they cite the spiritual experience it offers them as a reason for visiting.

The stay in objects of religious significance, for 50% of the respondents, is from 3 to 4 h; 31% spend 1 to 2 h, and the remaining 19% say they spend more than 4 h.

Decomposing the processed data reveals a picture where for 71% of respondents, religion is very important for choosing a tourist destination, and for another 26%, religion has an important role; for only 3% of respondents, the religious motive is not important when choosing a tourist destination. A total of 61% of respondents stated that they combine religious tourism with other types of tourism when traveling. The remaining 39% answered that when it comes to traveling with a religious purpose, they only visit religious sites.

When asked what their expectations were regarding the services and amenities that should be available in the three surveyed sites, almost half of the respondents (48%) indicated that the availability of places to eat was important to them. For 27% of respondents, it is important to have accommodation available on-site or nearby. An interesting point that requires additional attention is that for 15% of the respondents in these places, it is good to have outlets selling natural foods and cosmetic products.

When asked about the sources of information that they used or considered appropriate in relation to the specific sites and the religious ones, in general, 37% of the respondents indicated in the first place the travel agencies as the main source. In second place are online sources, indicated by 35% of respondents; 13% are informed by their relatives, and another 12% by their friends. Only 3% indicated that representatives of the religious community to which they belong influenced their trip (Fig. 5). At the same time, in 99% of cases, the respondents to the questionnaire were ready to recommend the relevant religious site to others.

Assuming that religious tourism combines a tourist trip with a visit to a religious site, various experiences are realized in connection with it, which rank the visitor in

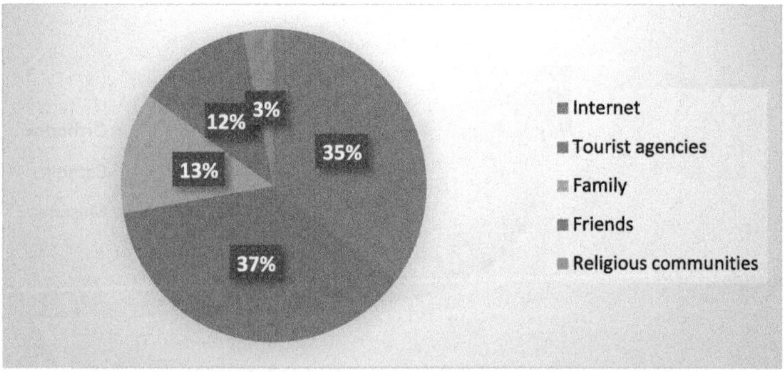

Fig. 5 Information provision of travel with a religious purpose, according to respondents (Authors' research)

terms of his motivational bases. Indirectly, this was also confirmed by the research carried out for the specific objects in the destination Bulgaria. Bearing this in mind, other confirmations were sought in specialized publications, strategic documents, and plans of destination Bulgaria. The review found that religious tourism is mentioned in the National Strategy for the Sustainable Development of Tourism in the Republic of Bulgaria as part of the cultural and cognitive tourism product (p. 54) according to Tourism.government.bg (2024). It clearly states that the lack of restrictions of a religious nature, the existing freedom of religion in the country, together with the support of the state, society, and church institutions, provide good opportunities for combining the religious with elements of the cultural-cognitive tourist product. In this sense, no explicit actions and measures are foreseen to support the development of religious tourism as an independent type. Therefore, specific recommendations should be made in connection with its sustainable promotion and the achievement of satisfaction among tourists who practice it. Against the background of several objective limitations arising from the lack of specialized scientific and applied research in the field of religious tourism as a destination in Bulgaria, as well as sufficient management experience, achieving success in its development could be realized through the implementation of targeted strategies for its promotion as a combination of culture, history, traditions, and religion.

6 Conclusion

The development of religious tourism is tempting for both already-developed tourist destinations, as well as for newly emerging ones. On the other hand, a significant online presence is critical for tourism businesses to continue to have access to the changing demand and global market. This is particularly important for the current and future market positioning in tourism and must be considered as a starting point for destination strategic planning. The region of South-West Bulgaria reveals significant potential for sustainable tourism development based on existing cultural and religious resources. The important thing is to capture the trend of modern orientation toward new types of tourist service to reach the minds of visitors to these sites. And since these visitors, in the general case of religious tourists, as well as in the specifics of the conducted research, travel with a well-defined and clear purpose, they have specific needs and interests that the usual tour guide talk is not able to satisfy. Therefore, considering the possibilities, the recommendation that is made is the religiously bound destinations to take advantage of the possibilities of interpretation.

Tourism interpretation is little known in Bulgaria and there is an objective need to clarify the issue, regarding the understanding and application of effective interpretation within religious tourism sites. On the other hand, the potential of Bulgaria as a destination for religious tourism exists, with interest from the domestic and international markets as well. The research conducted proves it, even in short-time intervals. Of course, the possibilities are much greater, but to achieve them, it is necessary to adapt with an open mind and collective efforts, to attract all interested parties.

References

Benea M, Petroman I (2007) Bazele turismului.Editura Eurostampa, Timişoara

Blackwell R (2007) Motivations for religious tourism, pilgrimage, festivals and events. In: Religious tourism and pilgrimage festivals management: an international perspective. Cabi, Wallingford, pp 35–47

Btsbg.org (2024). www.btsbg.org/100nto/stobski-piramidi. Accessed 25 Apr 2024

Bulgariatravel.org (2024) Rupite. www.bulgariatravel.org/bg/%d1%80%d1%83%d0%bf%d0%b8%d1%82%d0%b5/. Accessed 22 Apr 2024

Bywater M (1994) Religious travel in Europe. Travel Tour Anal 2:39–52

Coleman S, Elsner J (1995) Pilgrimage: past and present in the world religions. Harvard University Press

Fernandes C, McGettigan F, Edwards J (eds) (2003) Religious tourism and pilgrimage – atlas: special interest group (1st expert meeting, pp 27–42). Tourism Board of Leiria/Fatima

Global Religious Tourism Market (2021). www.dataintelo.com/report/global-religious-tourism-market?gad_source=1&gclid=Cj0KCQjw97SzBhDaARIsAFHXUWCm55WhSgU_W2CZTm7vjIc4Y4atArbkQ7c0ERH-D5-rJV1xIVeaR4waAvI-EALw_wcB. Accessed 23 Apr 2024

Heydari Chianeh R, Del Chiappa G, Ghasemi V (2018) Cultural and religious tourism development in Iran: prospects and challenges. Anatolia 29(2):204–214. https://doi.org/10.1080/13032917.2017.1414439

Iliev D (2020) The evolution of religious tourism: concept, segmentation and development of new identities. J Hosp Tour Manag 45:131–140. https://doi.org/10.1016/j.jhtm.2020.07.012

Jackowski A (2000a) Religious tourism—problems with terminology. In: Jackowski A (ed) Peregrinus Cracoviensis, Cracow: publishing unit. Jagiellonian University, Institute of Geography

Jackowski A (ed) (2000b) Selected research problems in the geography of pilgrimage. Peregrinus Cracoviensis 10. Jagellonian University, Kraków

Jusserand JJ (1920) English wayfaring life in the middle ages (XIVth century). TF Unwin Limited

López AM (2023). www.statista.com/statistics/1371225/worldwide-estimated-value-of-faith-based-tourism-market/. Accessed 20 Apr 2024

Lundegren H, Farrell P (1985) Evaluation for leisure service managers. Saunders College Publishing, New York, p.160 цит. по Ham, Sam H. (Ham, S. & Weiler, B). Interpretation Evaluation Tool Kit. Methods and tools for assessing the effectiveness of face-to-face interpretive programs, CRC for Sustainable Tourism Pty Ltd., 2005. www.sustain.pata.org/wp-content/uploads/2014/12/InterpEvalToolkit_Sep07.pdf. Accessed 10 Apr 2024

McCannell D (1973) Staged authenticity: arrangement of social space in tourism settings. Am J Sociol 79(3):589–603. www.travelstudies.com/sites/default/files/MacCannel%20Staged%20Authenticity.pdf. Accessed 25 Mar 2024

Merriam-Webster (2024). www.merriamwebster.com/dictionary/context. Accessed 25 Mar 2024

Nolan M, Nolan S (1989) Christian pilgrimage in modern Western Europe. University of North Carolina Press, Chapel Hill

Nolan ML, Nolan S (1992) Religious sites as tourism attractions in Europe. Ann Tour Res 19(1):68–78. https://doi.org/10.1016/0160-7383(92)90107-Z

NSI (2024). www.nsi.bg/bg/content/. Accessed 22 Mar 2024

Pravoslavieto.com (2024) Rozhenski monastery. www.pravoslavieto.com/manastiri/rozhen/index.htm, Accessed 21 Mar 2024

Projects 2014–2020interregerope.eu (2014) Swot Analysis. HARGHITA COUNTY COUNCIL, WP3 -Training needs and population wider attitude in Romania. www.projects2014-2020.interregeurope.eu/fileadmin/user_upload/tx_tevprojects/library/file_1645023600.pdf. Accessed 20 Mar 2024

Raj R, Griffin K, Blackwell R (2015) Motivations for religious tourism, pilgrimage, festivals and events. CABI Books. CABI International. https://doi.org/10.1079/9781780645230.0103

Rashid AG (2018) Religious tourism – a review of the literature. J Hospit Tour Insights 1(2):150–167. https://doi.org/10.1108/JHTI-10-2017-0007

Religious Tourism in the European Union: Resources And Trends (2014). www.hotelmule.com/html/86/n-3286.html. Accessed 20 Mar 2024

Riddick C, Russell R (1999) Evaluative research in recreation, park, and sport settings: searching for useful information. Sagamore Publishing, Champaign, Illinois, Цит. по Ham, Sam H. and Weiler, B. Development of a research-based tool for evaluating interpretation, Sustainable Tourism CRC, p 13

Rila Monastery (2024). www.rilskimanastir.org/. Accessed 5 Apr 2024

Rila Monastery Nature Park (2024). www.parkrilski-manastir.eu/en/content/tourists. Accessed 4 Apr 2024

Rila National Park (2024). www.rilanationalpark.bg/tourism/general-information/location/-2.html. Accessed 9 Apr 2024

Rinschede G (1992) Forms of religious tourism. Ann Tour Res 19(1):51–67. https://doi.org/10.1016/0160-7383(92)90106-Y

Ron AS, Timothy D (2019) "4. Promoting Christian tourism". Contemporary Christian travel: pilgrimage, practice and place, Bristol, Blue Ridge Summit: Channel View Publications. pp 64–81. https://doi.org/10.21832/9781845416652-006

Russell P (1999) Religious travel in the new millennium. Travel Tourism Analyst 5:39–68

Shackley M (2004) Tourist consumption of sacred landscapes: space, time and vision. Tour Recreat Res 29(1):67–75

Shackley M, Welton R (2008) Religious tourism in the European Union: resources and trends. Nova Science Publishers, New York

Stausberg M (2014) Religion and spirituality in tourism. The Wiley Blackwell companion to tourism, pp 349–360. https://doi.org/10.1002/9781118474648.ch28

Tarlow PE (2010) Religious and pilgrimage tourism. Destination World no. 33, viewed 16 June 2015. www.destinationworld.info/newsletter/feature65.html. Accessed 7 Apr 2024

The European Market Potential for Religious Tourism (2010). www.cbi.eu/market-information/tourism/religious-tourism/market-potential. Accessed 5 Apr 2024

The Examination of Master William Thorpe, Arber's Engl. Garner, 1407

Timothy DJ, Boyd SW (2003) Heritage tourism. Prentice Hall, Harlow

Timothy DJ, Olsen DH (eds) (2006) Tourism, religion and spiritual journeys (Vol. 4). Routledge, London

Tourism.goverment.bg (2018) Concept for touristic zoning of Bulgaria, p 31. www.tourism.government.bg/bg/kategorii/opik-2014-2020/koncepciya-za-turistichesko-rayonirane-na-bulgariya. Accessed 25 Apr 2024

Tourism.govemnent.bg (2024) The National Strategy for Sustainable Development of Tourism in the Republic of Bulgaria. https://www.tourism.government.bg/bg/kategorii/strategicheski-dokumenti/nacionalna-strategiya-za-ustoychivo-razvitie-na-turizma-v. Accessed 10 Apr 2024

Training needs and population wider attitude in Romania. file:///D:/Downloads/SWOT+religious+tourism+training+needs+in+Romania.pdf (30.04.2019). Accessed 8 Apr 2024

UNWTO (2016) Secretary-General, Taleb Rifai. www.unwto.org/archive/europe/press-release/2016-11-14/religious-tourism-catalyst-cultural-understanding. Accessed 15 Apr 2024

Vukonic B (1996) Tourism and religion. Pergamon Press, Oxford

The Religious Tourism in Greece: Current Status and Future Development

Evangelia Kasimati and Vincenzo Asero

Abstract Travel for religious purposes was essentially the first "leisure" travel recorded in various regions of the world, regardless of religion. The purpose of this chapter is to highlight the importance of religious tourism, as a special form of tourism, in the tourist development of Greece, to identify its current status, to determine the possibilities that exist for its further development and to provide policy proposals for its sustainable growth. The methodological approach is based on a qualitative analysis, taking into account secondary data and research work. Our findings contribute to the Greek tourism literature by providing valuable insights for the Greek religious tourism, especially since the body of such academic research for Greece is not adequate.

1 Introduction

The human movement, with the aim of searching for the sacred or the divine, constitutes an important psychological need for every individual, regardless of race, nationality, or religion, which entails multiple economic, cultural, and social consequences in the places of pilgrimage or religious destination. In the academic literature, this phenomenon is called religious tourism. Religion tourism, also known as pilgrimage tourism, can be defined as a form of travel where individuals or groups journey to particular sites searching and expressing for values and identities (Liutikas 2017). This type of tourism is driven by faith, spiritual beliefs, or the

The views expressed in this article are personal and do not necessarily reflect the views of the Bank of Greece. Any errors or omissions are the sole responsibility of the authors.

E. Kasimati (✉)
Economic Analysis and Research Department, Bank of Greece, Athens, Greece
e-mail: Ekasimati@bankofgreece.gr

V. Asero
Department of Political and Social Sciences (DPSS), University of Catania, Catania, Italy
e-mail: vasero@unict.it

M. Valeri (ed.), *Contemporary Religious Tourism*, Tourism, Hospitality & Event Management, https://doi.org/10.1007/978-3-031-79074-4_16

desire to experience and participate in religious rituals and practices, serving as motive for the journey. In other words, religious tourists are people whose main purpose of travel is to visit religious and holy places.

Travel for religious purposes has been an integral part of human history and continues to be significant in modern times, contributing to both spiritual enrichment and cultural exchange. It is not just reserved for followers of a specific religion. Today, 'modern' pilgrims are often motivated by less religious interests than in the past. Hence, the phenomenon draws people from a variety of religious backgrounds in addition to secular tourists who are curious about the cultural and historical value of religious locations. Therefore, religious sites are not only visited by the pilgrims but also visited by non-religious tourists since they have cultural, historical, and religious significance. In this regard, tourism scholars have found that religious tourists may visit sacred places for reasons, such as appreciation of nature and educational and cultural enrichment (Collins-Kreiner 2010; Ron 2009).

The market for religious travel remains a powerful force in the world of travel, continuing to shape the experiences of millions of pilgrims and tourists globally. Although it is difficult to measure, due to the absence of specific statistical data at both national and international levels, religious tourism market size was valued at USD 181.6 billion in 2023 and is expected to reach USD 321.6 billion by the end of 2030 with a CAGR (Compound Annual Growth Rate) of 8.4% (Verified Market Reports 2024). A significant portion of religious tourists are domestic travellers, but there are also a notable number of international visitors. The rising interest of people towards religious activities and pilgrimage to sacred places drives the market and major religious sites to attract millions of pilgrims each year. Key destinations include the Hajj to Mecca, the Vatican City, and Bible cities like Jerusalem, Bethlehem, and Nazareth in Israel.

Today, religious tourism with all its different components and types plays an important role in the context of global tourism because of its specific structural and functional characteristics. According to UNWTO, being key tourism destinations, religious heritage sites not only drive international tourism and economic growth, but also provide important meeting grounds for visitors and host communities, making crucial contribution to tolerance, respect, and mutual understanding between different cultures as well as in building cultural dialogue and peace. In addition, religious tourism can also be a powerful instrument for raising awareness regarding the importance of safeguarding cultural heritage and help preserve the important sites for future generation. Thus, knowing the motivations that lead millions of peoples to travel to sacred places is an essential condition for the sustainable development of destinations and for specific policies for protection and management of religious and sacred sites.

Sacred places or places of divine worship and religious rituals have been destinations for travellers since ancient times. In these places (temples, monasteries, places of apparitions or tombs of saints and places of religious holidays), the faithful people sought communication with the divine. Religious tourism, therefore, is one of the oldest special forms of tourism, since it flourished back to ancient times, both in Greece and in many other countries. In Greece there are many areas (e.g. Mount Athos, Meteora, Tinos, and Patmos) with religious attractions (monasteries, churches) that are an integral part of the national heritage and attract a large number of tourists. Also, some of the tourist attractions in Greece are included in the UNESCO's World Heritage List. Since Greece is filled with unique churches, monasteries, and religious sites, it is considered as one of the world's most popular destinations for religious tourism.

The purpose of this chapter is to highlight the importance of religious tourism, as a special form of tourism, in the tourist development of Greece, to identify its current status, to determine the possibilities that exist for its further development and to provide policy proposals for its sustainable growth. The methodological approach is based on a qualitative analysis, taking into account secondary data and research. Our findings contribute to the Greek tourism literature by providing valuable insights for the Greek religious tourism, especially since the body of such academic research for Greece is not adequate. Towards this direction, the structure of the chapter is as follows. The first section examines the theoretical framework of the religious tourism and gives a short overview of the existing literature. The following section presents the current status of the religious tourism in Greece. Finally, the implications, providing policy proposals for its future development and growth, and conclusions are presented in the last section.

2 Literature Review on Religious Tourism

Religious tourism includes all kinds of travel motivated by religion in combination with other kinds of motivation. Many authors (i.e. Bachimon and Thery 2021) affirm that the relationship between tourism and religion is particularly complex and takes several hybrid forms. In the literature of tourism studies, religious tourism is often described interchangeably as a cultural heritage tourism, cultural tourism, cultural heritage tourism, and spiritual tourism by different authors. Then, religious tourism is a broader term encompassing travel for religious purposes, including pilgrimages, missionary travel, religious conferences, and visits to religious sites. As suggested by Jackowski and Smith (1992), religious tourism is a relatively recent term, used to define travel that is motivated largely by a search for spiritual knowledge.

Globalization significantly impacts the religious and spiritual aspects of people's lives, including religious tourism and pilgrimages (Liutikas and Raj 2024). Today, pilgrimage must be defined in a holistic way, encompassing both traditional religious pilgrimage and modern secular journeys (Collins-Kreiner 2018). Thus, in modern pilgrimage tourism, sacred and secular aspects can both apply. According to Timothy and Olsen (2006), the term pilgrimage specifically refers to journeys undertaken for religious or spiritual reasons, often to sacred sites. Jackowski and Smith (1992) affirmed that pilgrimage takes two different forms depending on the motivations. In the first form, the driving force is the religious feeling, while in the second one is the quest for knowledge the area and the towns they visit. Therefore, the latter form embodies religious tourists. Some scholars observe that the difference between pilgrimage and tourism is due to the tourists' motivations (Lloyd 2015; Margry 2009). However, the boundaries between pilgrimage and tourism have become blurred (Collins-Kreiner 2018). In the same line, Hudman and Jackson (1992) stated that it is very difficult to differentiate clearly between aspects of pilgrimage that are touristic and those that are religious.

The literature often explores the continuum between the sacred and the secular in religious tourism. Many tourists may visit religious sites for cultural, historical, or architectural interests rather than purely spiritual reasons (Eade and Sallnow 1991). Liutikas (2021) highlights that pilgrims engage in diverse types of rites, rituals, and behaviours, including internal and external actions and use of attributes. The same author states that the primary distinction between a recreational journey and a values-rich journey lies in the traveller's motivation and inner disposition (Liutikas 2012). Then, in a religious pilgrimage, spiritual and religious motives take precedence. However, nowadays the interest in pilgrimages is often motivated by less personal reasons than in the past. Hence, the focus is now on travel formulas that activate tourist flows and trigger economic processes that require specific planning and management by the organizations involved, be they secular or religious. As Bachimon and Thery (2021) noted, religious events such as holidays and pilgrimages show a clear tendency to take on all the attributes of tourism and leisure. As a consequence, places of pilgrimages have become religious destinations with all the attributes of other touristic destinations, even they have not that specific status. In other words, the relationship characterizing tourism and religion has caused the 'touristification' of religion. According to Gauthier (2020), in the context of globalization and in the era of neoliberalism, religion has transformed into a commodity that exists alongside other goods and services.

Several scholars highlight the complexity and multifaceted nature of religious tourism and underscore the need for a balanced approach that respects the sacredness of religious sites while promoting sustainable development. Within this context, the UNWTO plays a significant role in promoting and supporting religious tourism as part of its broader mission to foster sustainable, responsible, and universally accessible tourism. Moreover, the UNWTO (2014) recognizes that the growing interest for religious tourism worldwide is not only beneficial for the tourism sector, but also crucial in building cultural dialogue and peace. Therefore, as stated in Bethlehem Declaration on Religious Tourism, the UNWTO (2015) works with member states, religious organizations, and the private sector to enhance and manage this segment, promoting entrepreneurship and developing proper dialogue between stakeholders in local communities and providing a balance between the needs of religious tourists and tourists visiting religious sites for other purposes. In a more general sense, a high level of partnership among different stakeholders is more and more crucial in the coordination between tourism policies and their related actions (Asero and Billi 2022). At the same time, effective management strategies are also essential to balance visitor numbers with the preservation of religious sites. This includes infrastructure development, crowd control measures, and the provision of adequate facilities (Timothy and Boyd 2003). Moreover, in recent years, UNESCO played a leading role in the development and promotion of religious and sacred heritage, as such approximately 20% of the properties inscribed on the World Heritage List have some sort of religious or spiritual connection, while many initiatives on heritage of religious interest are carried out.

As both a social and economic phenomenon, religious tourism has attracted the attention of academics and entrepreneurs due to the dynamics it generates. For academics, the interest lies in the study of the motivations, interests and spiritual or cultural needs awakened by religious places; for entrepreneurs, it represents an economic impact generated by visitors in host communities and creating new business initiatives and employment sources and additional income opportunities (Tobón and Tobón 2013). Following this interest, within the religious tourism literature, Shtudiner et al. (2018) evaluate the impact of religiosity, specifically Jewish manifestation, on the attitudes of communities towards tourism in a sacred city such as Jerusalem. Higgins and Hamilton (2020) study the Catholic pilgrimage site of Lourdes, the second biggest tourist destination in France and the third largest Catholic pilgrimage site in the world and explore the role of religious souvenirs beyond their market value. From another perspective, the case of pilgrimage at Lourdes has been investigated also in the study of Thomas et al. (2018). Chang et al. (2020) analyse the dimensionality of the pilgrimage experience as a reference in the destination planning process. The case of the Shikoku pilgrimage in Japan and the connection between slow tourism/travel and pilgrimage have been analysed by Olsen and Wilkinson (2016). Terzidou et al. (2018) explore the complexity of religious tourism motivations to sacred places within the Greek Orthodox context. Investigating the visitors' shopping behaviour (pilgrims and secular tourists) within Kykkos Monastery on the island of Cyprus, it has been observed that the consumption patterns reflect the different levels of religious motivation (Dimitrovski et al. 2020). By reframing and reconceptualizing Muslim pilgrimage, Luz (2020) analyses the complex relations among Islam, pilgrimage, and religious tourism and explores Islamic religious voyages. Following this path, this study explores religious tourism and seeks to contribute to the existing body of knowledge by providing a picture of the phenomenon in Greece.

3 What Is the Current Status of the Religious Tourism in Greece

Religious tourism in Greece is inextricably linked to the Orthodox faith given that Greeks, compared to other peoples, show high rates of religiosity: 81.4% declare themselves Orthodox Christians and more than seven out of ten declare themselves believers (Kapa Research 2018). In Greece there are a large number of religious monuments (churches and monasteries) that are part of the national cultural heritage and attract a significant percentage of visitors. The significant benefits arising from this type of tourism are lengthening the tourist season, increasing demand for local products and contributing to the preservation of tradition (Spilanis and Vayanni

2003). The motivation for the travel of tourists is either religious reasons (trips submitted by the religion of tourists) or cultural reasons, such as acquaintance with other cultures. The object of interest focuses on the religious monuments of a region, churches, monasteries, etc., while it may also include visits on the occasion of celebrations, festivals of some peculiarities (miraculous icons, relics of saints, etc.). In addition, the resilience of the Greek religious destinations such as the Mount Athos monastic community was strong in the COVID-19 era, despite the face of looming health risks (Tsironis et al. 2021).

The country's connection with this particular faith, already since the Byzantine years, has contributed decisively to the creation of a plethora of religious monuments in all regions of Greece, forming of a diverse mosaic of churches, monasteries, metochions (glebe), sketes (hermitages), shrines, and chapels, which constitute a network of religious resources suitable for the development of religious tourism. The sacred icons, frescoes, mosaics, and architectural works make up a religious wealth of unparalleled religious and historical value, which is visited by both believers and tourists with historical and cultural interests (Moira 2003, 2018).

On the other side, the Christian festivals are a pole of attractions and a reason for a movement for tourists (mainly domestic tourists) in the places where they take place (Lagos and Christogianni 2006). Indicative of this wealth is the enumeration by the World Tourism Organization of a total of 280 areas of religious interest in Greece, although only 6 of them have been designated by UNESCO as world heritage sites the Agion Oros, the paleochristian and byzantine monuments of Thessaloniki, the Meteora, the Mystras, the monastery of Daphne at Haidari, the Monastery of Saint Loukas in Boeotia, and the New Monastery of Chios on the homonymous island, the Monastery of Ages Ioannis the Theologian in Patmos and the cave of the Apocalypse in Patmos (Panagiou 2020).

However, there are many places of shrines (pilgrimages) which, although not included in the UNESCO monuments—mainly due to a lack of visibility and unorganized actions on the part of the states towards their recognition (Lagos 2011)—do not lack religious and historical value. Indicatively listed are the Holy Monastery of Ikosifoinissa in Drama, Panagia Soumela in Veroia, Panagia of Tinos, the Holy Monastery of Limonos in Mytilene, the Monasteries of Agia Lavra and the Great Cave in Kalavrita, Panagia Kosmosoteira in Evros, Arkadi in Rethymnon, etc. Antosidou (2015) mentions that the list includes at least 889 religious monuments registered in the Greek territory.

Churches, places intended to meet the needs of Christian worship, are a characteristic and integral element of the Greek landscape. From the simple, one-room chapels to the imposing cross-in-square churches, the churches follow specific architectural styles that were gradually established, from the first centuries of the spread of Christianity and onwards. Their decoration, painting, mosaic or other, always reflects the Orthodox doctrine following specific stylistic trends and established iconographic programmes. Like the churches, the monasteries too are closely linked to the history and tradition of Greece. As a monastery is defined as a set of all the buildings, in which monks or generally the members of a religious community live. The monasteries usually consist of a central church (the so-called Katholikon), chapels, monks' cells, hostels, the refectory (dining room), the galley, the library, etc.

The largest monastic community in Greece is located on Mount Athos, also known as the orchard of the Virgin Mary. The peculiarity of Mount Athos lies in the fact that it is a self-governing part of the Greek state, politically subordinated to the Ministry of Foreign Affairs and religiously under the jurisdiction of the Ecumenical Patriarchate of Constantinople. All 20 Athonite monasteries operate coenobitically, imposing a common liturgy, prayer, shelter, food, and work on the monks. The latter also elect the head of each monastery, the Abbot, who retains his title for life. Numerous relics (manuscripts, historical documents, chrysobulls, portable icons, etc.) have been treasured in the monasteries of Mount Athos, which is still inaccessible to the female population. Apart from these, however, the monasteries themselves are unique cultural monuments, not only for Orthodoxy and Greece, but for the whole world.

The majority of religious tourists in Greece (about 85%) are Greek Orthodox Christians, precisely because the connection between faith and religious monuments is particularly strong (Lagos and Christogianni 2006). Also, Greeks who live abroad visit religious monuments especially during the periods of the great Christian festivals, when they visit Greece and the feeling of nostalgia and connection with the roots is particularly strong (Lagos and Poulaki 2013). The largest percentage of foreign religious tourists, as it to be expected, comes from countries of the same faith, mainly from Russia.

Regarding the time of visit, the greatest movement of religious tourists is in summer (Polyzos 2017), as it is combined with holiday vacation and it is intensified due to religious holidays that, due to the climate, take place mainly during the summer months. Also, because religious trips are usually short duration, they usually take place during 'three-day' holidays (Lagos 2011). This seasonality could be bent, as there are religious festivals and holidays throughout the year and most religious pilgrimages are not related to a specific period. Besides, religious faiths as an incentive for the development of religious tourism do not have a specific time of event. Polyzos (2017) reports that for this reason the government seems to be strongly interested in this type of tourism, while religious tourism has been an important goal of the Greek tourism policy. Tourists travelling, either entirely for religious reasons or partially, spend some time to additional tourist attractions and therefore contribute to the local development of the areas they visit (Konstantinidis et al. 2021).

4 Policy Proposals for the Development of the Greek Religious Tourism

Regarding the development of religious tourism in Greece, it seems that the vast majority who choose a destination driven by religious motives are mainly domestic tourists (domestic religious tourism). This is explained by the fact that Greeks are orthodox Christians and the places of pilgrimage in Greece are holy places of Orthodoxy. The majority of Christians in Europe and throughout the world are catholic or protestant. This reality greatly reduces the possibility of attracting religiously motivated tourists from abroad (external religious tourism). Nevertheless, it is the significant number of foreign tourists who visit Mount Athos and Meteora every year. On the other hand, of course, there are also catholic churches and monasteries in Greece, due to the existence of small percentage of Greek citizens who are Christian Catholics, but the movements of foreign tourists for religious tourism in Greece are limited.

In order to develop religious tourism in Greece and attract foreign tourists, first, the existence of catholic churches and monasteries should be made known to these interested in Europe and the USA, so that they can be converted into pilgrimage centres that will attract catholic Christians to visit them. Since catholic pilgrims do not easily prefer a visit to Greece to satisfy their religious motivations, due to the abundance of pilgrimage sites in central and Western Europe, this could be made easier if it were combined with a trip to our country that included the satisfaction of other motivations and interests (holidays, cultural tourism, etc.).

Another suggestion is the creation of religious travel packages to the regions through which the Apostle Paul passed in Greece, who is particularly religiously accepted and worshipped by Catholics. Starting from Troy, the Apostle Paul travelled through the following regions: Samothrace, Neapolis, Kavala, Philippi, Amphipolis, Apollonia, Thessaloniki, Veroia, Athens, Corinth, Kefalonia, Rhodes, Patmos, Kalampaka, Delphi, Heraklion, and Ierapolis. In these areas, following the teachings of Apostle Paul, temples were built that are now considered monuments of Christianity with special historical value (GNTO 2003). These areas also have a plethora of religious and other attractions. A travel programme 'in the footsteps of the Apostle Paul' or 'in the monasteries of Mount Athos' could satisfy the religious needs of tourists from abroad, combining other forms of tourism (Urban Tourism: Samothrace, Kefalonia, Rhodes, Patmos). It is noteworthy that cruises are also organized for the purpose of pilgrimage in the footsteps of the Apostle Paul (GNTO 2003).

More generally, a tourism policy framework could be formed that would include a number of selected actions for the planning and management of tourism in areas with important religious monuments, which are prerequisite for the development of religious tourism. This framework may include:

- To facilitate and encourage those involved in the preservation and management of Orthodox tradition and ecclesiastical heritage to make their significance accessible to members of the local community and visitors.
- Facilitating and encouraging entrepreneurs to promote and manage religious tourism in a way that respects and promotes the culture and living traditions of local communities.
- Encouraging processes by formulating programmes in order to develop detailed and measurable objectives and strategies linked to the promotion and interpretation of sites with religious monuments and activities.

- The undertaking of initiatives, such as construction of a website, creation of promotional material, signage on the roads to the monasteries, use of digital tools by the church for the development of religious tourism at local level.
- The organizations of conferences of theological content, as well as speeches and lectures in conference centres near the ecclesiastical and monastic sites.
- The creation of walking areas on designated paths located in areas of outstanding natural beauty and connecting religious moments.
- The creation of enriched religious packages, which are marketed by travel agencies and tour operators.
- The programming of special information broadcasts on television networks in Greece and abroad.
- The possibility for guided tours of historical and religious character when visiting metochia, monasteries, churches, and other monuments of our ecclesiastical heritage.

Moreover, the percentage of Orthodox Christians living in other countries, who could potentially choose Greece as a destination for a travel to satisfy, among other things, their religious motivation, should not be considered insignificant. The Orthodox church counts 225 million believers worldwide and ranks third after the Catholic Church (with more than 1 billion believers) and Islam (with 900 thousand believers). The majority of the Orthodox belongs to the Russian Orthodox Church. Apart from the Greeks, orthodox Christians are Bulgarians, Serbs, many of the inhabitants of the eastern countries, a percentage of Americans (mostly immigrants from Europe), the majority of the Ethiopians, along with smaller groups around the world. Many of these eastern countries into the European Union will facilitate the access of religiously motivated tourists to Greece.

While exploring the potential and prospects for the development of religious tourism in Greece, the number of people of the Greek Diaspora that could be attracted to make trips with a religious purpose, exclusively or in combination with other specific forms of tourism in Greece, should not be neglected. It is noteworthy that in 2021 more than 5 million Greeks in about 140 countries of the world lived outside Greece.

5 Conclusions

Religious tourism is one of the oldest special forms of tourism, since it flourished back to ancient times, both in Greece and in many other countries. Travel for religious purposes was essentially the first 'leisure' travel recorded in various regions of the world, regardless of religion. The main motive that pushes people to make trips of this type is the religious element, which is often combined with other motives that cover their tourist need. Basically, religious tourism encompasses a wide range of spiritual sites and associated services, which are visited for both secular and religious reasons. Within this context, religious tourism and sacred places play an important role as a tool for socio-economic and cultural development of destinations.

Today, UNESCO and the international community define appropriate measures to preserve the values of religious and sacred places, which are the foundation of different cultures, aiming to prevent any gradual loss of cultural and religious traditions. Therefore, the main objective of many organizations is to understand the continuing nature of religious and sacred heritage, to preserve the spiritual value associated with a sacred place, having the capacity to protect its authenticity and integrity, including its particular spiritual significance. That requires specific policies for protection and management of the religious sites that take into account their diverse spiritual nature, developing proper dialogue between all stakeholders involved. In the same line, UNWTO emphasizes how effective partnerships, in sacred places, can benefit local communities and encourage their sustainable development and socio-economic empowerment in the long run.

In Greece there are many regions with religious sites that represent an integral part of the national heritage and attract a large number of tourists. Also, some of the tourist attractions in Greece are included in the relevant UNESCO list, such as Mount Athos, Meteora, the church of Agios Dimitrios in Thessaloniki, Daphni and Osios Loukas Monasteries and the New Monastery in Chios, as well as the monastery of Saint John the Baptist and the Cave of the Apocalypse in Patmos. For the better utilization of all these places of worship, the formation of a tourism policy that can develop Greece's religious heritage in a balanced way is required. This, beyond the economic benefits that will exist, is expected to contribute to the easing of tourist seasonality and to the expansion of Greece's multicultural identity, which is further estimated to strengthen Orthodoxy and attract travellers with religious motives. Those could form the basis for the organized development of this special form of tourism, giving an added value to the various religious local destinations. The development of a national strategic approach to religious tourism can build on the comparative advantage of Greece in this field by creating new tourist sites to attract travellers with special interests and intellectual pursuits (Chantziantoniou and Dionysopoulou 2017). On the other side, as COVID-19 refreshed the public's interest in pilgrimage, future studies are needed to investigate the pilgrimage under new conditions, in terms of mobility, and as part of wider interpretations of modern-day life (Tsironis 2022).

Our study is not free from limitations. It is a qualitative research work that comprises an examination of religious tourists, investigating only a part of the range of religious locations and forms of religious travel in Greece. For a more comprehensive understanding of the phenomenon, there is a need to research other religious places in Greece and to further enhance the research methodology applying statistical analysis derived by a questionnaire survey. Finally, as Terzidou (2020) mentions, future research in religious tourism concerning Greece could consider an in-depth exploration of the display and positioning of objects in Greek sacred places and their effect on religious tourists.

References

Antosidou R (2015) Religious tourism and marketing strategy: the case of the Greek reality. Postgraduate Dissertation, University of Macedonia, Interdepartmental Postgraduate Programme in Business Administration (in Greek)

Asero V, Billi S (2022) New perspective of networking in the DMO model. In: Valeri M (ed) New governance and management in touristic destinations. IGI Global, Hershey, PA, pp 105–118. https://doi.org/10.4018/978-1-6684-3889-3.ch007

Bachimon P, Thery H (2021) Tourism and religion. Via [En ligne] 20. https://doi.org/10.4000/viatourism.7949

Chang AY-P, Li M, Vincent T (2020) Development and validation of an experience scale for pilgrimage tourists. J Destination Market Manage 15:100400. https://doi.org/10.1016/j.jdmm.2019.100400

Chantziantoniou A, Dionysopoulou P (2017) The religious tourism in Greece: case study of Saint John Russian in N. Evia. J Tour Heritage Services Market 3(2):15–24. https://doi.org/10.5281/zenodo.1160590

Collins-Kreiner N (2010) Researching pilgrimage: continuity and transformations. Ann Tour Res 37(2):440–456

Collins-Kreiner N (2018) Pilgrimage-tourism: common themes in different religions. Int J Religious Tour Pilgrimage 6(1):7–17. https://doi.org/10.21427/D73428

Dimitrovski D, Ioannides D, Nikolaou K (2020) The tourist-pilgrim continuum in consumer behaviour: the case of international visitors to Kykkos monastery, Cyprus. J Tour Cult Change 19(5):568–586. https://doi.org/10.1080/14766825.2020.1828437

Eade J, Sallnow M (1991) Contesting the sacred: the anthropology of Christian pilgrimage. Routledge, London

Gauthier F (2020) Religion, Modernity, Globalisation. Nation-State to Market. Routledge, London

Greek National Tourism Organization - GNTO (2003) In the footsteps of Paul the Apostle in Greece. Athens

Higgins L, Hamilton K (2020) Pilgrimage, material objects and spontaneous communitas. Ann Tour Res 81:102855. https://doi.org/10.1016/j.annals.2019.102855

Hudman LE, Jackson RH (1992) Mormon, pilgrimage and tourism. Ann Tour Res 19(1):107–121. https://doi.org/10.1016/0160-7383(92)90110-B

Jackowski A, Smith V (1992) Polish pilgrim tourists. Ann Tour Res 19(1):92–106. https://doi.org/10.1016/0160-7383(92)90109-3

Kapa Research (2018) Easter, faith and religion in Greece. Kapa Research, Athens. Accessed 15 June 2021 (in Greek)

Konstantinidis C, Mystridis C, Tsagkalidou E, Rizopoulou E (2021) The religious tourism as a competitive advantage of the prefecture of Pieria, Greece. J Tour Hospit Manage 9(3):173–181

Lagos D (2011) Management and tourism policy, lecture notes for the course designing training in tourism and tourism development plan. University of the Aegean, Chios. (in Greek)

Lagos D, Christogianni P (2006) The development of religious tourism in Greece. In: Vogiatzis A (ed) Honorary volume of Maria Negreponti-Delivani. University of Macedonia, Thessaloniki. (in Greek)

Lagos D, Poulaki P (2013) The contribution of religious tourism to regional development. The case of Chios. 11th Conference of the Hellenic Society of Regional Development, Patras, 14–15 June 2013 (in Greek)

Liutikas D (2012) Experiences of valuistic journeys: motivation and behaviour. In: Sharpley R, Stone P (eds) Contemporary tourist experience: concepts and consequences. Routledge, New York, pp 38–56

Liutikas D (2017) The manifestation of values and identity in travelling: the social engagement of pilgrimage. Tour Manag Perspect 24:217–224

Liutikas D (2021) The expression of identities in pilgrim journeys. In: Liutikas D (ed) Pilgrims: values and identities. CAB International, Wallingford, pp 17–34

Liutikas D, Raj R (eds) (2024) Religious tourism and globalization. The search for identity and transformative experience. CAB International, Wallingford

Lloyd S (2015) Sacrilisation of secular pilgrimages as archetypal transformational journeys: advancing theory through emic and etic interpretations. Int J Tour Anthropol 4(1):25–45

Luz N (2020) Pilgrimage and religious tourism in Islam. Ann Tour Res 82:102915. https://doi.org/10.1016/j.annals.2020.102915

Margry JP (2009) Marian interventions in the wars of ideology: the elastic politics of the Roman Catholic Church on modern apparitions. Hist Anthropol 20(3):243–263

Moira P (2003) From the pilgrim to the religious tourist. Sociological approach. Tourist Scientific Rev 1(1):87–102. (in Greek)

Moira P (2018) Religious tourism. Faidimos, Athens (in Greek)

Olsen DH, Wilkinson G (2016) Are fast pilgrims true pilgrims? The Shikoku pilgrimage. Ann Tour Res 61:228–230. https://doi.org/10.1016/j.annals.2016.08.004

Panagiou A (2020) Case of study: the development of religious tourism in Epirus and the prospects of pilgrimage routes in the prefecture of Ioannina. Postgraduate Dissertations, Hellenic Open University, School of Social Sciences, Tourism Business Administration, Patra (in Greek)

Polyzos S (2017) The spatial and economic dimension of religious tourism in Greece. Res Project Series 23(1):1–48. (in Greek)

Ron A (2009) Towards a typological model of contemporary Christian travel. J Heritage Tourism 4(4):287–297. https://doi.org/10.1080/17438730903045548

Shtudiner Z, Klein G, Kantor J (2018) How religiosity affects the attitudes of communities towards tourism in a sacred city: the case of Jerusalem. Tour Manag 69:167–179. https://doi.org/10.1016/j.tourman.2018.06.003

Spilanis I, Vayanni H (2003) Sustainable tourism: utopia or necessity? The role of new forms of tourism in the Aegean Islands. University of the Aegean

Terzidou M (2020) Re-materialising the religious tourism experience: a post-human perspective. Ann Tour Res 83:102924. https://doi.org/10.1016/j.annals.2020.102924

Terzidou M, Scarles C, Saunders MNK (2018) The complexities of religious tourism motivations: sacred places, vows and visions. Ann Tour Res 70:54–65. https://doi.org/10.1016/j.annals.2018.02.011

Thomas S, White GRT, Samuel A (2018) To pray and to play: post-postmodern pilgrimage at Lourdes. Tour Manag 68:412–422. https://doi.org/10.1016/j.tourman.2018.03.021

Timothy DJ, Boyd SW (2003) Heritage tourism. Prentice Hall, Harlow

Timothy DJ, Olsen DH (2006) Tourism, Religion & Spiritual Journeys. Routledge, London

Tobón S, Tobón N (2013) Turismo religioso: fenómeno social y económico. Anuario Turismo y Sociedad xiv:237–249

Tsironis C (2022) Pilgrimage and religious tourism in society, in the wake of the COVID-19 pandemic: a paradigmatic focus on 'St. Paul's Route' in the Central Macedonia Region, Greece. Religions 13(10):887–902. https://doi.org/10.3390/rel13100887

Tsironis CN, Sylaiou S, Stergiou E (2021) Risk, faith and religious tourism in second modernity: visits to Mount Athos in the COVID-19 era. J Heritage Tourism 17(5):516–532. https://doi.org/10.1080/1743873X.2021.2007252

UNWTO (2014) Religious heritage and tourism: types, trends and challenges. International conference, Elche, Spain, 27–28 November 2014

UNWTO (2015) Bethlehem declaration on religious tourism as a means of fostering socio-economic development of host communities. ISSN electronic version: 2221-6987, vol 24, n 4. https://doi.org/10.18111/unwtodeclarations.2015.24.04

Verified Market Reports (2024). https://www.verifiedmarketreports.com/product/religious-tourism-market/. Accessed 15 July 2024

Strategic Management in Catholic Tourist Destinations and Their Territorial Network of Stakeholders

Noel M. Muniz, María J. Vázquez-García, Maribel Rodríguez, and Leonor M. Pérez

Abstract This work analyzes how, within the life cycle model for tourism planning, strategic management of Catholic tourist destinations is inherently related to its territorial network of stakeholders—key for the present and future management of religious tourism in regions. Strategic planning, networking management, and case analysis in religious tourism are addressed, identifying the intervention times and potential value generation in destination management based on the life-cycle model. This examination exposes how the weight of key actors in catholic religious tourism sites may accelerate certain stages of the development of tourist destinations, while other stakeholders, traditionally considered pivotal, could come to play a less decisive role given social transitions. The originality of this work derives from the purposeful scope of networking and stakeholder management in territories.

1 Introduction

This work aims to analyze how, within the life cycle model for tourism planning, strategic management in Catholic tourist destinations is inherently related to its territorial network of key actors (stakeholders), being crucial for the optimization of the present and future management of religious tourism in a region and, therefore, of the stages of growth, development, consolidation, and subsequent evolution of its initiatives. To this end, the recent history of strategic planning management and the formation of networks in religious tourism are examined, as well as the principles of the evolution cycle of a tourist destination (Valeri 2022). Likewise, different cases of Catholic religious tourism sites, located both in Spain and Latin America, are comparatively examined, identifying through maps of actors the intervention times and the potential generation of value in the territorial management of the

N. M. Muniz (✉) · M. J. Vázquez-García · M. Rodríguez · L. M. Pérez
Department of Statistics, Econometrics, Operational Research, Business Organization, and Applied Economics, University of Cordoba, Cordoba, Spain
e-mail: nmunizr@uco.es; r52vagam@uco.es; es3rozai@uco.es; lperez@uco.es

M. Valeri (ed.), *Contemporary Religious Tourism*, Tourism, Hospitality & Event Management, https://doi.org/10.1007/978-3-031-79074-4_17

destination based on the life cycle model of tourism planning. Finally, together with current trends in world religious tourism, the strategic territorial capacity of networking management to act as a competitive driver of changes in sustainable and inclusive development of this tourism model is analyzed.

Religious tourism attracts millions of people from all around the world; thousands and thousands of Muslims visit Mecca every year, while many Buddhists and Hindus make the pilgrimages to Bodh Gaya and Kumbh Mela, respectively.

Catholic devotees, for their part, flock by the thousands to sacred sites and places such as the Basilica of the Virgin of Guadalupe in Mexico, St. Peter's Basilica in the Vatican, or the Cathedral of Santiago de Compostela, where those who have decided to make the internal journey to Santiago end up.

In the absence of recent statistics, the evolution of religious tourism has tended to rise since the first decade of this century (UNWTO 2011). Various reports project growing global income for religious tourism. Thus, Future Market Insights (2024) estimates that this type of tourism will generate around $16 billion worldwide by 2024, potentially reaching $38 billion in 2034, while Coherent Market Insights (2024) considers that the tourism market will be valued at $1293.71 billion in 2024, reaching $2010.41 billion in 2031. This source currently assigns 24.1% of global market revenue to Catholic religious tourism.

In this sense, Catholic religious tourism appears to represent the largest segment of this tourism market, apparently mainly due to the existence of more than 1.3 billion people eager to make ancient pilgrimages, visit legendary churches or historic cathedrals, as well as places world-renowned for their Marian apparitions.

In Latin America and the European continent, there are places considered sacred by Catholicism. Many of them have been visited by their followers for decades and even centuries; however, not all have managed to build a mechanism, physical, logistical, and human, that, in an integrated manner, has led them to jointly optimize the potential of their latent capacities to enrich and increase the value of the overall experience of their visitors.

Along these lines, based on the theory of the evolution cycle of a tourist destination (Butler 1980; Valeri 2015), it would be valid to understand that those underdeveloped or forgotten destinations have experienced a downward trajectory due to different reasons.

Firstly, it is appropriate to consider that if the core religious reason that gives primary meaning to the sacred place is not solid enough to provoke acts of faith in its visitors, it will be complex to expect that the evolution of development and consolidation of the initiative will progress.

Similarly, if the Catholic Church and the local clergy do not recognize the site as consecrated, even though its population venerates it with intense devotion, its opportunities to grow and consolidate itself on a global scale will be truly scarce.

However, we will assume that both factors are fulfilled and can be used as foundations to give rise to a robust territorial religious tourism project. What other factors should influence the consolidation of the initiative?

Another limiting or empowering variable could be scarce or fully concerted strategic planning, respectively; in other words, territorial management in which the transformation processes have been or are being facilitators of the changes required to dynamize religious tourism activity. Undoubtedly, the organized human factor represents the main ingredient in materializing this opportunity for progress.

Based on this premise, religious tourism initiatives in the territories could decline or advance independently of the stage of the evolution cycle of a specific tourist destination, given that the organized management of the community would represent the main lever of change for the optimal development of the site. This multiplier effect, the result of planned community effort, is what we call strategic management based on the territorial network of stakeholders.

In this sense, this work analyzes how, within Butler's Life Cycle Model for Tourism Planning (1980), a strategic and planned vision of a Catholic religious tourism destination must be closely related to its territorial network of key actors (stakeholders), which is crucial for optimizing the management of present and future religious tourism in a region and, therefore, the stages of growth, development, consolidation, and subsequent evolution of its initiatives. To this end, the recent history of strategic planning management and the formation of networks in religious tourism are examined, as well as the principles of the evolution cycle of a tourist destination.

Likewise, different cases of Catholic religious tourism sites located in Spain and Latin America are comparatively examined, identifying through maps of actors the intervention times and the potential generation of value in the territorial management of the destinations based on the life cycle model of tourism planning.

Finally, together with current trends in global religious tourism, the strategic territorial capacity of network management to act as a competitive driver of change in sustainable and inclusive development of this tourism model is analyzed.

2 Theoretical Backgrounds

2.1 Strategic Planning in Catholic Tourism Destinations

Butler (1980) said that the tourism planning of a site considered implicit that strategically managed tourist areas would always remain attractive for their visitors, which was expressed more as a fallacy than a fact consummated by reality.

The fact is that cultural, political, or financial issues, according to Liutikas (2020), can shake the establishment of a new religious tourism product or the use of a newly created route, giving rise to risk in its success due to potential resistance of the context to the effective implementation of a program, strategy, or project.

For his part, Coccossis (2005) recognizes that religious tourism's development must be considered based on the characteristics of the destination, respecting the local community under a strategic intervention framework that contains principles, goals, objectives, and political measures.

In addition, the relevance of strategic destination planning is essential in terms of the chances of success of the tourism initiative. Indeed, several authors have pointed out the importance of effectively managing religious destinations and their visitors' itineraries (Nolan and Nolan 1989; Digance 2003; Olsen 2006; Schinde 2012).

However, according to Shackley (2012), only a small minority of sacred destinations manage their sites under the implementation of some strategic planning.

Certainly, a planning strategy could be affected by the need to plan specific initiatives in specific contexts and times, often facing adjustments and redistributions of the resources used or their management structures (Liutikas 2020). In this regard, sometimes the planning strategy in Catholic tourist destinations can falter due to reasons such as little or no shared planning in the community, which could be due to the sacred nature of the facilities, their poor management, the fragmentation of the authorities responsible for each site, and the limitation of economic resources and knowledge or experience (Liutikas 2020).

A valued tool for strategic planning in tourism, which, when applied to Catholic tourism, facilitates its management, is the Life Cycle Model of a tourist destination (Butler 1980), allowing comparison between more developed territories and the formulation of key strategic principles for effective strategic planning (Millán et al. 2012).

This effective strategic planning, according to McGettigan et al. (2011), on the one hand, is inherently related to the main motivation for traveling of the Catholic tourist and his universe of beliefs and faith and, on the other hand, with the motivations of the secular tourist who visits a religious site for cultural, leisure or wellness purposes, or to seek an intimate and emotional connection with the cultural heritage and tradition of the destination (Smith 1992). These varied motivations, characteristic of religious tourism, give rise to a diverse number of visitor subgroups that constitute this niche of the tourism industry (see Fig. 1).

This is why McGettigan et al. (2011) recommend, when proposing a planned strategy for a Catholic tourism destination, considering key factors such as developing a strategy through interregional cooperation, considering the current changing lifestyles of visitors, shorter and more frequent breaks for the typical tourist for this segment, and active learning from other innovative experiences in the market.

2.2 *Strategic Territorial Management and Stakeholder Networks in Catholic Tourism Destinations*

Collins-Kreiner (2016) points out that religious tourist destinations, in terms of strategic territorial management, behave as living entities whose existence depends on their capacity for change and adaptation. Liutikas (2020) argues that all concepts and actors of the destination are involved in a continuous process of transformation that integrally and consequently gives rise to efforts to adapt to the ever-changing environment of the sector.

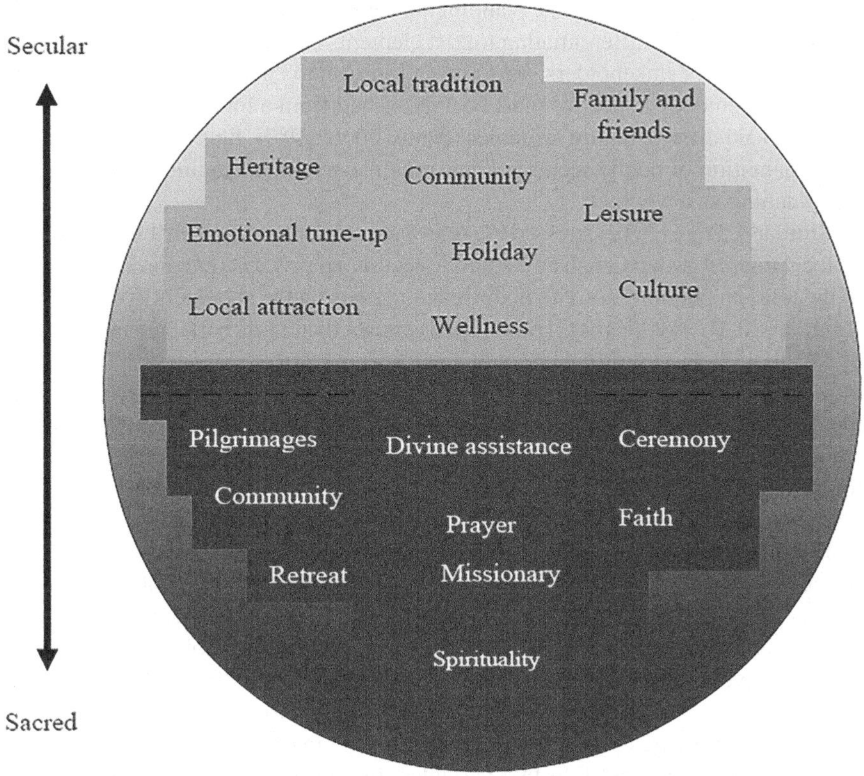

Fig. 1 Segments and motivations of religious tourism (McGettigan et al. 2011)

According to Uriely et al. (2003), one of these key actors is the collective of residents in religious destinations. As stated by this author, the study of the behavior of each stakeholder is essential to recognize possible processes of change in the strategic management of the religious tourist destination, such as attitudes—sometimes open, empathetic, and collaborative; and other adverse, hostile and aggressive reactions—of residents toward tourists, mainly due to the positive or negative strategic management of the tourist development of the territory and the acceptance or rejection of the residents to it (Butler 1980; Doxey 1975).

Despite this potential and relevant reaction of only one of the multiple and possible stakeholders of a religious tourist destination—in terms of its promotion and development, according to Butler (1980), both public entities and private organizations—also act as key actors in the territory—rarely or rarely take into consideration in their planning schemes the expected useful life of a tourist destination or its differentiating resources and/or attractions.

For this reason, Butler (2006) suggests the need to know how to include, as a transformative change lever, these key actors within the comprehensive strategic management of a tourist destination. To do so demands a change of attitude on the part of those responsible for the planning, development, and management of tourist areas, given that the differentiating tourist elements of the destination, as well as its interest groups or stakeholders, are finite and subject to change over time; therefore, their management perspective must be approached from a finite and non-renewable vision of said differentiating resources (Butler 2006, 1980); thus demanding that the implementation of public sector policies requires an understanding of management and planning as a whole.

Similarly, Page (2019) stressed the relevance of planning sustainable tourism and its local interest groups, such as the public sector and private companies, since much of the tourism planning activity in the territory rests on the shoulders of these agents (Liutikas 2020). For his part, Trono (2015) argues that, in the strategic management of a Catholic tourist destination, only a marked and active management in synergy of the set of key actors of the destination—be they called the church, local governments, and other interested parties—can channel the management of religious tourism in their territories and take the perceived value of the experience lived by their tourists to greater levels.

Along these lines, Liutikas (2020) recognizes that the goal of successful strategic management in Catholic tourist destinations, such as pilgrimages, is based on forging structures and relationships between key territorial actors that can promote this specific tourist aspect. This same author points out that planning management effectiveness of the Catholic tourist destination should include variables such as a transparent intervention framework that guarantees cooperation between the stakeholders involved.

More specifically, Evans (2015) singled out the interest groups of a destination by classifying them based on two variables: the power of the stakeholders in the territory and their ability to influence the destination's network of stakeholders and the interest of said key groups or desires to transform the reality of religious initiatives for the better (Liutikas 2020).

Undoubtedly, the stakeholder of greatest interest in the strategic management of religious tourist destinations is the tourist itself. In analyzing this main actor, Afferni and Ferrario (2016) observe the project management and planning process for a specific destination, considering the types of visitors and their frequencies.

Mainly due to the emerging development of the destination, these authors established as the cause of these adverse results, firstly, the lack of true strategic planning that involved all the interest groups involved—neither for its valorization nor for the development of its tourist potential; secondly, the functioning of many institutions and organizations that were poorly coordinated with each other; and finally, the divergence of interests between representative actors of public and private entities that in the long run hindered successful and shared planning of the destination.

Therefore, managing key actors in a territory is not only a matter of involving and linking them but also of knowing how to identify and collectively reconcile their interests. The management of Catholic tourist destinations becomes even more complex when different interests are present and unable to coexist in the proposed stakeholder network.

For this reason, Afferni and Ferrario (2016) affirm that local actors should focus their management and strategic planning efforts on a destination toward revaluing the experience lived by the tourist and toward strengthening the network of local actors involved and shared planning among all of them; otherwise, the chances of success could be significantly reduced.

Finally, McGettigan et al. (2011) use a methodological model proposed by the Irish Western Development Commission (WDC 2006) to study the strategic management of Catholic tourist destinations in Ireland. This model proposed the identification, mobilization, improvement, and value addition to the resources that can be used in the territory. It also involved collaborative work between private, community, and government stakeholders for project management and implementation with strategic potential to generate knowledge and innovation, seeking their successes (McGettigan et al. 2011).

The most interesting thing about this contribution is that in establishing its principles, it sets out seven steps for its implementation in which it makes abundantly clear the capital importance of the strategic management of the territorial stakeholder network, in this case, in the planning of Catholic tourist destinations: (a) Involve interest groups in the processes of association, consultation, and management; (b) jointly recreate a shared strategic framework; build a pool of data; (c) information and knowledge for the determination of strategic priorities; (d) develop a strategic plan with political, budgetary, and execution recommendations; (e) constitute a steering group of key stakeholders capable of facilitating the materialization of the actions present in the plan; (f) implement the elements of the plan; and (g) ensure the continuity of the implementation of the transformations proposed in the plan (McGettigan et al. 2011).

2.3 *Principles of the Evolution Cycle of a Catholic Tourist Destination*

To develop this section, it is essential to focus on the Tourism Area Life Cycle (TALC) Model proposed by Butler (1980), which recognizes the concept of change in the development of the tourist area considered, using curves to show its popularity and decline, while describing special characteristics and potential future events.

This model is still considered one of the most important in tourism research (Xin et al. 2013), maintaining its value for assessing possible implications related to tourism resource management (Alipour et al. 2017).

Regarding his model, Butler (1980) states that although a lasting progression of the tourist destination in question can be conceptualized, it must be considered that not all destinations will respond equally to the principles that govern each of the stages proposed in his model as clearly as others (Butler 1980). The Tourism Area Life Cycle Model is divided into different phases or stages of the potential evolution of the destination: (a) Exploration Stage; (b) Participation Stage; (c) Development

Stage; (d) Consolidation Stage; (e) Stagnation Stage; (f) Decline Stage; and (g) Rejuvenation Stage. Briefly, and in the words of Butler himself (1980), they could be summarized for this work as follows:

(a) Exploration stage (Butler 1980):

> …is characterized by small numbers of tourists: individual travel arrangements and following irregular visitation patterns. They can also be expected to be non-local visitors attracted to the area by its unique or considerably different natural and cultural features. At this time, there would be no specific facilities provided for visitors. The use of local facilities and contact with residents are likely to be high, which may be a significant attraction to some visitors. The physical fabric and social milieu of the area would be unchanged by tourism, and the arrival and departure of tourists would be of relatively little significance to the economic and social life of the permanent residents. …

(b) Involvement stage (Butler 1980):

> …As numbers of visitors increase and assume some regularity, some residents will be entering this new stage, beginning to provide facilities primarily or even exclusively for visitors. Contact between visitors and locals can be expected to remain high and, in fact, increase for those locals involved in catering to visitors. As this stage progresses, some advertising specifically to attract tourists can be anticipated, and a basic initial market area for visitors can be defined. A tourist season can be expected to emerge, and adjustments will be made in the social pattern of at least those residents involved in tourism. Some level of organization in tourist travel arrangements can be expected and the first pressures put upon governments and public agencies to provide or improve transport and other facilities for visitors…

(c) Development stage (Butler 1980):

> "…It will reflect a well-defined tourist market area, shaped in part by heavy advertising in tourist-generating areas. As this stage progresses, local involvement and control of development will decline rapidly. Some locally provided facilities will have disappeared, being superseded by larger, more elaborate, and more up-to-date facilities provided by external organizations, particularly for visitor accommodation. Natural and cultural attractions will be developed and marketed specifically, and these original attractions will be supplemented by man-made imported facilities. Changes in the physical appearance of the area will be noticeable, and it can be expected that not all of them will be welcomed or approved by all the local population…". "…Regional and national involvement in the planning and provision of facilities will almost certainly be necessary and, again, may not be completely in keeping with local preferences. The number of tourists at peak periods will probably equal or exceed the permanent local population. As this stage unfolds, imported labor will be utilized, and auxiliary facilities for the tourist industry (such as laundries) will make their appearance. The type of tourist will also have changed, as a wider market is drawn upon …"

(d) Consolidation stage (Butler 1980):

> "…the rate of increase in numbers of visitors will decline, although total numbers will still increase, and total visitor numbers exceed the number of permanent residents. A major part of the area's economy will be tied to tourism. Marketing and advertising will be wide-reaching, and efforts will be made to extend the visitor season and market area. Major franchises and chains in the tourist industry will be represented, but few, if any, additions will be made. The large numbers of visitors and the facilities provided for them can be expected to arouse some opposition and discontent among permanent residents, particularly those not involved in the tourist industry in any way, and to result in some deprivation and

restrictions upon their activities...". "...The resort cities will have well-defined recreational business districts, and, depending upon the length of time involved, old facilities may now be regarded as second rate and far from desirable..."

(e) Stagnation stage (Butler 1980):

...the peak numbers of visitors will have been reached. Capacity levels for many variables will have been reached or exceeded, with attendant environmental, social, and economic problems. The area will have a well-established image, but it will no longer be in' fashion. There will be a heavy reliance on repeat visitation and conventions and similar forms of traffic. Surplus bed capacity will be available, and strenuous efforts will be needed to maintain the level of visitation. Natural and genuine cultural attractions will probably have been superseded by imported 'artificial' facilities. The resort image becomes divorced from its geographic environment. New development will be peripheral to the original tourist area, and the existing properties are likely to experience frequent changes in ownership. The type of visitor can also be expected to change towards the organized mass tourist...

(f) Decline stage (Butler 1980):

...the area will not be able to compete with newer attractions and so will face a declining market, both spatially and numerically. It will no longer appeal to vacationers but will be used increasingly for weekend or day trips if it is accessible to large numbers of people. Property turnover will be high, and tourist facilities will often be replaced by non-tourist-related structures as the area moves out of tourism. This latter factor, of course, is cumulative. More tourist facilities disappear as the area becomes less attractive to tourists, and the viability of other tourist facilities becomes more questionable. Local involvement in tourism is likely to increase at this stage, as employees and other residents are able to purchase facilities at significantly lower prices as the market declines. The conversion of many facilities to related activities is likely. Hotels may become condominiums, convalescent or retirement homes, or conventional apartments since the attractions of many tourist areas make them equally attractive for permanent settlement, particularly for the elderly. Ultimately, the area may become a veritable tourist slum or lose its tourist function completely...

(g) Rejuvenation stage (Butler 1980):

...although it is almost certain that this stage will never be reached without a complete change in the attractions on which tourism is based. Two ways of accomplishing this goal can be seen at present. One is the addition of a man-made attraction. Obviously, though, if neighboring and competing areas follow suit, the effectiveness of the measure will be reduced. An alternative approach to rejuvenation is to take advantage of previously untapped natural resources. The development of new facilities becomes economically feasible and simultaneously serves to revitalize the older summer holiday trade. As new forms of recreation appear, it is not impossible that other tourist areas will find previously unappreciated natural resources to develop. In many cases, combined government and private efforts are necessary, and the new market may not be the allocentric section of the population (which would suggest a recommencement of the complete cycle) but rather a specific interest or activity group. Ultimately, however, it can be expected that even the attractions of the rejuvenated tourist area will lose their competitiveness. Only in the case of the truly unique area could one anticipate an almost timeless attractiveness, able to withstand the pressures of visitation. Even in such a case, human tastes and preferences would have to remain constant over time for visitors to be attracted. Artificial attractions may also be able to compete effectively over long periods by adding to their attractions to keep in tune with contemporary preferences...

3 Methodological Approaches

3.1 Case Study: Comparative Analysis of Catholic Tourist Destinations in Europe and Latin America

For the development of this section, we have relied on the analysis carried out by Millán et al. (2012) on the evolutionary state of six Catholic tourist destinations—three European and three Latin American—to subsequently propose territorial networks of stakeholders that would potentially facilitate the strategic tourist management of the destination.

Likewise, the digital tool *Smaply* has been used to propose a map of stakeholders for these Catholic tourist destinations, identifying those individual and collective interests that would most frequently come into play in the interactions of these key actors.

The summary table of the evolutionary state of the six Catholic tourist destinations would be the following:

The proposed stakeholder map for these Catholic tourist destinations, identifying those individual and collective interests that would most frequently come into play in the interactions of these key actors, would be as follows (Fig. 2):

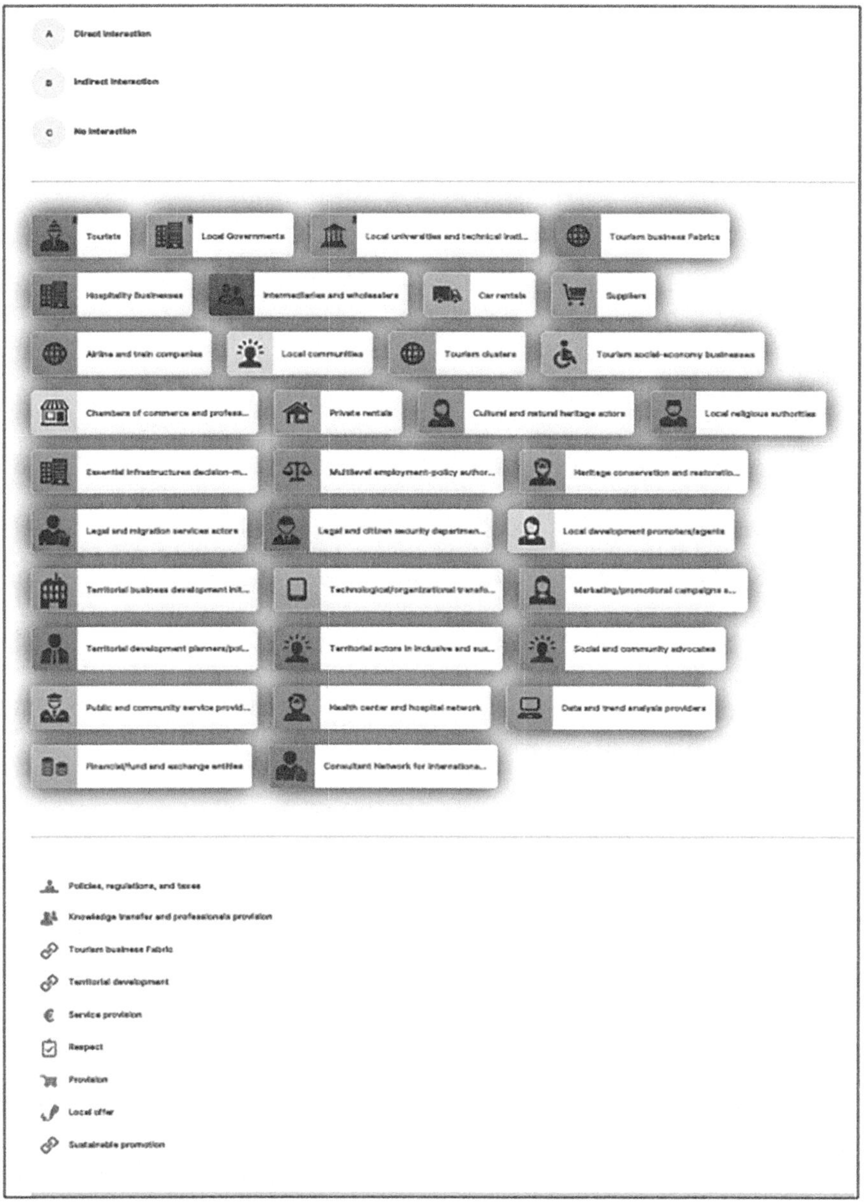

Fig. 2 Proposed stakeholder map for Catholic tourist destinations

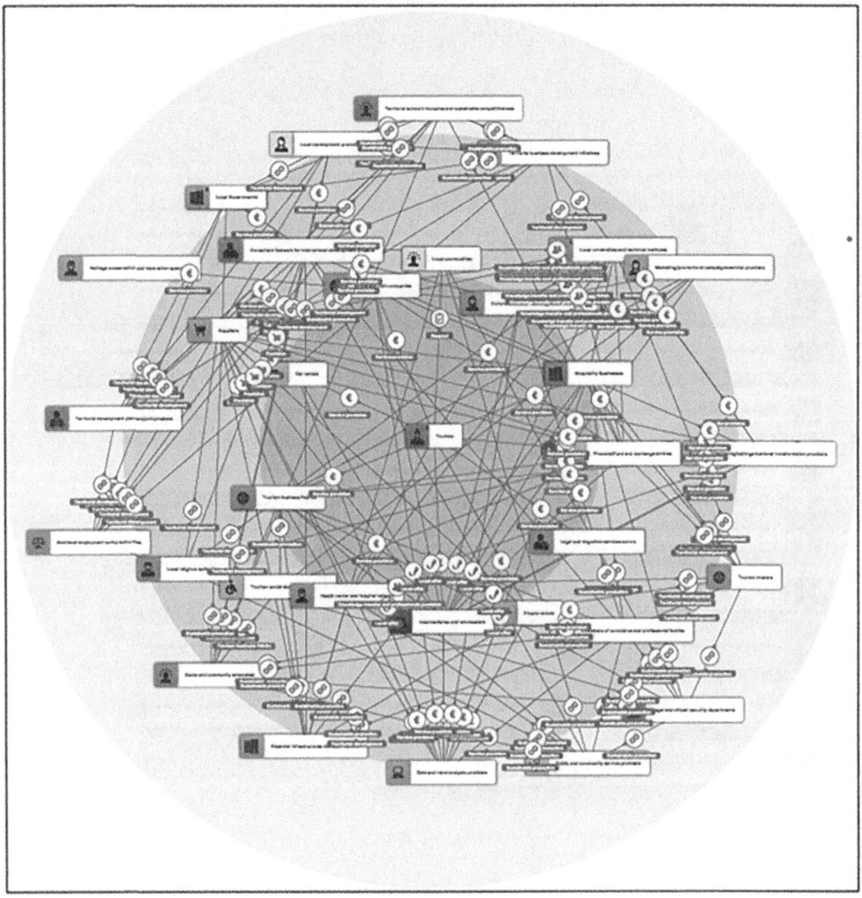

Fig. 2 (continued)

4 Network Interpretation Results

This section is addressed by establishing an order of explanation for the scale of possible influence on the tourist (primary stakeholder) of the different members that make up the network of stakeholders or key actors in the strategic management of a Catholic tourist destination. Therefore, the proposed levels of influence are First level or direct interaction with the Catholic religious tourist; Second level or indirect interaction with the Catholic religious tourist; and Third level or little or no interaction with the Catholic religious tourist.

Likewise, the explanation of said influence is described in a comprehensive, connected, and joint manner, as is the behavior of a network that aims to be active, dynamic, and collaborative to optimize the possibilities of success in the strategic management of a Catholic tourist destination. Finally, having expressed the orientation of the interactions of this network, the possible interconnections of its potential stakeholders are related to the different cases of Catholic religious tourism included in Table 1 (methodological approach) and their specific stage of evolution according to the model proposed by Butler (1980), which has also been described in detail in the section on the Theoretical Framework. In this sense, they are considered as Catholic tourist destinations in the development stage:

- Involvement: Banneux (Belgium), Knock (Ireland), and Santa Ana de Guadalupe (Mexico).
- Development: Talpa de Allende (Mexico).
- Consolidation: Lourdes (France), Fátima (Portugal), and San Juan de los Lagos (Mexico).

Table 1 Evolutionary state of the six Catholic tourist destinations (Millán et al. 2012)

Destination	Evolutionary state
Lourdes	It reached a consolidation phase in the early 2000s and was about to enter a period of stagnation, but the last two visits by the Popes (in 2004 and 2008) and the celebration of the jubilee year (in 2008) have translated into a new boost for religious tourism
Fatima	It has entered a consolidation phase and will probably soon approach a stagnation situation
Banneux and Knock	After the state of exploration and a certain participation of the local population in the creation of services for pilgrims, there are a number of conditions that do not exist, preventing progress toward an advanced stage of development. In both cases, the ecclesiastical approval of the apparitions has been weaker and there has not been a continuous dynamism in the construction of religious buildings
San Juan de los Lagos	Religious tourism has progressed to a state of maturity, but it is not fully mature, as the destination has great potential for this type of tourism to continue growing. Its economy depends heavily on religious tourism, but there is a lack of greater involvement and interaction between lay and church leaders. San Juan de los Lagos is not geared toward tourism. Some services are provided to pilgrims, but many basic services are not covered. The economic impact generated by religious tourism would be enhanced if basic services for religious tourists were covered and the development of the hospitality industry was encouraged by promoting the destination in other countries, attracting foreign tourists who tend to spend more at the destination. The participation of the local population in covering these services with business initiatives of all kinds is very important, even offering alternative activities to religious tourists themselves. In San Juan de los Lagos, there has been a debate for several years about the possibility of developing a suitable space for religious tourists in a more strategic location outside the town center. However, the main opponents are the local residents who live around the parish because of the effects on the drop in sales and the decrease in the cost per square meter of the premises and land located in that area (Martínez 2009). Likewise, the Church is also opposed. Nevertheless, this idea should be taken up again, perhaps trying to come up with a new proposal that would favor the ecclesiastical authorities, the municipal government and the merchants, letting them know the benefits that it would bring. The construction of a large basilica in this area would be a great opportunity for the development of religious tourism

(continued)

Table 1 (continued)

Destination	Evolutionary state
Talpa de Allende	A number of conditions have impeded progress toward an advanced stage of development. Accessibility has now improved somewhat, and some entities have undertaken initiatives on the development of tourism in general and religious tourism in particular. The Pilgrim's route project will cover some basic services for pilgrims; however, the tourist infrastructure for accommodation and catering is still insufficient. As in San Juan de los Lagos, local initiatives to cover these services and the promotion of the destination in other countries would generate a greater economic impact. Religious and business leaders should work together to promote this type of tourism
Santa Ana de Guadalupe	Its relatively short period of time as a religious tourism destination does not allow us to explore its evolutionary cycle, but it can be concluded that, after the state of exploration and local initiatives for pilgrim services, there is a commitment to the development of religious tourism and the opportunity is being taken to build a medium-sized basilica. Likewise, the hotel and hostel capacity should be increased, as well as the number of shops selling artisanal and gastronomic products. This, together with a promotion of the destination in other countries, would increase the economic impact of religious tourism

4.1 First Level: Direct Interaction

4.1.1 The Tourist

Despite its geocentric nature, when it comes to examining a network of interactions linked to the tourism industry, the tourist does not remain an agent who is only a receiver of influences; on the contrary, the tourist decisively influences the tastes, tendencies, and behaviors of other potential visitors through his/her comments and virtual interactions in the different social networks that today make up the impressive wave of information that we all enjoy when deciding our next destination to visit.

On the other hand, the tourist influences the other stakeholders of the Catholic tourism network of territory by expressing his/her specific desires, needs, and demands on how he/she wants to be treated and served, what type of goods he/she requires to enjoy his/her experience to the fullest, what expectations he/she intends to fulfill with his/her visit, etc. Using this information, the remaining network of stakeholders adjusts its value proposition and transforms itself organizationally to be more attractive and competitive. Therefore, in cases that are in the participation stage, it would be highly advisable for the nascent network in the territory to invest efforts in promoting active and constructive management in virtual social networks so that these channels of interaction between Catholic tourists allow the transmission of religious and faith feelings that have moved repeat visitors to the destination; and also so that such constructive management facilitates the studied and planned growth of the potential demand for goods and services in the territory.

Likewise, in cases in the development stage, it would be important to direct such active management toward factors more related to the need for visitors to comply with standards of respect for traditions, the natural environment, and the life of the community; the message at this stage should be more focused on the need to behave in harmony with the spirituality of the destination and to create awareness about the indissoluble link between said spirituality of the destination and the sustainability of the environment.

Finally, in the cases defined in the consolidation stage, it would be advisable to strengthen the messages described above but interpret how to use said interactions to put them at the service of knowledge management and innovation; that is, for example, in knowing how the tourist is reacting or would react to enriching their religious experience—increasing the value of their perception—by making use of new technologies such as virtual or simulated realities: augmented reality and immersive virtual reality.

4.1.2 Airlines and Railway Companies, Hospitality and Entertainment Companies, Financial/Fund and Exchange Entities, Legal and Immigration Service Actors, Car Rental Companies, Intermediaries and Wholesalers, Private Rentals, Social Economy Tourism Companies, and the Tourism Business Network

All these organizations have as a shared characteristic the provision of goods and/or services directly to visitors of the Catholic tourist destination. Therefore, the influence of the tourists on this group described above is more than foreseeable. However, we must take into account two additional dimensions of influence: the first is that exerted by the same group on its members through the destination's tourism value chain through vertical economic interactions (integration) and horizontal interactions (complementarity), which in turn are infinitely multiplied thanks to "Business to Business" (B2B) platforms where deals are closed almost instantly and in real time.

The other dimension is that exerted by the organizations and institutions that make up the second level of influence and indirect interaction. In this sense, companies that provide goods and/or services directly to visitors will demand from key actors in the territory, but with indirect interaction with the visitor, the necessary support for the provision of their goods and services in terms of quality, quantity, time, form, etc., as well as the fluidity and efficiency of the processes involved in the said provision.

The level of influence will also be manifested through the requirement of the latter for a higher level of involvement as non-direct complementary agents of their services and/or through the demonstration of high standards of quality, knowledge, innovation, and sustainability in the strategic management of the resources that support the meaning of the Catholic tourist destination: structures, heritage, traditions, security for the enjoyment of the experience, etc.

4.2 *Second Level: Indirect Interaction*

4.2.1 Local Communities and Religious Authorities of the Territory

Both actors play a crucial role in the dynamics of the network of stakeholders of a Catholic tourist destination. Thus, in the case of local communities, the bidirectional influence of the tourist is indirect but sensitive and increasingly noticeable. It is indirectly bidirectional because the visitor influences the economic, social, and environmental context of the local community, being able to contribute respect, wealth, and culture or, on the contrary, putting excessive stress on the resources, capabilities, and social peace of its inhabitants; and because the community is an essential factor in making the visitor live a unique experience with its kindness, proactivity, desire to communicate the goodness and heritage of the destination, and above all, in our case, the perfect way to experience the sacredness of the place. For this relationship to remain positive, the local community must be involved from the beginning in the strategic management of a religious tourist destination, asserting its weight in the decisions made by the dominant interest groups, maintaining strong vigilance and constructive pressure on the political development proposals promoted by local or multi-level governments, and finally, contributing to creating and strengthening awareness in the community and among the members of the network of stakeholders about the importance of designing and implementing sustainable tourism development models from their origins, thus building continuous, permanent, and stable proposals, from their foundations, for their Catholic tourist destinations.

If the indirect interaction reaches negative levels of rejection, certain negative manifestations by the community may become evident, such as public expressions of protest, impolite and unfriendly treatment, and scarce collaboration and help, until reaching what we know as "tourism phobia," with social expressions such as protests, incidents, and acts of vandalism. But we must not go that far, the territorial network must work on these multidirectional relationships of direct, indirect, and little or no influence (since it would involve various actors) from the beginning, and if, in any case, we reach that point, this will be a clear example of how poorly the territorial network of stakeholders has grown and been formed relationally.

For their part, the religious authorities of the territory are one of the central axes of the territorial network of stakeholders for Catholic tourist destinations for several indirect interrelated reasons. Firstly, the religious experience will generally be accompanied by activities and acts of faith, such as pilgrimages, ceremonies, retreats, etc. These, to be administered and conducted, will require the accompaniment of the spiritual leaders of the destination (Shekhar and Valeri 2023). Likewise, the religious experience will be even more enriched with the more doors, mysteries, and facilities the clergy and their community are willing to open and show to their visitors. Finally, just as in the case of local communities, religious authorities maintain an indirect two-way relationship with Catholic religious tourists: tourists seek to strengthen their faith by living a sublime spiritual experience, while religious authorities fulfill their mission of disseminating God's message on earth to their faithful visitors.

4.2.2 Territorial Heritage Stakeholders (Cultural, Natural, etc.), Network of Health Centers and Hospitals, Providers of Public and Community Services (Police, Cleaning, etc.), Suppliers of Products and Consumer Goods, and Territorial Stakeholders in Inclusive and Sustainable Competitiveness

These interest groups that interact indirectly with Catholic religious tourists have a common pattern: on the one hand, the provision of goods and/or services (inputs) to the companies that generate the goods and services—known as "principals"—that will be consumed directly by the visitors; and on the other hand, the provision of services that may or may not be used by tourists on some occasions, such as the use of public services, such as health services for reasons of illness or police services for reasons of citizen insecurity, or the creation of social and environmental awareness in the case of institutions that ensure the sustainable conservation of resources and their sustainable use.

Whatever the type of service, interactions with Catholic religious tourists have an indirect behavior since they are not part of the network that provides those components that were pre-established to be consumed and experienced by visitors when the travel packages or online contracts were purchased. In other words, they are services that tourists expect when visiting a territory but are not directly included in the travel package purchased to visit the destination.

This approach takes us to the next circle of influence, which is that of this group of stakeholders, with those belonging to the third level of influence or with little or no interaction.

The interactions between this second group of the second level (indirect interactions) and the groups of the third level (explained in the next paragraphs) are again bidirectional and direct; that is, the police and security forces, as well as the institutions belonging to the health network, will need strategies, budgets, policies, etc., which must come from entities such as local governments or multi-level governance with supra-regional powers; and in turn.

Local police forces and public hospitals/clinics will have to make government institutions aware of the vital importance of having the necessary policies/strategies and resources promptly to serve not only the resident population of the destination but also the waves of visitors who at certain times of the year will surely invade the city (in the good sense of the word) hoping to enjoy a pleasant and safe environment.

4.3 Third Level: Little or No Interaction

As explained in the previous section, the institutions that make up this third level of influence maintain a series of dynamic interactions of a direct nature with the members of the second level of influence and practically none with the tourist.

The stakeholders of this group can be divided into two subgroups: (1) Local, regional, territorial, national, and supranational government institutions that have a close relationship with the formulation of policies and decision-making in the territories and organized social pressure bodies with direct influence on the formulation of territorial policies and decision-making. We will now briefly discuss each of them.

4.3.1 Local Governments, Multi-level Governments, Decision-Makers on Essential Infrastructure, Public Institutions of a Legal and Citizen Security Nature, and Social and Community Advocates

Their main role in interaction is to respond to, and above all to anticipate, the demands of all the interest groups in the Catholic religious tourism network of territory in terms of standards, regulations, protection and security policies, etc. Therefore, they do not interact directly with tourists but with the sub-network of other services not contracted but expected in the territory by visitors.

Their ability to resolve, but above all to anticipate potential or real problems will depend on how close the interrelations become with the key agents of the first and second level of influence since these will be essential in communicating the possibilities of taking advantage of the opportunities of the environment and of strengthening its strengths or, on the contrary, of influencing those factors that are weakening the destination itself to be able to work now to counteract and apply measures to the threats of the context.

For this communication to be fluid, this subgroup must facilitate the necessary channels of expression to the community. That is, to enable, while being protected by territorial regulations and strategy, citizen participation at all levels while creating or strengthening social awareness in the community of the need to manage a network of stakeholders for the Catholic, sustainable, and inclusive tourist destination at the service of the well-being of the community.

4.3.2 Local Universities and Technical Institutes, Heritage Conservation and Restoration Specialists, Technological/Organizational Transformation Providers, Marketing/Promotion Campaign Service Providers, Data and Trend Analysis Providers, Network of Consultants for International Development Projects, Territorial Business Development Initiatives, Local Development Promoters/Agents, Tourism Clusters, and Chambers of Commerce and Professional Associations

This last group shares the generation and management of knowledge and innovation as a common characteristic. Therefore, all of them are authentic levers of technological change in the territories, and this is fundamental for the renewal, adjustment, and transformation of the tourist offer that surrounds a Catholic religious destination: universities, providing new professionals with improved ways of managing companies/institutions and their resources, while providing updated knowledge to the network; or providers of technological transformation in organizations linked to

Catholic religious tourism, who not only offer their technological tools but also guide organizations—from a utilitarian, cost, and value creation perspective—on how to implement these technologies, efficiently optimizing their internal resources and capabilities, while promoting communication and organizational learning among their members. Other service providers such as marketing and promotion, data management for decision-making, the achievement of highly specialized projects for access to international funds, territorial initiatives for tourism business development—e.g., incubators, innovation centers, technology parks, etc., for the implementation of local tourism development projects, for the joint/collaborative articulation and management of the effort in a territorial network or cluster, and the professional and collegiate organization of interest groups belonging to the destination's network of stakeholders, are essential to renew and add value to the experience lived by the visitor to the destination, always supported by the fundamentals of this niche: faith and religious experience.

Therefore, their interactions mainly with the interest group of the first level of influence are monumental and are bidirectional since the former (direct providers of goods and services to Catholic tourists) can guide the latter in the need to transform their service offerings, adding value through the implementation of knowledge and technology; or vice versa, that third level organizations know how to identify the potential of these new technologies and knowledge and, on the other, adapt them to the needs and capacities of the requesting organizations.

In this sense, for this sub-network of generators and providers of innovative knowledge to emerge and grow, develop, multiply, and consolidate, the dynamic and continuous interaction of stakeholders belonging to levels of influence two and three is necessary: religious authorities will eventually need their services to maximize the experience of their parishioners, while local communities could find a response to their demands through the application of new ways of communicating to visitors the urgent need to live the religious experience sustainably, in harmony and with respect for the heritage and local community.

A similar relationship would be expected between this subgroup of level three and the key actors of the second subgroup of level two: actors linked to territorial heritage will also require knowledge and technology to promote, maintain, and protect in perfect form the heritage of the Catholic tourist destination; the territorial health network can rely on this third group by specifying the use of new equipment and diagnostic tests, or renewed action protocols—based on scientific knowledge—in the face of new health emergencies; or the police security forces, for their part, will demand new means of aerial surveillance (e.g., more sophisticated drones and means to block their frequencies) and non-invasive recognition of the individual (e.g., facial recognition based on artificial intelligence), while public cleaning services will request the transfer of greater knowledge related to green waste transformation processes and implementation of new processes that involve both the Circular Economy and compliance with the Sustainable Development Goals (SDGs) of the territory.

The interactions of this subgroup of the third level of influence with the first subgroup also of level three will be essential to guarantee the robustness of knowledge management and innovation in the territory. The companies that generate and provide knowledge and innovation can emerge spontaneously and reactively to the demanding context of the Catholic religious tourist destination, but in no case will it be sufficient to sustainably maintain the levels of value perception in the Catholic tourist.

This perception of value must be conceptualized as a living, flexible entity with the capacity to be adapted and enriched if it is to remain competitive in the Catholic religious tourism market.

It is at this point where key territorial actors linked to the formulation and implementation of development policies must include as a priority the aid and promotion of the growth, multiplication, and consolidation of this sub-network. The effort, therefore, activated by these providers must be accompanied, in a planned and strategic manner, by policymakers, government officials, and decision-makers in the territories, making available to them regulations, resources, tools, and procedures to accelerate their strengthening and evolution.

Finally, the stakeholders themselves that form this sub-network of provision of knowledge and technological services must interact intensely in a collaborative manner, generating cohesion and social capital among themselves, envisioning potential vertical and horizontal integrations in the value chain of its services, identifying new forms of "servitization" and monetization points in its activities, and contributing to the materialization of increasingly smart and sustainable Catholic tourist destinations.

5 Conclusions

Religious tourism is a growing type of tourism, attracting millions of people around the world every year, with Catholic religious tourism representing the largest segment within this type of tourism (Coherent Market Insights 2024; Future Market Insights 2024). However, in many cases, these sacred destinations have not been or are not being managed optimally. Thus, it is necessary to investigate the strategic management of these destinations (Ambrosio and Pereira 2007; Shackley 2012). Specifically, the objective of this work focuses on analyzing how, within the life cycle model for tourism planning, strategic management in Catholic tourist destinations is inherently related to their territorial network of stakeholders, being crucial for the stages of growth, development, consolidation, and subsequent evolution of their initiatives, and therefore, being key in the optimization of the present and future management of religious tourism in a region. The results of this study show the relevance of strategic management, based on the territorial network of stakeholders, in the planning and evolution of Catholic religious tourist destinations. The tourism life cycle model (Butler 1980) has been especially useful in understanding the evolutionary dynamics of destinations so that the identification, as well as the activation, of the relevant stakeholders in each of these phases, is crucial for the competitiveness and sustainability of destinations in the long term. The ability to generate value in a Catholic destination depends, to a large extent, on the quality and effectiveness of the collaboration networks between stakeholders.

This study allows us to reflect on the role played by the different stakeholders in the strategic management of Catholic destinations, especially when established as a complex and interrelated network. Through a multi-level approach, the impact of stakeholders on this type of destination is considered based on the degree of interaction with religious tourists, promoting a better understanding of the network of stakeholders involved, as well as a more detailed exploration of how to optimize the interaction between them to maximize the perceived value of the Catholic religious tourist destination.

The originality of this work derives from the purposeful scope of networking and stakeholder management based on the perspective of territorial strategic management as a driver of transformative change in the sustainable development of religious tourist destinations.

Finally, this study presents several limitations. The analysis carried out in this work entails certain limitations, given its geographical representativeness and cultural nature. However, the results obtained could potentially be extrapolated and scalable to other territories and destinations since they furnish some learned lessons regarding territorial management for religious destinations. That is why it is recommended to perform further research in this promising tourism realm.

References

Afferni R, Ferrario C (2016) Religious tourism and Italian Sacred Mounts: experiences of networking and co-operation at a UNESCO site. Int J Rel Tour Pilgrimage 4(1):1–16. https://doi.org/10.21427/D7NM7Z

Alipour H, Olya HG, Forouzan I (2017) Environmental impacts of mass religious tourism: from residents' perspectives. Tour Anal 22(2):167–183. https://doi.org/10.3727/108354217X14888192562285

Ambrósio V, Pereira M (2007) Case study 2: Christian/Catholic pilgrimage—studies and analyses. In: Raj R, Morphet ND (eds) Religious tourism and pilgrimage festivals management. CABI, Wallingford, pp 140–152

Butler RW (1980) The concept of a tourist area cycle of evolution: implications for management of resources. Can Geogr 24(1):5–12. https://doi.org/10.1111/j.1541-0064.1980.tb00970.x

Butler RW (2006) The tourism area life cycle: Vol. 2. Conceptual and theoretical issues. Channel View Publications, Clevedon

Coccossis H (2005) Operational management of cultural and heritage sites. In: Sigala M, Leslie D (eds) International cultural tourism: management, implications and cases. Elsevier, pp 54–65

Coherent Market Insights (2024) Religious tourism market size and share analysis – Growth trends and forecasts (2024–2031). https://www.coherentmarketinsights.com/industry-reports/religious-tourism-market. Accessed 10 Jun 2024

Collins-Kreiner N (2016) The lifecycle of concepts: the case of 'Pilgrimage Tourism. Tour Geogr 18(3):322–334. https://doi.org/10.1080/14616688.2016.1155077

Digance J (2003) Pilgrimage at contested sites. Ann Tour Res 30(1):143–159. https://doi.org/10.1016/S0160-7383(02)00028-2

Doxey G (1975) A causation theory of visitor-resident irritants: methodology and research inferences. In: Travel and tourism research association, sixth annual proceedings. University of Utah, Salt Lake City, pp 195–198
Evans N (2015) Strategic Management for Tourism, Hospitality and Events. Routledge
Future Market Insights (2024) Faith-based Tourism Market Outlook from 2024 to 2034. https://www.futuremarketinsights.com/reports/faith-based-tourism-sector-overview. Accessed 10 Jun 2024
Liutikas D (2020) Management of pilgrimages in Lithuania: resources, stakeholders and projects. Int J Religious Tour Pilgrimage 8(2):8. https://doi.org/10.21427/j8w8-cw12
Martinez R (2009) Un acercamiento al turismo religioso en Los Altos de Jalisco. Revista Científica Electrónica de Ciencias Humanas 13(5):47–66
McGettigan F, Griffin C, Candon F (2011) The role of a religious tourism strategy for the west and northwest of Ireland in furthering the development of tourism in the region. Int J Bus Global 7(1):78–92. https://doi.org/10.1504/IJBG.2011.040847
Millán G, Pérez LM, Cárdenas RM (2012) Life-cycle stages in religious tourism development: a comparison of case studies. Cuadernos de Turismo 30:241–266
Nolan LM, Nolan S (1989) Christian pilgrimage in modern Western Europe. Chapel Hill, London
Olsen DH (2006) Management issues for religious heritage attractions. In: Timothy DJ, Olsen DH (eds) Tourism, religion and spiritual journeys. Routledge, pp 104–118
Page JS (2019) Tourism management. Routledge
Schinde K (2012) Policy, planning, and management for religious tourism in Indian pilgrimage sites. J Policy Res Tour, Leis Events 4(3):277–301. https://doi.org/10.1080/19407963.2012.726107
Shackley M (2012) Managing sacred sites. Cengage Learning EMEA, Andover
Shekhar A, Valeri M (2023) Modelling the effect of spiritual tourism motivators on spiritual tourism. Tourism 71(3):634–646. https://doi.org/10.37741/t.71.3.13
Smith VL (1992) Introduction: the quest in the guest. Ann Tour Res 19(1):1–17. https://doi.org/10.1016/0160-7383(92)90103-V
Trono A (2015) Politics, policy and the practice of religious tourism. In: Raj R, Griffin K (eds) Religious tourism and pilgrimage management. An international perspective. CABI, pp 16–36
UNWTO (2011) Religious tourism in Asia and the Pacific. UNWTO, Madrid
Uriely N, Israeli A, Reichel A (2003) Religious identity and residents' attitudes toward heritage tourism development: the case of Nazareth. J Hospit Tour Res 27(1):69–84. https://doi.org/10.1177/1096348002238881
Valeri M (2015) Sustainability development and competitiveness of Rome as a tourist destination. Tour Hospit Manage 21(2):203–218. https://doi.org/10.20867/thm.21.2.7
Valeri M (2022) New governance and management in tourist destinations. IGI Global Publishing, Hershey PA
WDC (2006) The Western Development Commission's Model of Rural Development. Irish Western Development Commission; May 2006. https://westerndevelopment.ie/wp-content/uploads/2020/09/reports_Models-of-rural-development.pdf
Xin S, Tribe J, Chambers D (2013) Conceptual research in tourism. Ann Tour Res 41:66–88. https://doi.org/10.1016/j.annals.2012.12.003

Identity Deconstruction Through Pilgrimage in Lhasa

Bob Yi-Chen Duan, Yuetian Zhang, Jingjing Dai, and Emily Ma

Abstract The existing literature has focused less on pilgrims' experiences in religious tourism, rather considering general tourists and pilgrims together. A better understanding of religious tourism and pilgrims has consequently been hampered. This study looks into Han Chinese pilgrims' experiences in Lhasa, Tibet, and discovers that religious identity changes due to individual and contextual factors after visiting religious places. The findings fill the gaps in the literature that only show favourable (positive) correlations between religious tourism and identity.

1 Introduction

Most religious tourism involves sightseeing and relaxation (Iliev 2020), which has not differentiated it from mass tourism or normal holidays. Only a segment of the market is purely for religious purposes, such as spiritual uplifting or to undertake a pilgrimage trip. This type of tourism is personal and religious. It can be considered in terms of identity (religious identity), such as self-identity, social identity, and self-congruity theory (Duan et al. 2020). Thus, a social psychology approach matches religious tourism. However, the role of identity in religious tourism has

B. Y.-C. Duan (✉)
School of Vocational and Technical Education, Yunnan Normal University, Yunnan, China

Griffith Institute for Tourism, Griffith University, Brisbane, QLD, Australia
e-mail: bob.duan@alumni.griffithuni.edu.au

Y. Zhang
School of Hospitality and Tourism Management, Macau University of Science and Technology, Macau, China

J. Dai
Office of Domestic and Foreign Affairs, Ningxia University, Ningxia, China

E. Ma
Department of Hospitality and Events, School of Hospitality and Tourism Management, University of Surrey, Surrey, UK

M. Valeri (ed.), *Contemporary Religious Tourism*, Tourism, Hospitality & Event Management, https://doi.org/10.1007/978-3-031-79074-4_18

been under-researched. Extant studies have indicated that religious tourism is positively linked with one's religious identity (e.g. Collins-Kreiner 2020). However, religious tourism and its identity have received limited research attention. In terms of the types of religions, Christianity leads the way, followed by Buddhism, Judaism, Islam, and Hinduism. However, Islam has been the most prominent religious tourism study in recent years (Collins-Kreiner 2020). Meanwhile, research on Buddhism has drastically declined, and updated studies are missing. Furthermore, most extant literature treats pilgrims and general tourists together in their sample (e.g. Kim et al. 2020), including Buddhism (Xie et al. 2020). Thus, the findings are mixed with general tourism motivation and experience (e.g. sightseeing and relaxation). Only a few studies are aimed at specific devout individuals (pilgrims), and a deeper understanding of this segment is needed.

Further, this study's context has a few distinguishing features (Han Chinese pilgrims travel to Tibet). Firstly, China has had a hiatus with its own traditional religious beliefs (e.g. religion; the Cultural Revolution; heavily censored access to information). Consequently, religion is more secularised and commercialised. The local authorities have tried to enhance patriotism through Chinese culture and heritage (such as that represented by religion). Thus, the background of religious tourism is different from other destinations (e.g. Italy). A struggle to move from folklore to a deeper understanding of the Buddhist faith is noticed in this market. Secondly, this chapter investigates Han Chinese pilgrims visiting Tibet. Existing studies have indicated that religion is not a motivation for religious tourism, and Tibet tends to be a tourist destination rather than a religious destination, even though most of its tourist attractions are based on Tibetan Buddhism (e.g. Cui et al. 2014; Xie et al. 2020). Thus, predominantly one-voice-only literature exists. In addition, a focus on religious believers (Han Chinese) visiting Tibet for pilgrimage is missing within the literature. Given the facts, the Han Chinese believed that Tibet had a spiritual role for its believers. Hence, this chapter aims to understand the religious experience of Han Chinese pilgrims in Lhasa, Tibet. Further, how does religious tourism shape their identity?

2 Literature Review

2.1 Religious Tourism

Religious tourism (RT) refers to tourism activities at religious sites (Heydari Chianeh et al. 2018). In this vein, tourists' motivation and experience are based on religion. However, existing studies have the counterintuitive finding that non-religious motivation and experience dominate the field: fun-seeking, sightseeing, and relaxation experiences or curiosity are examples of the non-religious attributes of religious tourist sites (Kim et al. 2020; Wong et al. 2013b). To illustrate, non-Buddhists found relaxation from their daily burdens in Buddhist temples in Los

Angeles (Choe et al. 2015); most Chinese religious tourists were more interested in sightseeing than the religion (Wong et al. 2013b); female tourists who visited Italian holy places were focused on socialisation (with locals and other tourists), while male visitors were engaged with the discovery of local places (Abbate and Di Nuovo 2013). Indeed, Amaro et al. (2018) claimed that religious motivation was the least prominent motivation present along the Camino de Santiago, a Catholic pilgrimage route in Spain. These critical findings outline two distinct "religious tourists": general/secular and pilgrim.

General tourists are not adherents of a religion or self-identified believers. Even though scholars have identified the motivation and experience of non-believers, a further investigation of their underpinning mechanisms is lacking. For example, compared to other kinds of tourism (e.g. mountains, lakes, beaches) that offer similar relaxation or fun experiences, why do general tourists choose a sacred site? Instead, the destination management perspective (DMP) has gained the most research attention. In this chapter, DMP refers to RT destination's overall offerings. One of the approaches in DMP emphasises the infrastructure of RT, which involves recognising tourism products. This includes identifying service quality, hotel/restaurant facilities, and transportation (Hung et al. 2015; Krešić et al. 2013). Similarly, in any tourism activity, religious tourists were satisfied by both facilities (e.g. souvenir shops, accommodation) and soft skills (e.g. courtesy and friendlessness; Kim et al. 2020). In other words, this approach explores RT in the same way as other types of tourism, disregarding the distinct character of RT regarding hospitality infrastructure.

Focussing on the destination experience from a tourist perspective is the second-most popular approach. This approach is prevalent in this field, as it has practical roles (e.g. economic benefits), and a widely existing scale of experience in other tourism fields may apply to RT. Besides the aforementioned experiences, the rural environment and the destinations' culture/heritage/historical events/artistic values are also common findings. Egresi and Kara (2018) illustrated that 93% of religious tourists were interested in cultural values at Italian sacred sites. Cathedrals in the UK are seen as tourist attractions rather than merely religious affiliations (Gutic et al. 2010; Kim et al. 2020). Sacred Sites in the UK attracted 3,087,441 tourists in 2023, and Westminster Abbey and St. Paul's Cathedral in London ranked 17th and 18th among Britain's most popular destinations, respectively (Group Leisure & Travel 2024). Due to the location of some holy places in remote areas, perspectives from rural tourism or ecotourism have also been considered. The countryside's lifestyle, natural environment, and peaceful feeling were highlighted in this case (Collins-Kreiner 2016; Dafuleya et al. 2017). Furthermore, in a Buddhist setting, RT has promoted the countryside's income and reduced the wide disparity between urban and rural areas (Wong et al. 2013b). Further, Buddhist-themed hotels have distinct experiences that reflect Buddhist culture in accommodation settings (Hung et al. 2015), in which the Buddhist-themed hotels/guesthouses may also apply in the countryside to bring economic benefits.

Studies that only focus on devotees or religious believers are rare. Instead, studies that look at mixed groups of the secular and the religious or studies of religious tourists' general tourism habits are more common. For this particular segment (pilgrims), reducing psychological distance or bringing spiritual connectivity with one's faith was the key motivation to visit and consequently satisfy the spiritual fulfilment experience (Buzinde et al. 2014; Kruger and Saayman 2016). Identity was considered in this context: self-identity with religion was the measure of distinguishing between general religious tourists and pilgrims (Nyaupane et al. 2015). This links back to general social-psychology identity theories (e.g. social identity theory; Burke and Stets 2022), such as people's identification as Christian/Buddhist in this setting. More specifically, religious identity is the key push element for pilgrims visiting the Camino de Santiago (Kim et al. 2019). During trips, they (pilgrims) also fostered self-identity with a religion, increasing religious belongings, and gaining a deeper understanding of their identity after visiting sacred places (Damari and Mansfeld 2016; Olsen 2012; Schott 2006). More recently, the rise of identity politics (research politics purpose through identity-building in tourism activities) in RT is the trend (Collins-Kreiner 2020). However, Kim et al. (2020) pointed out that a pure pilgrim experience is overlooked and that their identity construction and exploration call for further examination.

2.2 Chinese Religious Tourism

Among the religions, Chinese authorities show a preference towards Buddhism, Daoism, and Confucianism over others (e.g. Islam and Christianity), as they believe these three are closer to Han Chinese culture. Buddhism is the most popular religion in China, including Han or Chinese Buddhism, Tibetan Buddhism, and Theravada Buddhism (Meulenbeld 2019). The State's Administration for Religious Affairs licenses and controls all religious sites.

China's contemporary RT arose due to efforts towards the restoration of traditional culture (including religion). A former crackdown on conventional culture during the Cultural Revolution massively disrupted religious beliefs, architecture, and heritage within China (Wong et al. 2013b). Thus, the present generation of Chinese has lost traditional ties with past religious beliefs. After the revival, pioneering studies were employed in Taiwan, emphasising tourism surrounding Mazu (a Chinese sea goddess; Yeh et al. 2009), and more RT research followed up in mainland China (e.g. Hung et al. 2017; Wang et al. 2016; Wong et al. 2013b; Wong et al. 2016). Regardless of the location (Taiwan or mainland China), most RT research has mixed general religious tourists and pilgrims (e.g. Wong et al. 2013b). The existing literature on Chinese RT has consistent findings that most/all tourists are interested in experiences other than those of the religion itself. These experiences are similar to those mentioned in research on other RT destinations, such as relaxation and interest in culture and heritage (Wong et al. 2013b; Wang et al. 2016). Socialisation needs with friends and family, sightseeing, cuisine, and shopping experiences were also identified as relevant to this segment (Kim et al. 2020). Cui et al. (2014) claimed that sightseeing is the Han Chinese's spiritual fulfilment/

pilgrimage in Tibet. Xie et al. (2020) supported and put forward that the Han Chinese perceived Tibet as a tourist destination rather than a religious destination, even though most of its tourist attractions were based on Tibetan Buddhism (e.g. Potala Palace, Jokhang Temple, Sera Monastery, and Drepung Monastery). Aiming for the specific RT experience of pilgrims in this market has been absent. A few close-to-equivalent studies are Buddhist monks and nuns protecting their culture (Wong et al. 2016) and their perceptions towards tourism (Wong et al. 2013a).

In sum, RT and perhaps all tourism activities involve multiple actors and perspectives, such as from an ecotourism view (e.g. countryside lifestyle, landscape/sightseeing, overtourism; Hung et al. 2017), from a heritage tourism foundation perspective (e.g. cathedrals in the UK), a hedonist experience (e.g. discovery, socialisation, shopping), and from the point of view of hospitality infrastructure (e.g. accommodation, service quality). Further, most extant studies regarding RT motivation and experience have not been aimed at pilgrims. Thus, a deep understanding of pilgrims was missing in the literature. The research context has two major gaps: firstly, research on Buddhist tourism is somewhat outdated, and most of the extant literature mixed general religious tourists with pilgrims; secondly, existing studies have focused on Han Chinese sacred sites; only a few studies have examined the RT experience for Han Chinese in culturally different Tibet.

3 Method

Purposive and snowball sampling were used in this chapter (Bakkalbasioglu 2020). Pilgrims were defined in this chapter as self-identified Buddhist believers. Thus, the sample was Han Chinese or non-Tibetan Chinese who visited Lhasa, Tibet, in the past 12 months. In-depth interviews were carried out. Data was collected from 12 interviewees, and data analysis was guided by the grounded theory approach. This involved open, axial, and selective coding (Qureshi and Ünlü 2020).

4 Results and Discussion

4.1 Religious Authenticity

Religious authenticity is one of the goals that this type of tourist pursues at an RT destination. At the same time, this is why some of them found Lhasa or Tibet would have but has rarely existed in mainland China (i.e. Han Chinese territories in China/Han China). One informant said:

> The last time we went to a (Buddhist) temple, we felt like we were being ripped off. The temple and the travel agency teamed up and took us to a temple. A monk came out and asked each family member to go into a room with him. The monk said that your family

> should give 1,000 yuan (Chinese currency; 138 US$). Some business families seemed to give more. Everyone gave at least 1,000 yuan.

Similar comments were commonly found among informants. Another informant narrated her experience as follows:

> I was scared to death at that time. That (Buddhist) temple was a formal and large temple that had existed in history… Each of us entered a private room, and the monk told me that my family would have a disaster in March next year, and I had to pay for a ritual, and the fee was not low.

Social media and official media have also mentioned these phenomena, such as the commercialisation of temples in China. In addition to the local government's need to stimulate regional economic development, there is more involvement of private capital for profit (BBC 2017). In this case, the pilgrims escape mainland China (Han China) to avoid the commercialisation of religions. The pilgrim believed Tibet was the last lost and pure land for religion. Their comments are:

> It has been spread online and has become more and more amazing. Tibet has become a complex… the purest place… What cleanses the soul?

> It is the last pure land in the world… I went to Tibet with a smile because I was happy and had fulfilled a lifelong wish. I left Tibet crying with mixed feelings, knowing that a beautiful dream was about to end. I didn't want to wake up, but I had to.

> I think Tibet is the last pure land, a dream, where people believe in religion and are kind and simple.

Another phenomenon is that overtourism has led religious sites to become overcrowded in mainland China. One of the side effects is that the peaceful feeling at sacred sites has disappeared. This peaceful feeling falls under the theme of emotional experiences that general religious tourists and pilgrims are both concerned about (Kim et al. 2020). For pilgrims, this feature is particularly essential, as the former step contributes to spiritual fulfilment and also because the outside peace stimulates inner peace (Kruger and Saayman 2016). The informant said:

> Religious places are all scenic spots. Some temples are always too crowded, with people like chopsticks (people squeezed together), making it difficult to walk. The place is not calming… The religious places, I imagine, are quiet and peaceful.

There is a high barrier to entry for a pilgrim to Tibet or even simply travelling to Lhasa. For example, it is usually more expensive to visit Lhasa from faraway, rich areas in China (physical distance). This means more time and money, one to be healthy (high altitude sickness), and maybe know about the ethnic minority cultures. Those are the common thresholds for visiting Lhasa. Some regions in Tibet also require special permission for Chinese people to travel. Thus, pilgrims to Tibet seem to avoid overtourism and also escape the commercialisation of temples in China.

4.2 *Barriers to Integration*

Informants found more challenges in their RT. Firstly, local Tibetan Buddhism is a "new religion" for them, even for Han Chinese who are also self-identified Buddhists. They (the Han Chinese) are not familiar with or even know the local religious ceremonies, rituals, theories, and rules. A feeling of detachment from local Buddhism is the common feedback. In this case, they have a relatively far psychological distance between religion and the religion that trends to be others' faith. The informants said:

> Sometimes I don't understand why we do these things, and I don't think some of the actions are good. I think some actions are wasteful... Scattering those coloured papers is bad for the environment. Those papers have chemicals and heavy metals, which are completely polluting the environment.
>
> I am particularly disgusted by the scattering of coloured paper. I know it's a local religious event, but many tourists also do these activities. They hurt the environment. I think we should not promote this ceremony, and we should not use these toxic papers. We cannot sacrifice everything for tourism... I believe that in the most primitive times, these papers should have no burden on nature.

Another informant noticed other phenomena and commented:

> I understand putting gold on Buddha statues, but I don't understand why so many people keep doing this over and over again. It would be better if the money was used to help people in need. This could include starting a cancer research centre, doing research, or feeding people who don't have enough food.

Language barriers are a strong impediment for tourists and pilgrims to understand the local culture (e.g. religion). The informants mentioned that those who can speak Mandarin Chinese are normally young Tibetans, and they are not even familiar with their own culture and religion. The tourist guides can speak Mandarin, but they just need money, and they introduce the local culture as second-hand information that they copied from the Internet. Thus, the language is an ostensible obstruction, but the genuine predicament is that the current generation of Tibetans has lost the connection with its past (e.g. religion). One informant mentioned:

> The locals don't know as much as I do... I want to know about religion, history, and culture. The locals seem to be like me, an outsider, and I, an outsider, know more than them.

The former two factors (cultural distance and language barriers) lead to superficial understandings of the locals and their culture. In other words, an authentic destination is fading away, and this authentic here is the religious authentic.

4.3 Misunderstanding of Self and Religion

A few of the informants even questioned the meaning of certain religious rituals, for example, the aforementioned 'putting gold on Buddha statues'. Another informant mentioned she noticed one pilgrim made her feel 'sick'. She explained:

> One person said that she brought a lot of pearls, crystals, gems, and other things with her on this pilgrimage. I don't remember how much she spent on this pilgrimage. I don't like this… Because it seems that since she spent a lot of money, the people in the temple, including other pilgrims, have to be nicer to her. And we ordinary pilgrims seem to be so unimportant at this time. Our existence is still meaningless. Especially when other people treat her so well, my religious concept is impacted.

Some informants also commented that the local lamas (Tibetan lamas) were just watching how much money tourists/visitors gave them and that it was the only thing they (lamas) cared about and were concerned about. One informant said:

> I feel like they (lamas) are not really chanting. They are just looking at the merits (money) you give them… They even secretly check their phones. And I saw monks eating meat and smoking. I don't think monks should be like this. They seem to be more secular than us… It's really sad to say this. I have always thought that Tibet is the final pure land. It should be clean (close to the truth, real, without deception) … I feel like I have been deceived. I have doubts about religion.

This informant is not the only one; others have similar comments:

> Why do religions tell us to be good people? Look at the monks nowadays. They are not monks. They are actors. They perform what you want to see. In the end, they just treat religion as a business. Look at some rich people or officials. They also believe in Buddhism. They do bad things and have no moral conscience. Then, what is the meaning of my belief in religion? I can't find any reason to believe in religion anymore.

Some religious rituals seem to be associated with complex feelings.

> I feel very conflicted when talking about this from a personal religious perspective. I personally feel very conflicted. Prostrating (walking from home to the religious destination) is a waste, and at the same time, I also reason about why we should do this. I feel very conflicted. I should not criticise this behaviour, but I really feel that the influence of religion on us should not be like this… Maybe I am too simple and naive.

Understanding about oneself can evolve, and the informants said:

> I have always thought that I am a very devout Buddhist, but I was shaken by this pilgrimage. Religion is too difficult if I don't meet certain requirements… For example, you have to chant sutras every day, eat vegetarian food, and be a good person. I can be a good person, but there are some requirements that I really can't meet.

> If I cannot meet certain requirements, I cannot call myself a Buddhist. I am hesitant, entangled, and in pain. If the correct expression is powerless, I have no way to hope that everything is just as I imagined (for religion and myself). I am lost. I know what to do, but I don't want to.

The perception towards the locals from the informants also shook their religious beliefs, and one informant said:

> The locals are not pious in their religion either, and they seem to be using religion to make money. Some believers seem to be just like the Han Chinese people, using religion to reach their own goals. For example, I want to be rich and marry this person. These rituals with a purpose are not religions. They are not real religions.

In addition, positive feelings about religion or the self are also revealed, and one informant said:

> During this trip, I was deeply impressed by a temple. Windchimes were hung on several corners of the temple's main hall. The wind blew and the chimes rang. Listening to the sound of the bells, I stepped onto the stairs step by step. The half-soul floating outside the body was summoned into the body by the bells. My heart was calm, and I didn't have any thoughts… I joined the group of people walking around the temple, walking in circles around the main hall. The sound of the bells was nice to hear. The cool wind blew across my cheeks. The fresh air filled my lungs. The saliva I swallowed was also slightly sweet, and my heart was quiet. After three circles, I left slowly, thankful for this beautiful moment and for giving me a deep memory of myself.

4.4 Identity Deconstruction

The identity deconstruction of the pilgrims has two pathways. Firstly, they questioned their religious identity and that identity started to change, in a way close to deconstruction. Secondly, identity deconstruction (identity-destabilising) led to two further trials (see Fig. 1). Each is discussed below.

As previous studies indicated, the pilgrims in this chapter originally searched to reinforce their religious identity (Collins-Kreiner 2020; Kim et al. 2020). Firstly, the existing studies have all proven that RT increases religious identity, and no opposite case has been found. Some of the findings are similar to research (Damari and Mansfeld 2016; Olsen 2012; Schott 2006) that shows that pilgrims have learnt more about religion from an identity perspective. However, the new finding in this chapter, which has not been examined or explored in the literature, is the deconstruction

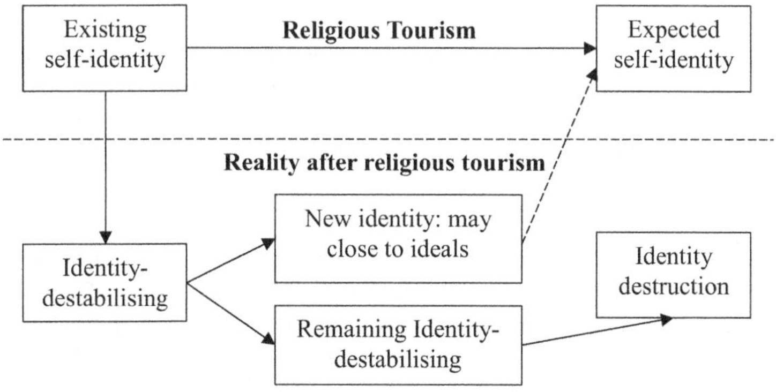

Fig. 1 Identity deconstruction framework

of identity. The emotional and physiological distance and cognition all together shaped the informants' attitudes, values, and beliefs, the constituents of identity. For example, some people do not agree with certain religious beliefs and practices. They already have their expected idea of Buddhism, which is different from what they have experienced and seen. In the meantime, a positive role model or a few Buddhists have the ideal behaviour that exists that the pilgrim wishes to witness. The underlying reason could be that Han Chinese found it hard to learn local Buddhism due to cultural differences and language barriers. The lost connection with the past in the current generation of Tibetans has further deteriorated the potential to learn. In this vein, this group of pilgrims cannot be satisfied only by the physical experience at the sacred sites, including the intangible feeling (e.g. peaceful) and tangible architecture. These factors (feeling peaceful in sacred sites) are proven to contribute to the RT experience and also to the building of intensifying religious identity (e.g. Damari and Mansfeld 2016; Kruger and Saayman 2016). In other words, the pilgrims in this chapter need both religious and educational experience. While the local tour guide somewhat filled this need, as the informants mentioned, all the religious information is copied from the Internet. Hence, authentic religious information is revealed as a weak RT experience and further leads to a change in religious identity.

Secondly, during the identity-destabilising process, two possible outcomes are expected. One is to rebuild the religious identity. This reconstructed identity may be closer to the ideal identity in the long term. The second outcome is that identity remains destabilised. From a postmodern perspective, identity is fluid and changes all the time (Onega and Ganteau 2020). Thus, theoretically, the two assumptions are practical. Subsequent studies on these pilgrims, testing these assumptions, are worthwhile in the long term.

5 Conclusion

A neglected research area (i.e. Han Chinese pilgrimage to Tibetan Buddhist destinations) was studied. Furthermore, this research uses an identity perspective to examine the religious tourism experience. In addition, identity deconstruction has received little research in the context of RT. Within this study, the pilgrims found that Buddhist identity was not what they had imagined, including feelings of doubt, and therefore found the identity hard to maintain. For part of the pilgrims who had not reached their ideal identity (e.g. uplifting), their identity began destabilising. Meanwhile, the religious identity of some of the pilgrims has been enhanced, which is consistent with existing studies. Religious authenticity is the pilgrims' search during the RT and has been revealed as currently a weak experience. Partially due to language barriers and cultural differences, the current generation has lost connection with its past in Tibet. Compared with general RT tourists, pilgrims have a higher demand for religious sites. Educational experiences, such as through communication, are one of the missing experiences. As mentioned, religious information (e.g. culture, religion) has been promoted by tour guides using online information, which did not fulfil the needs of the pilgrims. All these former steps (e.g. learning a local religion) are the basis for later spiritual fulfilment. Furthermore, recreational and aesthetic experiences seem to be less important for this segment of tourists.

It is suggested that this religious identity be considered in the context of an identity theory in further research, for example, social identity theory or other theories. The pilgrim identity can also be investigated as a long-term project, such as using a time series of data. The sampling method, size, and location also may have a bias; thus, generalisability calls for verification.

References

Abbate CS, Di Nuovo S (2013) Motivation and personality traits for choosing religious tourism. A research on the case of Medjugorje. Curr Issues Tour 16(5):501–506

Amaro S, Antunes A, Henriques C (2018) A closer look at Santiago de Compostela's pilgrims through the lens of motivation. Tour Manag 64:271–280

Bakkalbasioglu E (2020) How to access elites when textbook methods fail: challenges of purposive sampling and advantages of using interviewees as "fixers". Qual Rep 25(3):688–699

BBC (2017) "Don't take over my faith": China strictly prohibits commercial capital from intervening in Buddhism and Taoism. https://www.bbc.com/zhongwen/simp/chinese-news-42106196

Burke PJ, Stets JE (2022) Identity theory: revised and expanded. Oxford University Press

Buzinde CN, Kalavar JM, Kohli N, Manuel-Navarrete D (2014) Emic understandings of Kumbh Mela pilgrimage experiences. Ann Tour Res:1–18

Choe J, Blazey M, Mitas O (2015) Motivations of non-Buddhists visiting Buddhist temples. Curr Issues Tour 18(1):70–82

Collins-Kreiner N (2016). The lifecycle of concepts: the case of 'Pilgrimage Tourism'. Tourism Geographies, 18(3), 322–334

Collins-Kreiner N (2020) A review of research into religion and tourism Launching the Annals of Tourism Research Curated Collection on religion and tourism. Ann Tour Res 82:102892

Cui QM, Xu HG, & Yang Y (2014). Secular Pilgrimage; Tourist Experience in Tibet. Tourism Tribune/Lvyou Xuekan, 29(2)

Dafuleya G, Gyekye AB, Oseifuah EK, Nethengwe T, Sumbana F (2017). Religiously motivated travel and rural tourism in Vhembe District of South Africa. Tourism Economics, 23(4), 911-918

Damari C, Mansfeld Y (2016) Reflections on pilgrims' identity, role and interplay with the pilgrimage environment. Curr Issues Tour 19(3):199–222

Duan B, Arcodia C, Ma E, Hsiao A (2020) The xiao zi effect on motivation to visit wineries from an identity perspective: the case of China. J Hosp Tour Manag 45:359–369

Egresi IO, Kara F (2018) Residents' attitudes to tourists visiting their mosques: a case study from Istanbul, Turkey. J Tour Cult Change 16(1):1–21

Group Leisure & Travel (2024) The most popular tourist attractions in 2023 were... https://www.groupleisureandtravel.com/news-and-views/the-most-popular-tourist-attractions-in-2023-were/10541.article

Gutic J, Caie E, Clegg A (2010) In search of heterotopia? Motivations of visitors to an English Cathedral. Int J Tour Res 12(6):750–760

Heydari Chianeh R, Del Chiappa G, Ghasemi V (2018) Cultural and religious tourism development in Iran: prospects and challenges. Anatolia 29(2):204–214

Hung K, Wang S, Tang C (2015) Understanding the normative expectations of customers toward Buddhism-themed hotels: a revisit of service quality. Int J Contemp Hosp Manag 27(7):1409–1441
Hung K, Yang X, Wassler P, Wang D, Lin P, Liu Z (2017) Contesting the commercialization and sanctity of religious tourism in the Shaolin Monastery, China. Int J Tour Res 19(2):145–159
Iliev D (2020) The evolution of religious tourism: concept, segmentation and development of new identities. J Hosp Tour Manag 45:131–140
Kim H, Yilmaz S, Ahn S (2019) Motivational landscape and evolving identity of a route-based religious tourism space: a case of Camino de Santiago. Sustainability 11(13):3547
Kim B, Kim S, King B (2020) Religious tourism studies: evolution, progress, and future prospects. Tour Recreat Res 45(2):185–203
Krešić D, Mikulić J, Miličević K (2013) The factor structure of tourist satisfaction at pilgrimage destinations: the case of Medjugorje. Int J Tour Res 15(5):484–494
Kruger M, Saayman M (2016) Understanding the Zion Christian Church (ZCC) pilgrims. Int J Tour Res 18(1):27–38
Meulenbeld M (2019) Confucianism, Daoism, Buddhism, and Chinese Popular Religion. In: Oxford Research Encyclopedia of Asian History
Nyaupane GP, Timothy DJ, Poudel S (2015) Understanding tourists in religious destinations: a social distance perspective. Tour Manag 48:343–353
Olsen DH (2012) Negotiating identity at religious sites: a management perspective. J Herit Tour 7(4):359–366
Onega S, Ganteau JM (2020) Introduction: transcending the postmodern. In: Transcending the postmodern. Routledge, pp 1–19
Qureshi HA, Ünlü Z (2020) Beyond the paradigm conflicts: a four-step coding instrument for grounded theory. Int J Qual Methods 19:1609406920928188
Schott SB (2006) Pilgrims, seekers and history buffs: identity creation through religious tourism. In: On the road to being there. Brill, pp 297–327
Wang W, Chen JS, Huang K (2016) Religious tourist motivation in Buddhist Mountain: the case from China. Asia Pac J Tour Res 21(1):57–72
Wong CUI, McIntosh A, Ryan C (2013a) Buddhism and tourism: perceptions of the monastic community at Pu-Tuo-Shan, China. Ann Tour Res 40:213–234
Wong CUI, Ryan C, McIntosh A (2013b) The Monasteries of Putuoshan, China: sites of secular or religious tourism? J Travel Tour Market 30(6):577–594
Wong CUI, McIntosh A, Ryan C (2016) Visitor management at a Buddhist sacred site. J Travel Res 55(5):675–687
Xie YJ, Wei YD, Jia YN, Yang K (2020) Phenomenon of marginal men and their experience in pilgrimage tourism fields in Tibet. Tourism Tribune/Lvyou Xuekan 35(6):50–65
Yeh S-S, Ryan C, Liu G (2009) Taoism, temples and tourists: the case of Mazu pilgrimage tourism. Tour Manag 30(4):581–588

Micro-Pilgrimage as a Ritual to Commemorate the Fighters for Freedom in Romania

Cernicova-Buca Mariana

Abstract The study is the first attempt to deal with the "1989 Revolution routes" as instances of micro-pilgrimage and to identify the secular and religious values incorporated into the remembrance rituals. Yearly, quasi-religious pilgrimages are organized in Timisoara, along the places of sacrifice where locals died in the clash with repressive forces defending Ceausescu's regime. The "route of freedom" was developed through monument-building, ritualic commemoration, and media representation. This practice is challenged by a city-hall initiative to offer a competing, tourist-friendly route partially overlapping with the existing one, with other commemorating visuals and stories and a strong digital component, but lacking the religious load. The competition can result in the disentangling between participants in 1989 Revolution and local authorities and a change in the commemorating rituals.

1 Introduction

Public manifestation of religious beliefs was vigorously discouraged in communist Romania for more than four decades during the twentieth century. Some of the denominations were banned (like the Greek Catholic church or the Jehovah's Witnesses); others had their level of religious organization downsized following the dispositions of political authorities, enforcing official atheism. Confessional schools, many churches and monasteries, and religious charities and associations were abolished or forbidden. At that time, only monasteries in the northern part of Romania, having external mural paintings and standing out as attractions in the region, were preserved and included in cultural heritage tourism routes for domestic and for international visitors. Otherwise, pilgrimages as religious-spiritual journeys were not possible and participation in religious celebrations could bring unwanted

C.-B. Mariana (✉)
Faculty of Communication Sciences, Politehnica University of Timisoara,
Timisoara, Romania
e-mail: mariana.cernicova@upt.ro

M. Valeri (ed.), *Contemporary Religious Tourism*, Tourism, Hospitality & Event Management, https://doi.org/10.1007/978-3-031-79074-4_19

attention to the attendees. Against such a social and political context, Romanians perceived the dramatic fall of communism in December 1989 in the key of the David and Goliath clash. For more than a week unarmed, defenseless civilians protested Ceausescu's rule and regime, considered among the harshest communists in the European East. Lives were lost, more than 1100 people falling victim to the attempts of the army to suppress the protests. The protest began ten days before Christmas in Timisoara, the largest city in the Western part of Romania, a multinational, multi-confessional city, the sparkling event being the eviction of a Reformed priest, in whose defense the local population gathered, irrespective of their denominational affiliation. Public prayers and lit candles helped people feel like a community and accompanied the protest until the upturn of the regime. One week later, a mass demonstration called by the communist ruler to demonstrate support of his authoritarian regime turned into an anti-communist manifestation and the uprising spread rapidly. The communist regime fell by Christmas 1989 and the newly installed authorities quite soon declared a National Day of Mourning, in January 1990, to commemorate "the martyrs of Timisoara" (Rusu 2020). Terms such as "victims" or "casualties" were rebuked by the surviving participants in what is officially called "the Romanian Revolution of 1989." In the following years, cities that had a substantial number of martyrs persuaded the Romanian Parliament to pass laws declaring them "martyr cities". Timisoara and Bucharest were the first to obtain such titles, in 1990. Plaques at the entrance to these cities mark the title. In addition, in Timisoara, all the names of martyrs are attributed to street names, to preserve the memory of the people whose lives were lost in December 1989 (Crețan and Matthews 2015).

The present study focuses on the celebrations organized around the topic of martyrdom in Timisoara, with direct reference to the deaths of civilians during the 1989 Revolution. Commemorative events are organized yearly and presented as "pilgrimages," in a ritualistic way that will unfold later in the present research. After the fall of communism, in December 1989, praying in public, inviting priests to perform religious service on various occasions, and displaying signs of religious affiliation became part of public life. Churches and monasteries were restored, religion was introduced as an obligatory subject matter, and religious observances were recommended as public holidays (based mainly on the Orthodox calendar, since the census shows that up to 90% of Romanians practice East Orthodox Christian religion). Statistics show that the number of church buildings more than doubled after the fall of communism, most religious denominations arguing that places of worship were undersized for the needs of the population (Iftimiu 2016). Participation in pilgrimages, religious processions, and religious events increased exponentially over the years, the main denominations organizing their own pilgrimage agencies, thus competing on the respective segment with the regular tourism agencies which offered religious-spiritual tourism as part of their regular tours (Giușcă et al. 2018; Sorea et al. 2023; Apostol 2011; Bănică 2014). This trend fits the global growth of pilgrimage in the twenty-first century (Reader 2007), increasing the debate over the nature and motivations of such journeys. As Ian Reader points out, the modern pilgrim engages in a personalized "spiritual search" which may preserve religious traditions but may also be detached from religious traditions and presupposed meanings (Reader 2007). The pilgrimages organized to commemorate the martyrs of the revolution in Romania, although having most features pertaining to spiritual voyages, represent a special and under-researched case.

2 Pilgrimage, Thanatourism, and "Dark Tourism"

Pilgrimage as a practice occurred thousands of years ago, but scientific literature shows that as a research topic it is relatively new to the realm. As Noga Collins-Kreiner states, it stirred the interest of sociologists, anthropologists, philosophers, and geographers starting with the second half of the twentieth century (Collins-Kreiner 2020). Her review of the state-of-art shows that while some decades ago the term "pilgrimage" was used restrictively in connection with religious motivations and meanings, in the twenty-first-century scholarship a larger, holistic approach is adopted, treating pilgrimage as a "phenomenon with religious and secular foundations," encompassing sites that can emerge from both religious and secular contexts. She also believes that "the current literature is finding it increasingly difficult to differentiate among religious pilgrims, secular pilgrims, dark tourists, heritage tourists, and pilgrimage, as they are all recognized as part of the growing phenomenon of tourist interest in sites that add meaning to life" (Collins-Kreiner 2020). George Greenia, seeking to offer the key elements for a workable definition of pilgrimage, lists among the most salient features the following issues: the adoration/admiration of a physical location (place) as a site invested with symbolic meaning, the journey (most often treated as transformative), the performance of ritualized behaviors during the journey or at the site, the expectation that visiting the site can create an enduring memory, and the sense of community (*communitas*) that animates the participants in the journey (Greenia 2018). Peter Jan Margry defines pilgrimage as "a journey based on religious or spiritual inspiration, undertaken by individuals or groups, to a place that is regarded more sacred or salutary than the environment of everyday life, to seek a transcendental encounter with a specific cult object for the purpose of acquiring spiritual, emotional, or physical healing or benefit" (Margry 2008). Griffin and Razaq systematize the topic of pilgrimage tourism, mapping the field along key features to be considered, such as factors influencing the growth of religious or spiritual tourism, the breadth of religious tourism products, destination sites, motivations of the pilgrims, features of the travelers, and scale of the tourism industry offering such experiences (Griffin and Razaq 2017).

The debate whether secular pilgrimages should be treated in the same way that "real" (faith-based) pilgrimages fuels ample debates. Griffin and Razaq list secular pilgrimage in the typology of pilgrimages, since a transformative experience "that impacts an individual's belief system" occurs during such travels (Griffin and Razaq 2017), although critics of such theories deplore that even "religiously motivated travel is not religiously motivated enough" (Zwissler 2011). However, many authors agree with Alphonse Dupront in the belief that "there is no pilgrimage without a sacred place" (Dupront 1987; Eade and Sallnow 1991). In this case, authors such as Kim Knott deepen the debate about "sacred" and "sacralization (Knott 2005)." Other authors are more concerned whether pilgrimage and thanatourism should be discussed in terms of interchangeability, since the latter indicates visitation of "dark" places, where dramatic events took place and most religious sites tell the story of suffering, violent death, or drama (Collins-Kreiner 2016; Light 2017; Olsen and Korstanje 2019; Berger and Luckmann 1967; Seaton 2002). Duncan Light (Light 2017) dives into the motivations of tourists who visit "dark places," while Philip Stone (Stone 2009) launches the idea that dark tourism or thanatourism offers the moral guidance and paces that religion ceases to provide, under the pressure of secular trends in society. The American urbanist and geographer Edward Soja offers a different view, shifting the emphasis toward the idea that the so-called third spaces blend the real space and the imaginary space in an interplay of elements that create both the space and its interpretation (Soja 2003). In such a key, the difference between religious and secular loses importance, leaving room for the focus on the interplay between the dynamics of the physical elements of the space as socio-transmitters of meanings and narratives that sustain the mental construction (or re-construction) of the events tied to the space (Candau 2005).

Despite the debates on the motivations triggering the attraction of pilgrimages, religious or sacred, many scholars converge in believing that these journeys help define or affirm collective identities. Participants in the endeavor repetitively absorb the narratives accompanying these journeys and contribute to the perpetuation of the collective memory of the narratives (Halbwachs 2010). The pilgrims become part of the performance at the destination sites and share narratives and interpretations that consolidate the attractiveness of the place. A wide body of research develops as inherent elements of analyzing the spiritual journeys (labelled in such a variety of terms, depending on the focus of the analysis) three defining elements: people (as participants in the production of meaning and collective memory), places (as loci that allow for the actualization of narratives), and practices (ritualized and repetitive, meant to create the sense of community and emphasize solemnity, spirituality, experience of something unique and memorable).

In the last decade of the twentieth century, former communist countries found themselves in the position to reinvent or rediscover forms of rituals and public manifestation of emotions and beliefs that would clearly signify a discontinuity with the communist ones. Many of them engaged in a ritualistic removal of the signs of the near past and an equally symbolic filling in the (public) space with new memories and narratives (Rusu 2020; Czepzyński 2008; Crețan and Matthews 2015; Light 2000; Palonen 2008; Light and Young 2015; Forest and Johnson 2011; Liu et al. 2021). Romania fits into this pattern, in the attempt to obliviate the traces of unwanted past (Light 2000; Rusu 2020), and create "places of amnesia" (Candau 2005; Viejo-Rose 2015) at no longer acceptable reference points or produce new destinations of public admiration. Two main possibilities seemed at hand: reviving the visitation of traditionally known places or creating new ones, responding to the needs of society for anchors building identity and sense of community. Thus, Romanians fervently recovered the pilgrimage sites of the East Orthodox Christian religion, both internally and abroad (Bănică 2014; Apostol 2011). For the Catholic church, by far, the Marianic pilgrimage to Şumuleu Ciuc recently became the largest South-Eastern European destination of the kind (Munteanu and Cocean 2019; Ungureanu 2023). However, Romanians discovered the attractiveness of esoteric places, the thrill of dark tourism, and the passion for secular sites that fill in the desire for connectedness and communion (Rabontu and Vasile 2014; Munteanu and Cocean 2019; Light 2017).

From the wealth of destinations, practices, and types, the present research focuses on the specificity of newly developed places of remembrance that yearly become destinations of micro-pilgrimages. Micro-pilgrimages are described in scientific literature as small-scale journeys to particular items that can trigger the spiritual experience such as relics, altars, roadside chapels, and crosses, even images (Abel 2009; Wigley 2016; Liutikas 2018; Bailey 2022). By contrast to the grand-scale pilgrimages (macro and mezzo) (Liutikas 2018), the micro-pilgrimages are typically of local or small regional significance (Gábor 2014), and the number of visitors is not remarkably high. The research is based on participatory observation over three decades, supplemented by individual interviews with the main organizers of the pilgrimages and with the monitoring of the media, reflecting the events. The micro-pilgrimage under scrutiny is of a secular nature, but the organizers invited a strong religious component and, lacking other models of organizing public commemorative rituals, appealed to the East Orthodox cult of martyrs (Anikin 2021; Stan and Turcescu 2000), adding to it testimonials/narratives of participants in the commemorated events, candle vigils, and wreath-laying.

3 The Production of "Secular Shrines" in Timisoara

Many of the participants in the uprising that put an end to the communist regime in Romania, recognized as freedom fighters for the revival of pluralistic democracy, formed nongovernmental organizations according to their shared experiences during December 1989: arrested and imprisoned during the revolution, successors of martyrs, memorial association, etc. Each association annually organizes commemorative events on a designated day, starting with 16 December (Grama 2005). The pattern for the "commemorative-anniversary" sequence of remembrance ceremonies was established as early as 1990, when these associations negotiated with local authorities to have grieving days for the first part of the sequence and anniversary, happy celebrations in memory of the victory over the oppressive regime. The first anniversary (in December1990) had as a focal point the inauguration of a symbolic memorial complex at the Heroes' Cemetery in Timisoara, specially built for the occasion, with the construction being personally supervised by the prefect of the region. Priests were invited to perform the religious consecration of the memorial complex and the lighting of the eternal flame. At the time, the first pilgrimage was announced, to a suburban area of Bucharest (Popeşti-Leordeni), at the only human crematory in Romania. The communist authorities ordered the bodies of 47 protesters shot during the revolution to be burned, in an attempt to hide the fact that human lives had been lost. The ashes were scattered in the vicinity. The truth regarding the fate of these victims created a great shock in public opinion and the memorial served as homage to the dead, recognized as martyrs for the cause of freedom. Pilgrimages were organized from sympathizing cities to Timisoara, to the symbolic graveyard and memorial in the Heroes' Cemetery, and subsequently from Timisoara to Bucharest, to Popeşti-Leordeni, where the ashes of the dead were thrown into a sewer, to erase the memory of the crime. This second segment is promoted under the title "Following the traces of the ashes."

Fearing that the memory of the 1989 sacrifice in the name of freedom can be trivialized, minimalized, or lost, the association Memorial to the Revolution actively lobbied the local administration of Timisoara to approve locations for the erection of monuments commemorating the most dramatic events of December 1989, thus contributing to the production of spaces (Lefebvre 1991) as social transmitters of heroic narratives. "A vice-mayor was appointed to support us and ease the administrative aspects of the project"—remembers Traian Orban, the first president of the Memorial Association (personal communication, July 15, 2024). The initiators of this project invited famous sculpturers to pitch in and deployed intensive fundraising efforts to bring the idea to life. Overall, the Memorial Association succeeded in creating a route of 14 monuments dedicated to Timisoara fighters for freedom in the 1989 Revolution. The number of monuments is symbolic: it deliberately appeals to the Christian memory of 14 stations on the Via Dolorosa in Jerusalem. The imagery in the monuments varies: some are symbolic representations of the martyrs' fountain, freedom's bell, or the crying church; others are terrifying, like the target man, the winner, the martyrs, or appeal to the Christian representations such as Pietà (only having the man supporting the dead body of a woman), the Crucifixion, or the Memorial of Student Martyrs (Pelerinaj 2024). The ceremonies, organized each year, are, however, a vivid illustration of the way secular and religious initiatives intertwine and aim to appeal to the religious, pious sentiments of the audience. To create a solemn atmosphere and instill respect for the commemorative events, the organizers (associations of revolutionaries jointly with local authorities) invite representatives of the Metropolis of Banat to officiate the memorial service along the route, at each monument, while the audience holds lit candles and other signs of grief. Rememorating the actions of the protesting civilians at the site is orally transmitted by one of the participants in the 1989 events. Representatives of the authorities (municipality, parliament, political parties) usually participate in the celebration in a VIP position, but without speeches (reserved for solemn sessions held in indoor settings). The organizers tour all the monuments. Attendance varies from year to year. Initially, the tours were designed as a pedestrian procession from one monument to the next, culminating in the Heroes' Cemetery, which attracted thousands of participants, linked by the recent memory of the dramatic events of December 1989. Later (as the route also became longer), the pedestrian procession was replaced by coach travelling. Ultimately the organizers realized that the local community becomes increasingly more absorbed by other preoccupations and that, apart from organizers, members of the associations of freedom fighters and journalists, the audience dropped to insignificant numbers. It became increasingly difficult to ensure significant participation in the commemorative activities at each monument. The reduced participation was deplored by the organizers after the twentieth anniversary of the 1989 Revolution (Mocioalcă 2010), and to ensure that each of the monuments (which are scattered in all districts of the relatively large city) receives attention, the Memorial Association partnered with schools for the yearly micro-pilgrimages. Schools in the vicinity of memorials to martyrs of 1989 were invited to "adopt" a monument and care for it. Also, yearly, students from schools participating in the project initiated by the Memorial Association represent the main part of the audience. Both church representatives and leaders of Memorial Association believe that this "participation in the pilgrimage" contributes to the religious and civic education of the youngsters, who develop, in this way, a sense of belonging to Timisoara and a pious attitude toward freedom fighters, ensuring, at the same time, intergenerational transmission of the significance attributed to the monuments (personal communication, Traian Orban, the first president of the Memorial to Revolution Association, 15 July 2024, Ionel Popescu, Pr., eparchial vicar at Banat Metropolis, 18 July 2024). The ceremonies end with wreath-laying and candlelight vigils. Media coverage of these events is detailed, maintaining the specific language and references for religious events, accepting the term "pilgrimage" in the program of memorial activities and in the description of the march (Hojda 2018). Eventually some local tourist agencies incorporated into their offer walks entitled "Timisoara – the 1989 Revolution Tour," but not the whole route is included in the package and a significant part of the monuments remain outside the scope of the tourist attention. On the thirtieth anniversary of the revolution, the tourist agencies offer organized tours of the sites, at the same time as the pilgrimage, but without merging the two. At least for the time being, the pilgrimage is perceived as having an educational, identity-building, and community-building value (Gino Rado, current president of the Memorial to Revolution Association, 15 July 2024. Ionel Popescu, Pr., eparchial vicar at Banat Metropolis, 18 July 2024), although its future is uncertain because of the fluctuating interest of potential participants in the celebrations and because of its proximity to winter holidays. At the same time, the city officials moved away from the revered, pious attitude toward the places of sacrifice, presented in a religious, deeply spiritual key, and intend a different approach of preserving/promoting the memory of the 1989 events. By the end of 2022, in preparation of the Timisoara–European Capital of Culture program, the municipal officials accessed a 4.3-million euro project funded by the European Union under the #NextGenerationEU temporary financial mechanism to develop "The 1989 Timisoara Revolution Route" (The 1989 Timisoara Revolution Route 2024).

The concept of this new route is based on Robert Musil's warning that "there is nothing in this world as invisible as a monument" (Musil 2006). The Austrian philosophical writer argued almost a century ago that monuments, once ingrained into the geography of a city, become part of the scenery and that passers-by can overlook their presence without the "slightest notion of whom they are supposed to represent" (Musil 2006). The new project promises to create a route linking the landmarks of the December 1989 Revolution in Timișoara which is media-sustained and (more) coherent visually, thematically, and conceptually. The intention of the promoters is to use the project to improve the quality of the public space and integrate new technological affordances (such as Augmented Reality, Virtual Reality, New Media Art), thus "allowing a sensorial and emotional experience of the place" and a sustainable commemoration of the 1989 Revolution (The 1989 Timisoara Revolution Route 2024). The project was under public scrutiny via the dedicated site of the city hall, in the section for public consultations, and the initiators responded to the objections regarding the fact that the new route is shorter and overlaps only partially with the one developed by the Memorial to Revolution association. The project (proposed to be finalized by the commemorative celebration of the 1989 Revolution to be held in December 2024) incorporates, besides the intervention in the urban space with a special type of pavement (turquoise granite and bronze), memorial plaques, urban furniture, and illumination, to develop a web app leading to a dynamic archive of audio and visual archives related to the revolution and the city. The intended internet platform will be accessible both directly online, remotely, and on-site, with the developers explaining that the differences occur in the type of visitor experiences. A Braille version of testimonials also accompanies the new objects, populating the Timisoara Revolution Route. Although interesting and modern in concept, the new project has a strong potential to compete with existing consecrated pilgrimage-like commemorations. The narratives behind the two routes are similar, but the visitation proposals are quite different in the unfolding of the contemplation of the created "places of memory" as social transmitters of the narratives (Candau 2005) regarding the heroic clash between civilians, longing for freedom, and an oppressive regime, defending its privileges at the gunpoint. The pilgrimage feature of the December commemoration might be lost in the competition. But this is also a consequence of the disentangling between the agendas of associations of participants in the revolution, who operate in the manner of nongovernmental organizations, and the authorities, which have the administrative and symbolic power to develop the infrastructure that supports or hinders the perpetuation of physical elements in the city's "sites of memory."

4 Conclusions

This research places under scrutiny competing visions on a recent development around the heritagization of commemorating fighters for freedom in post-communist Romania, on the example of Timisoara. The "sites of memory" are physically marked by monuments that motivate stops for ritualized storytelling and religious service during annual micro-pilgrimages commemorating the sacrifice of freedom fighters. The annual December pilgrimage to the places of sacrifice combines secular and religious to trigger emotional engagement of the audiences, to consolidate the memory of the 1989 Revolution in the local community, and to elicit reverence and admiration from audiences. The route and rituals have been assiduously cultivated as a civil society initiative, with freedom fighters inviting religious authorities to support and bring solemnity to commemorative events. Alternatively, the local authorities encourage a secular approach closer to the touristic interpretation and use of the route, only partially overlapping with the consecrated one described above. The new route, heavily financed from 2024, proposes new narratives on the 1989 events in Timisoara. It promises to enhance the chances of multisensory experiences, but it will do so at the expense of the spiritual dimension of the travel. Until the new route is in place, it is early to anticipate which version will appeal more to the local population. A cautious conclusion, however, can lead to the idea that secular micro-pilgrimages are fragile and, unless supported by the stakeholders, can prove to be short-lived.

References

Abel M (2009) Within, around, between: micro pilgrimage and the archivolted portal. Hisp Res J 10(5):385–416. https://doi.org/10.1179/146827309X12541437923748

Anikin DA (2021) Traumatization of the past and martyrological thinking in the Soviet Union and the Post-Soviet Space. Folklore: Electron J Folklore 83:47–62

Apostol M-S (2011) Pilgrimage phenomenon in the new context of Post-Communist Romanian Media. ESSACHESS - J Commun Stud 4(8):127–143. Available at https://www.essachess.com/index.php/jcs/article/view/131

Bailey AE (2022) Micro pilgrimages: a new post-secular trend? Religions 13:665. https://doi.org/10.3390/rel13070665

Bănică M (2014) Nevoia de miracol. Fenomenul pelerinajelor în România contemporană. Polirom, Romania, Iaşi

Berger P, Luckmann T (1967) The social construction of reality: a treatise in the sociology of knowledge. Doubleday, Garden City, NY

Candau J (2005) Anthropologie de la mémoire. Armand Colin, Paris

Collins-Kreiner N (2016) Life cycle of concepts: the case of pilgrimage tourism. Tour Geogr 18:322–334

Collins-Kreiner N (2020) Pilgrimage tourism-past, present and future rejuvenation: a perspective article. Tour Rev 75(1):145–148. https://doi.org/10.1108/TR-04-2019-0130

Creţan R, Matthews PW (2015) Popular responses to city-text changes: street naming and the politics of practicality in a post-socialist Martyr City. Area 48(1):92–102. https://doi.org/10.1111/area.12241

Czepzyński M (2008) Cultural landscapes of post-socialist cities representation of power and needs. Ashgate, Abingdon, UK

Dupront A (1987) Du sacré. Croisades et pèlerinages. Images et langages. Éditions Gallimard, Bibliothèque des Histoires, Paris

Eade J, Sallnow N (1991) Contesting the sacred: the anthropology of Christian pilgrimage. Routledge, London and New York

Forest B, Johnson J (2011) Monumental politics: regime type and public memory in post-communist states. Post-Sov Aff 27(3):269–288. https://doi.org/10.2747/1060-586X.27.3.269

Gábor B (2014) Places of pilgrimage and pilgrimage routes as community-creating places. In: Gavrilović D, Nikolić V (eds) Cult places on the border. Yugoslav Society for the Scientific Study of Religion, Niš, Serbia, pp 7–14

Giuşcă M-CG, Gheorghilaş A, Dumitrache L (2018) Assessment of the religious-tourism potential in Romania. Hum Geogr – J Stud Res Hum Geogr. https://doi.org/10.5719/HGEO.2018.122.6

Grama S (2005) In between places of remembrance and realms of memory: the 15-years commemoration of the Romanian Revolution in Timisoara. Philobiblon. Transylv J Multidiscip Res

Humanit 10–11:311–342. https://www.philobiblon.ro/sites/default/files/public/imce/doc/2005/philobiblon_2005_10-11_16.pdf
Greenia GD (2018) What is pilgrimage. Int J Relig Tour Pilgr 6:2, Article 3. Available at: https://arrow.tudublin.ie/ijrtp/vol6/iss2/3
Griffin K, Razaq R (2017) The importance of religious tourism and pilgrimage: reflecting on definitions, motives and data. Int J Relig Tour Pilgr 5(3)., Article 2. Available at: https://arrow.dit.ie/ijrtp/vol5/iss3/
Halbwachs M (2010) La mémoire collective. Albin Michel, France, Paris
Hojda D-M (2018) Pelerinajul – itinerar inițiatic în formarea identității naționale în conștiința elevilor, Memorial 1989. Buletin științific și de informare 1(21):174–180
Iftimiu A (2016) Only 37.9% of the places of worship built in Romania after 1989 belong to the Romanian Orthodox Church. Available at https://basilica.ro/en/only-37-9-of-the-places-of-worship-built-in-romania-after-1989-belong-to-the-romanian-orthodox-church/. Accessed 15 July 2024
Knott K (2005) Spatial theory and method for the study of religion. Temenos - Nordic J Study Relig 41(2). https://doi.org/10.33356/temenos.4778
Lefebvre H (1991) The production of space. Blackwell Publishers, Oxford
Light D (2000) An unwanted past: contemporary tourism and the heritage of communism in Romania. Int J Herit Stud 6(2):145–160. https://doi.org/10.1080/135272500404197
Light D (2017) Progress in dark tourism and thana-tourism research an uneasy relationship with heritage tourism. Tour Manag 61:275–301. https://doi.org/10.1016/j.tourman.2017.01.011
Light D, Young C (2015) Public memory, commemoration and transitional justice: reconfiguring the past in public space. In: Stan L, Nedelsky N (eds) Post-communist transitional justice: lessons from twenty-five years of experience. Cambridge University Press, New York, pp 233–251
Liu Y, Dupre K, Jin X (2021) A systematic review of literature on contested heritage. Curr Issues Tour 24(4):442–465
Liutikas D (2018) Catholic pilgrimage in Europe: contemporary traditions and challenges. In: El-Gohary H, Edwards DJ, Eid R (eds) Global perspectives on religious tourism and pilgrimage. IGI Global, USA, pp 84–103. https://doi.org/10.4018/978-1-7998-2457-2.ch014
Margry PJ (2008) Secular pilgrimages: a contradiction in terms. In: Margry PJ (ed) Shrines and pilgrimage in the modern world: new itineraries into the sacred. Amsterdam University Press, Amsterdam, The Netherlands, pp 13–46
Mocioalcă S (2010) Pelerinaj la monumentele dedícate Eroilor Martiri din Timișoara, Memorial 1989. Buletin științific și de informare 1(6):189–190
Munteanu G, Cocean P (2019) The sacred sites of Transylvania. Revue Roumaine de Géographie 63(1)
Musil R (2006) Posthumous papers of a living author. Archipelago Books, New York
Olsen DH, Korstanje M (eds) (2019) Dark tourism and pilgrimage. CABI, Oxfordshire, UK. https://doi.org/10.1079/9781789241877.0000
Palonen E (2008) The city-text in post-communist Budapest: street names, memorials, and the politics of commemoration. GeoJournal 73(3):219–230. https://doi.org/10.1007/s10708-008-9204-2
Pelerinaj la Monumentele dedicate Eroilor. Available on https://memorialulrevolutiei.ro/pelerinaj-la-monumentele-dedicate-eroilor-2/. Accessed 1 July 2024
Rabontu CI, Vasile MC (2014) Dark tourism - new form of tourism in Romania. Annals - Economy Series, Constantin Brancusi University, Faculty of Economics, vol 6, pp 269–273, December
Reader I (2007) Pilgrimage growth in the modern world: meanings and implications. Religion 37(3):210–229. https://doi.org/10.1016/j.religion.2007.06.009
Rusu M-S (2020) The politics of mourning in post-communist Romania: unravelling the thanatopolitics of grievable deaths. Mortality 25(3):313–331. https://doi.org/10.1080/13576275.2019.1682983
Seaton AV (2002) Thanatourism's final frontiers? Visits to cemeteries, churchyards and funerary sites as sacred and secular pilgrimage. Tour Recreat Res 27(2):73–82. https://doi.org/10.1080/02508281.2002.11081223

Soja E (2003) Writing the city spatially[1]. City 7(3):269–280. https://doi.org/10.1080/1360481032000157478
Sorea D, Defta M, Popescu IM (2023) Journeys to significant places in orthodoxy as a source of sustainable local development in Romania. Sustainability 15(7):5693. https://doi.org/10.3390/su15075693
Stan L, Turcescu L (2000) The Romanian Orthodox Church and post-communist democratisation. Europe-Asia Stud 52(8):1467–1488. https://doi.org/10.1080/713663138
Stone P (2009) Chapter 4. Dark tourism: morality and new moral spaces. In: Sharpley R, Stone P (eds) The darker side of travel: the theory and practice of dark tourism. Channel View Publications, Bristol, Blue Ridge Summit, pp 56–72. https://doi.org/10.21832/9781845411169-005
The 1989 Timisoara Revolution Route (2024). https://tm89tour.primariatm.ro/en/about. Accessed 31 July 2024
Ungureanu L (2023) The dynamics of religious tourism. Case study: Romania. J Environ Manag Tour 14(2):582–598. https://doi.org/10.14505/jemt.v14.2(66).27
Viejo-Rose D (2015) Cultural heritage and memory: untangling the ties that bind. Cult Hist Digital J 4:e18. https://doi.org/10.3989/chdj.2015.018
Wigley E (2016) The Sunday morning journey to church considered as a form of 'micro-pilgrimage'. Soc Cult Geogr 17(5):694–713. https://doi.org/10.1080/14649365.2016.1139168
Zwissler L (2011) Pagan pilgrimage: new religious movements research on sacred travel within Pagan and new age communities. Relig Compass 5(7):326–342. https://doi.org/10.1111/j.1749-8171.2011.00282.x

[illegible]

An Overview of Religious Tourism with Special Reference to Shakti Peeths in Odisha, India

Monalisha Dash, Suneel Kumar, and Jyoti

Abstract This chapter explores about religious tourism in Odisha, India, focusing on the significance of Shakti Peethas, which are sacred pilgrimage sites dedicated to goddess Durga. Odisha, a state with a rich religious heritage, draws numerous pilgrims to its holy sites, contributing to both cultural exchange and economic growth. The research aims to explore, highlight these Shakti Peethas, and promote them as major tourist attractions. Shakti, or the "Great Divine Mother," represents the divine feminine force in Hinduism, with Shakti Peethas being revered shrines where devotees worship Durga in her various forms. Odisha houses over ten Shakti Peethas, each associated with unique supernatural events. The study suggests ways for the community to support religious tourism, benefiting both local culture and the national economy.

1 Introduction

One of the earliest methods of demonstrating religious devotion is the pilgrimage. It is a crucial part of the tourism business and among the first types of tourism. Onions (1983) described pilgrimage as a journey taken as an act of devotion by a pilgrim, usually covering a considerable distance to a sacred location. Stated that a pilgrimage is a planned march of a group of people, usually associated with a religious event. Vukonić (1996) witnessed a pilgrimage as a planned visit or tour wherein there are religious reasons for visiting a location and where religious rituals are

M. Dash · Jyoti
Department of Hospitality & Tourism, Galgotias University,
Greater Noida, Uttar Pradesh, India
e-mail: jyoti@galgotiasuniversity.edu.in

S. Kumar (✉)
Department of Commerce, Shaheed Bhagat Singh College, University of Delhi,
New Delhi, India
e-mail: suneel.kumar@sbs.du.ac.in

M. Valeri (ed.), *Contemporary Religious Tourism*, Tourism, Hospitality & Event Management, https://doi.org/10.1007/978-3-031-79074-4_20

practiced, "since the time of the ancient mythologies up to the present, all the religions in the world have made hope concrete by proclaiming to believers that a visit to a holy place would relieve all or some of their troubles, whether spiritual or worldly.

Visits to holy places for the sake of purification, redemption, fulfillment of vows, healing, or something else are called pilgrimages." Like all other identical industries, the tourism sector has been adapting to the shifting socioeconomic conditions. It is therefore clear that there is a constant hunt for prospective markets. Tourism and pilgrimage have a close relationship. But for a very long time, the relationship between "tourists," "religions," and the "religious" settings they visit on vacation has been disregarded (Collins-Kreiner 2010). The question that arises is why tourists visit sites of worship in order to learn about their significance and how they comprehend that their visit will provide answers to crucial concerns regarding the similarities and differences between tourism, religion, and spirituality. It is especially drawn to those who engage in religious activities or who experience spirituality in a religious setting. Odisha is one of India's most popular tourist destinations because of its illustrious two-millennia-old histories. Many Shakti Peeths, which are extremely significant both religiously and culturally, are located in the eastern Indian state of Odisha. These holy locations generate interest from foreign travelers hoping to witness India's rich spiritual legacy in addition to thousands of pilgrims from all throughout the country. The Shakti Peeths of Odisha offer a distinctive perspective for examining the dynamics of religious tourism because of their rich mythology, historical significance, and colorful ceremonies.

More significantly, the state is one of the most popular religious destinations in the nation because it is home to an excessive number of temples and sacred locations. The state's economy benefits greatly from the increased number of pilgrims that come here each year. Many religious sects, festivals, and cultures are worshipped in Odisha. It is a known truth that people who travel to holy sites for religious reasons have deep attachments to their beliefs. Rather than being referred regarded as tourists, some visitors who visit Indian would rather be called as pilgrims than tourists. Their religious beliefs serve as their driving force, and their travel goals could include obtaining wisdom, finding tranquilly, curing specific problems, performing rituals, fulfilling wishes, or curing illnesses. Tourism not only generates employment for the locals but also provides an opportunity to interact with people from different parts of the country and world. The Puri district was the main draw for tourists because of its countless tourist attractions. In Puri town alone, there were 17.59 lakhs domestic and 14,627 foreign visitors. On 2006, the number of visits crossed 50 lakh. In the same fiscal year, the capital city welcomed almost seven lakh tourists. Large numbers of tourists visited other locations, such as Chandipur in Balasore, Ratnagiri and Udaygiri in Jajpur, and Vedvyas in the districts of Sundargarh. Additionally, locations like Dhabaleswar in Cuttack, Kapilas in Dhenkanal districts, Harishanakar in Bolangir, and Nrusimhanath in Baragarh proved to be popular tourist spots.

Hotel occupancy rates in pilgrimage centers other than Bhubaneswar and Puri are as low as 50%, according to data available with the Tourism Department. It is attested to by Konark's dismal 33% hotel occupancy rate. Hindu texts from the Middle Ages mention 51, 52, 64, and 108 Shakti Peethas, of which 18 are acknowledged as Astadasha Maha (major). The exact number of these shrines' changes depending on the Puranas. The Shakti Peethas have many legends explaining their origins; the most well-known one is connected to the tale of the goddess Sati's death. Lord Shiva, engulfed in sorrow, travelled the cosmos carrying Sati's body and thinking back on their shared past. Lord Vishnu utilized his Sudarshana Chakra to split Sati's body into 18 pieces, each of which fell on the earth, devoting these spots as sites of worship for humanity. This relieved him of this enormous duty.

2 Literature Review

A substantial portion of the service industry, which plays a major role in India's economic growth, is the tourism sector. More quickly than in years past, the tourism sector in Odisha is expanding. Tourism encompasses a wide range of subfields that are frequently pursued in accordance with the objectives and conditions of the activity. These days, travel to places of health (medical tourism), places of worship (religious/spiritual tourism), sports, education, and agriculture are all popular niche markets (Das, September 2013). Any tourism has a great impact on socio economic factor of the state. By attracting tourists to local events, it promotes cross-cultural exchanges between the two groups. Both directly and indirectly, these tourist destinations have a major impact on the local economy. When event tourists spend their money on local products and services, it supports local productions and distributes the benefit more broadly throughout the community and economy. Both the guest and the host should benefit from it. Extensive research has demonstrated the beneficial impacts of tourism on the local economy, culture, and society. Nevertheless, greater efforts are required to effectively present local customs and traditions to tourists.

In this postmodern, high-tech era, people typically do not appear to lose their philosophical curiosity for which they travel all around the world to meet their quest. Research on spirituality is prevalent now in the fields of business, health, and finance and numerous branches have acknowledged the impact of religious tourism support in uplifting spiritual mindset and thoughts (Kumar et al. 2022a, b). Someone travels to a fascinating tourist location with the goal of venturing new things. Globally, spirituality has been expanding. Consequently, travel related to this activity has become a significant innovation for the modern tourism sector. Research in the field of spiritual tourism is continually developing. Notably, there has been a notable increase in the literature on spiritual tourism in recent years (Kumar et al. 2022a, b). India is among the world's most spiritually advanced nations and has been the spiritual center of the globe for more than 4000 years. India has always been seen as the home of temples and Vedas, where people can enhance and nourish their spirituality and enable all religions to grow and thrive (Mukherjeel 2020). The concept of spiritual tourism is not new, although it is generally a recent one. A traveler who goes to a fascinating tourist location with the goal of growing spiritually in line with their own beliefs is known as a spiritual visitor. This can happen for any number of reasons, including religious obligations that are essentially sacrosanct, nonreligious, spiritual, or religious.

Highlighting about the western world, it has been found that the majority of Western theology and philosophy of religion has avoided discussing goddesses, or "the Goddess," with the notable exception of those fields where feminists have made significant contributions. Hindu goddesses have drawn the interest of certain feminists, mostly because to the inclusion of strong and self-reliant individuals among them. In reality, Śhakti, which means "power, ability, strength, might, effort, & energy" in Sanskrit, is one of the most often used titles for the Hindu goddess. With the relationship between power and female divinity, it seems sense to wonder if the popularity of goddesses has helped or will help women in Hindu communities become more powerful (Burley 2021). Concerning the flow of tourists into Odisha, the state has observed that, except for 2015, which saw a decline in visitor numbers, there has been a consistent increase in both domestic and foreign visitor numbers over the past years. The number of people visiting Odisha has generally been rising steadily. 2017 was Odisha's best year ever in terms of foreign visitor arrivals. The months of October through December and January through March are the busiest for tourists visiting Odisha; April through September are the slowest, a time primarily due to weather patterns and seasonal preferences. The year 2013 marked the peak of the change in the overall amount of money coming in as well as the amount of money coming in from both domestic and foreign sources through tourism expenditures. Due to the several cyclonic storms that struck Odisha in 2015, the state was no longer/less a popular travel destination for international visitors, resulting in a negative growth or change in the amount of money coming in through foreign tourist expenditure (Behera 2019). Necessarily to gain better socioeconomic result and to promote spiritual tourism, the religious places that revolve around spirituality, and the local community who are into hospitality offering practice to the tourist, it is very important to understand the value of the tourist visiting those sacred places. Understanding the value of "knowing your customer" is crucial when it comes to promoting spiritual tourism.

2.1 *Objective of the Study*

The following are the study's objectives:

1. To explore inclinations and prospects of religious tourism with special reference to Shakti Peeths in Odisha
2. To analyze the socioeconomic impact of religious destinations with respect to Shakti Peeths for developing as a tourism product

2.2 Research Methodology

This study is based on primary and secondary sources of information including a wide range of research studies on religious tourism, particularly as it relates to Odisha's spirituality, books on the subject, numerous websites, newspapers, travel magazines, and other publications that may be useful for learning more about religious and spiritual tourism, and the principal method of data gathering for this research project direct communication with local people at the several Shakti Peeth destinations visited, including different parts of Odisha.

3 Religious Tourism Potential: An Overview of Shakti Peeths in Odisha

Odisha's potential to become a religious tourism destination is evident in its compassion, as the state is home to over 250 religious temples, shrines, and Shakti Peethas, making it a melting pot of diverse cultures and religions. Therefore, any of these holy locations might be thought of as the hub for the development of religious tourism. Since many of the aforementioned locations are well-served by road, train, and lodging options ranging from low-cost to five-star hotels, it is necessary to analyze their potential as a significant tourist destination from a statistical viewpoint. The research analysis has surfaced that Odisha is rich in culture so are the places of holy abodes. Since Shakti Peeths are located across various locations, some of them are well approachable while others are not or less approachable. Nevertheless, those holy shrines are well connected through various means of transport. The cities have beautiful home stays and budgeted hotels which caters tourist stay. According to data produced by the state tourism department, Odisha has seen an increase in domestic visitor numbers by 110% and overseas visitor numbers by 875% since 2021. In 2022, there were 78,67,909 domestic tourists, up from 37,42, 221 in 2021. While there were just 2269 international visitors to the state in 2021, there were 22, 121 in 2022. When all of this is considered, the percentage of tourists is rising over the previous year.

In Hinduism, Shakti is revered as the goddess of supreme, ultimate, and eternal reality of endless existence. She is seen as the source of creation itself, the embodiment of its vitality, and the force that directs everything. She is the basis of everything that exists. Though it accepts the presence of men, Shaktism emphasizes the divine feminine. The worship of Shakti has a long history in Odisha. As the embodiment of the divine feminine creative force, "Shakti," also known as the "Great Divine Goddess," has been revered throughout the nation since the Vedic era. Power goddess Shakti is considered to be the whole embodiment of Adi Shakti, the beginning and the end. The nine avatars of Navdurga, also known as Shakti, are worshipped all over the nation during Navratri and Durga Puja. How the 51 Shakti Peethas, where the goddess is revered as incarnations of Dakshayani-Sati, subsequently known as Parvati or Durga, came to be these locations are spread out throughout the subcontinent. According to the research from primary and secondary sources it has been found that Odisha is home to various Shakti Peeths. Twelve well-known Shakti shrines are taken for the research analysis which is popular among inhabitant/local tourist of those regions.

3.1 Maa Bhagabati Temple

One of the most well-known Shakti Peetha in Odisha is Banapur Bhagabati. It is 105 km from Bhubaneswar and 12 km from Barakul of Chilka. The main deity of this hallowed temple is goddess Burga, also known as Bhagabati in the local dialect; the idol shows the demon Mahisasura's cut head along with eight arms. The ruling goddess of Banapur is Maa Bhagabati. It is thought that the Gajapati Maharaja of Puri constructed both the current temple and the Jagamohana. This temple is significant because Banapur is associated with the Vajrayana school of Buddhism, as seen by the numerous Buddhist pictures found there. One more attraction which is at Banapur in Shaila dam which is only 2 km from the temple which encourage picnic, excursion, and weekend activity.

3.2 Maa Charchika Devi Temple

The eight-armed goddess Chamunda, sometimes referred to as Maa Charchika Devi locally, is the ruling deity. She wears a garland of human skulls and is perched on a human body in a prostrate position. She is holding four right hands, which represent Khadga, Shula, Katarina, and Varadamudra; her four left hands are holding a severed head, a blood-cup, "damru," and a blood-soaked finger from the leftover hand. Located on top of a small hillock called Ruchika Parvata next to the banks of the Renuka river, this temple is situated in the Odisha district's Banki, a small town. It is 52 km from Cuttack and 53 km from Bhubaneswar. This holy abode is not far off from Cuttack Railway Station and 60 km from Biju Pattanaik International Airport. Many hotels are around that support the stay of local visitors to the temple.

3.3 Maa Katakachandi

An ancient shrine to the goddess Chandi, the patroness of Cuttack, Orissa, is called the Katak Chandi Temple. Near the Mahanadi River's banks is where the temple is situated. Maa Katak Chandi is revered as "The Living Goddess" by the people of Cuttack. In the center of the historic city resides and controls the goddess known by the traditional name Maa Katak Chandi. She carries Paasha (a noose), Ankusha (a goad), Abhaya (a gesture that dispels fear), and Varadha (a boon) in her four hands. Every day, Utkala Brahmin Sevayatas worship her as Bhuvaneshvari Mahavidya, the universe's queen. This temple is very famous as like Durga Puja in Odisha, and located very close to the travel transition points like railway station and airport.

3.4 *Maa Ramachandi Temple*

The Ramachandi temple is situated 7 km before Konark on the Marine Drive Road connecting Puri and Konark and is the temple dedicated to the goddess "Ramchandi" at the mouth of the Kushabhadra River. It is commonly thought that Ramchandi is the most compassionate "Chandi" that exists and that she is the presiding deity of Konark. Compared to the Sun temple at Konark, it is undoubtedly older. One of Puri's most well-known Saktapithas is the temple dedicated to Ramachandi. The scenery is magnificent, with a dense growth of Casuarina plantation surrounding the river mouth of "Kushabhadra," a gorgeous Chandi sitting on a lotus in a little temple partially hidden by sand dunes, and the boundless Bay of Bengal stretching to eternity. This place is also a beautiful picnic spot which attracts visitors to the seashore. Boating option in the river is also help thronging crowd to the sacred Shakti Peeth.

3.5 *Maa Mangala in Kakatpur*

This temple, which dates to the fifteenth century, represents the old Kalinga Shakti Peethas. The holy river Prachi's eastern bank is where the temple is located. This temple is one of the Shakti's incarnations. There is a very old connection between Jagannath shrine in Puri and this shrine. The wooden icons of Jagannath, Balabhadra, and Subhadra from the Jagannath Temple in Puri are replaced every 12 to 19 years at the Nabakalebara ceremony. Priests from the Puri temple pray to Mangala in the Kakatpur temple for divine guidance. In their dreams, the goddess shows up and tells them where the three sacred Daru Bramha trees, which are used to make idols of the gods.

3.6 *Maa Hingula Temple*

Devi Hingula was highly revered by King Nala of the Vidarbha area in Western India. The King of Puri contacted him and asked for assistance. Devi Hingula was needed to provide fire for the temple kitchen so that he could begin preparing "Mahaprasada" for Lord Jagannath. After giving her approval, the goddess transformed into fire and travelled to Puri. Given that goddess Hingula is associated with Agnikona, or the southeast direction, it should come as no surprise that Odisha is located in this direction in India, as is Puri in Odisha, Gopal Prasad Gada in Talcher's southeast, Hingula Peeth in Gopal Prasad Pada's southeast, and the temple of Hingula in the southeast of the priest's house.

3.7 Shri Samaleswari Temple

In the western part of Orissa, India, there is a strong religious presence associated with the patron deity of Sambalpur, Shri Samaleswari. On the banks of the Mahanadi River, people have worshipped the mother goddess Samaleswari, also known as Jagatjanani, Adishakti, Mahalaxmi, and Mahasaraswati, since ancient times. The inhabitants of Sambalpur hold a shared belief that the goddess Samaleswari bestows strength upon all devotees who visit her temple from all over the world, empowering them to feel feelings of terror, affection, amazement, and veneration. Three major festivals are celebrated annually in front of the goddess: Nuakhai, Ashwin Navaratri, and Chaitra Navaratri, during which farmers present the goddess with their first harvest before utilizing it for their own consumption.

3.8 Maa Vimala Temple

Maa Bimala Temple, which is located inside Lord Jagannath's Puri temple grounds on the southwest corner next to the Rohini Kund, is another significant Shakti Peetha. It is said that this is where goddess Sati's feet (Pada Khanda) fell. Despite its diminutive size, the temple is built in the likeness of the temple of the Lord. Maa Vimala is regarded as Lord Jagannath's Tantric Consort. It is a deeply held belief that before worshipping Jagannath at the main temple, pilgrims should pay their respects to the goddess Bimala. Lord Jagannath's Prasad is reoffered to goddess Bimala on a golden platter each day. At that point, the Prasad is consecrated and made into Mahaprasad for the devotees' eating. Nothing can be touched without the permission of Maa Vimala in the temple. The one occasion of the year when the goddess is served special foods (nonvegetarian) is during Durga Puja. At the Vimala Temple, Durga Puja is observed for 16 days, during which mother is dressed in a variety of outfits. Vijaya Dashami brings everything to a close.

3.9 Tara Tarini Temple

Located around 28 km from Brahmapur City in the Ganjam district of Odisha, near Purushottampur, on the bank of the Rushikulya River, lies the Adi Shakti temple known as Tara Tarini Mandir. It is thought that the pieces of Maa Sati's body that were cut by Lord Vishnu's Sudarshan chakra during the Daksha yajna fell into the

Shakti Pithas. Where Maa Sati's breasts fell is considered to be the location of Tara Tarini Mandir. The temple is reached by 999 steps on the hill's front side, and there is a vehicle-accessible pucca road on the hill's backside that makes it easier for devotees to approach the Pitha. Kendujhargarh Railway Station is situated approximately 50 km away from this temple, while Harichandanpur Railway Station is the closest railway station, at a distance of roughly 17 km. This temple can be reached by bus from Bhubaneshwar (80 km) and Keonjhar district headquarters (50 km). These cities and this temple are well connected by road. Along with being a spiritual and cultural hub, Taratarini Temple serves as a springboard for exploring some of the other neighboring sights that are well worth seeing, such as Gopalpur Beach, a lovely and tranquil beach that is roughly 45 km away from the temple. It is a well-liked tourism destination where guests may unwind and replenish themselves. The beach's golden sand, blue waves, refreshing breeze, and picturesque vistas can all be enjoyed, as well as water sports and pastimes including sailing, boating, and surfing.

3.10 Maa Biraja Temple

Maa Biraja Temple is the one of the Ashtadasha Shakti Peethas and the 11th Shakti Peeth. The navel region of Devi is said to have fallen here. The current temple was constructed in the thirteenth century. Devi Durga, also known as Girija or Viraja, is the primary idol. This goddess is special because the idol is portrayed as Mahishasuramardini, the demon Mahishasura's chest being speared with one hand while the other is used to drag the tail. Her feet are shown as one on a lion and the other on Mahishasura's breast, which is shown as a water buffalo. Ganesha, a lingam, and a crescent moon are depicted on the idol's crown. The temple is quite vast and contains shrines dedicated to Shiva as well as other deities. The Skanda Purana refers to it as Lingams in Jajpur Adi Shankaracharya is credited for installing the Lord Shiva shrine. An additional noteworthy feature of this temple is a well close to Girija Devi's main temple, where devotees offer ritualistic offerings to their ancestors (Pinda pradhaan). People think that Kasi will directly receive these Pindas. The well appears to be between four and five feet deep, but it has never dried out. One can choose from a range of places to stay. There are OTDC government hotels, bed breakfasts, homestays, and star-rated hotels among the options, as well as eco tour nature camps with cutting edge Swiss tents and amenities.

3.11 Sarala Temple

Jhankad, the sacred site of goddess Sarala, is recognized as one of the most spiritually advanced manifestations of Shaktism dating back thousands of years. Sarala is a composite of the three main Hindu cults: Tantric, Vaishnavite, and Vedic. She is revered as a hybrid of the goddess Durga and Saraswati. It is one of Odisha's most well-known Shakta shrines. The primary idol is a stone carving of an eight-armed woman holding her right foot atop a lion in the Mahinsha Mardini position. It was constructed in the eighth century by the fabled Bhauma Kings. According to mythology, the location is connected to the song of Sage Yamadagni, Parshuram. Consequently, this location holds great historical significance as the hub of the Parshuram cult. Parshuram, the sixth incarnation of Lord Vishnu, is said to have inscribed the goddess with his arrow and worshipped her for her strength and knowledge, according to the Purana. The earliest epic poet of Odisha, Adikavi Sarala Das, who lived in the fifteenth century AD, is also linked to it.

3.12 Bhattarika Temple

Situated at the Sasanga village near Badamba, in Athgarh, it is one of the well-known Hindu pilgrimage sites in Odisha, the Bhattarika Temple. Built sometime between the sixth and sixteenth centuries BC, the Bhattarika Temple is thought to have existed. Storytelling suggests that Parshurama erected this Shakti Peetha at Bhattarika. At the end of his arrow, he engraved an image of the goddess Bhattarika. After losing the fights against Saharasjuna, he found the goddess Durga to be the source of his powers. The non-Brahmin community that worships the goddess Bhattarika cooks fish as part of their daily ritual and presents it to her. As a result, the fisherman community in the area regards the Temple of Goddess Bhattarika as a sacred place, with the goddess being associated with both fishing and navigating. It is located on the bank of river and 82 km from Cuttack town. During winter and spring seasons it is seen here that a large number of tourists come for picnic. There is also a viewpoint of the bank of Mahanadi River so one can enjoy hilly mountains besides the river. The place is attracting thousands of visitors and pilgrimages for its panoramic picturesque as this temple.

4 Conclusion and Future Research

"Orissa" is another name for Odisha. Strong religious convictions, colorful cultural celebrations, and historic temples are what set the stunning coastal state of Odisha apart. The investigation into the socioeconomic effects of holy locations—with a particular emphasis on Shakti Peeths—discloses the great potential that these locations have to offer as sustainable tourism offerings. Spiritually significant sites known as Shakti Peeths draw millions of pilgrims each year, generating significant economic benefits for the surrounding towns. The study suggests that the exploration of these locations as tourist attractions not only strengthens local economies by creating jobs, enhancing infrastructure, and improving services, but also promotes cultural preservation and a sense of community. Politicians and other interested parties can encourage responsible, inclusive tourism that strikes a balance between the preservation of cultural and religious integrity and economic rewards by leveraging the intrinsic spiritual appeal of Shakti Peeths. But in order to optimize the positive effects and reduce any possible drawbacks, a calculated strategy is required. This entails thorough planning, community engagement, environmentally friendly procedures, and ongoing assessment. It will be commercially and socially beneficial to grow Shakti Peeths as tourism goods if integration of local customs, culture, and environmental preservation are prioritized in tourism strategy. The study's conclusion emphasizes the transformative potential of religious tourism in promoting socioeconomic development and calls for a careful and comprehensive approach to the growth of the tourism industry. Shakti Peeths are potential pillars for sustainable tourism because of their rich spiritual legacy, which can propel regional economic growth and improve social harmony. Religious places with reference to Shakti Peeths, the holy abodes of powerful feminine in Odisha, are not only associated with its mythological belief but also with many cultural beliefs. Throughout the year tourist throngs in to Odisha to experience the beauty of it.

Indian tourism can draw visitors from across the globe. Travel for religious purposes is only one of the many travel opportunities that Odisha offers. The absence of essential infrastructures such as transportation and communication networks, clean drinking water, restrooms, lodging options, and entertainment and leisure spaces could potentially impact religious sites. So, it is mandatory to implement the above basic tourist requirement to any of the potential religious site. To manage both domestic and foreign guests, tour guides lack the necessary education and experience. When tourists ask for their assistance, tour guides are not prepared to offer more beneficial services like counseling or recommendations. Certain religious tourism sites lack adequate railroad connectivity, making them inconvenient for visitors and travelers. If a connectivity point is there, then it is not convenient for visitors to travel such a great distance from the religious site because it is far from the connectivity point. The density of railroads is quite low. A well-planned event can prevent safety and health-related challenges that may arise from large gatherings at a religious site so deploying disaster management troops and using crowd management systems will support to maintain the crowd. Safety for guests to religious sites will be guaranteed by promoting these behaviors. New activities for visitors' amusement are essential for any tourism site to expand. A number of the most popular activities that bring thousands of visitors each year include water sports and adventurous sports (Mohanty 2017). The religious location should prioritize attracting water and adventure tourists in addition to the usual tourist destinations, as this will draw visitors for longer periods of time and strengthen the local economy.

References

Behera RR (July 2019) Trends and patterns of tourism in Odisha. Int J Res Soc Sci
Burley M (2021) The nature and significance of the Hindu Divine Mother. Relig Stud

Collins-Kreiner N (2010) Researching pilgrimage: continuity and transformations. Ann Tour Res
Das DS (September 2013) Growth and prospects of Odisha tourism. Odisha Rev
Https://en.wikipedia.org/wiki/Tara_Tarini_Mandir#Sankranti_Mela
Https://odishabytes.com/shakti-peethas-in-odisha-their-significance-forshaktas/#google_vignette
Https://odishatour.in/charchika-temple-banki-cuttack
Https://ommcomnews.com/odisha-news/know-the-major-12-shakti-shrines-of-odisha
Https://timesofindia.indiatimes.com/city/bhubaneswar/arrival-of-foreign-domestic-tourists-to-odisha-records-875-110-growth-in-2022/articleshow/103257114.cms
Https://timesofindia.indiatimes.com/travel/destinations/all-about-18-astadasha-shakti-peethas-holy-abodes-of-shakti/photostory/104467781.cms
Https://trends.google.com/trends/explore?Cat=67&date=today%205-y&geo=IN-OR&gprop=youtube&q=Tourist%20visit%20in%20odisha,tourist&hl=en
Https://www.guidedtirth.com/en/travel-category/astadasha-maha-shakti-peethas/
Https://www.orissatourism.org/Https://www.speakingtree.in/allslides/shakti-peethas-in-odisha
Kumar S, Gupta S, Asthana S (2022a) Determinants of spiritual tourism consumption: a hierarchical approach. Int J Relig Tour Pilgr 10(2):9
Kumar S, Valeri M, Sageena G (2022b) Spiritual tourism transition: an ethos of hope during COVID-19 crisis from Indian perspectives. In: Tourism risk. Emerald Publishing Limited, pp 281–301
Mohanty S (2017) Plight of Odisha tourism: a conceptual study. Int J Adv Educ Res 2(6):251–254
Mukherjeel DS (2020) Religious to spiritual tourism - an era of paradigm shift in India. J Xi'an Univ Archit Technol
Onions CT (1983) The shorter Oxford dictionary on historical principles, 3rd edn. Book Club Associates with Oxford University Press, Oxford
Vukonić B (1996) Turizam i Religija (Tourism and religion) (trans: Matešić S). Elsevier Science, Oxford

The manufacturer's authorised representative in the EU is Springer Nature Customer Service Centre GmbH, Europaplatz 3, 69115 Heidelberg, Germany. If you have any concerns regarding our products, please contact ProductSafety@springernature.com

Printed and bound by CPI Group (UK) Ltd, Croydon, CR0 4YY
30/09/2026
02225589-0012